Virus Infection and Infectious Diseases: Unraveling Mechanisms, Innovations, and Therapeutic Strategies

Virus Infection and Infectious Diseases: Unraveling Mechanisms, Innovations, and Therapeutic Strategies

Guest Editor

Jung-Hyun Lee

MDPI

Basel • Beijing • Wuhan • Barcelona • Belgrade • Novi Sad • Cluj • Manchester

Guest Editor
Jung-Hyun Lee
Department of Life Science
University of Seoul
Seoul
Republic of Korea

Editorial Office
MDPI AG
Grosspeteranlage 5
4052 Basel, Switzerland

This is a reprint of the Special Issue, published open access by the journal *International Journal of Molecular Sciences* (ISSN 1422-0067), freely accessible at: www.mdpi.com/journal/ijms/special_issues/0Q6QLF7Q0A.

For citation purposes, cite each article independently as indicated on the article page online and as indicated below:

Lastname, A.A.; Lastname, B.B. Article Title. *Journal Name* **Year**, *Volume Number*, Page Range.

ISBN 978-3-7258-4090-8 (Hbk)
ISBN 978-3-7258-4089-2 (PDF)
https://doi.org/10.3390/books978-3-7258-4089-2

Contents

About the Editor

Jung-Hyun Lee

Dr. Jung-Hyun Lee is an Assistant Professor in the Department of Life Science at the University of Seoul. She obtained her Ph.D. in Molecular Virology and Cancer Biology from Friedrich-Alexander-University Erlangen-Nürnberg in 2014. Prior to her current position, she held research roles at institutions including Ulm University, the University of Chicago, and Cleveland Clinic Florida. Her research focuses on virus–host interactions, extracellular vesicle-mediated immune modulation, and the engineering of extracellular vesicles for therapeutic applications. Her work aims to elucidate the molecular mechanisms of viral infections and develop novel antiviral strategies.

Preface

Viral infections continue to pose major global health challenges as evidenced by recurrent pandemics and the continuing burden of chronic and emerging infectious diseases. This reprint of the Special Issue "Virus Infection and Infectious Diseases: Unraveling Mechanisms, Innovations and Therapeutic Strategies", published in the *International Journal of Molecular Sciences*, aims to provide a comprehensive overview of recent advances in virology and infectious disease research. This collection brings together experimental studies and reviews, investigating the molecular and cellular basis of viral pathogenesis, host–pathogen interactions, immune modulation and therapeutic innovations, including antiviral drugs, vaccines and diagnostic tools.

The motivation for this volume is rooted in the need to integrate basic mechanistic insights with translational research to better understand, diagnose and treat infectious diseases. We have attempted to highlight both established and novel viral systems and to reflect the interdisciplinary nature of the field, from molecular virology to clinical application.

This reprint is intended for researchers, clinicians, and students in virology, immunology, infectious disease medicine and translational biomedical sciences. The contributing authors are international experts who provide a diverse and up-to-date perspective on their respective fields.

We are deeply grateful to all the authors for their valuable contributions and to the reviewers for their constructive feedback. We also extend our sincere thanks to the *IJMS* Editorial Team for their professional support and dedication throughout the publication process.

Jung-Hyun Lee
Guest Editor

International Journal of
Molecular Sciences

MDPI

Article

Early Intervention in Herpes Simplex Virus-1 Replication in Vitro with Allenic Macrolide Archangiumide

You Li [1,†], Jia-Qi Hu [2,†], Wen-Hai Feng [1], Changsheng Wu [2,*] and Li Gao [3,4,*]

1 State Key Laboratory of Animal Biotech Breeding, Frontiers Science Center for Molecular Design Breeding, Ministry of Agriculture Key Laboratory of Soil Microbiology, Department of Microbiology and Immunology, College of Biological Sciences, China Agricultural University, Beijing 100193, China
2 State Key Laboratory of Microbial Technology, Institute of Microbial Technology, Shandong University, Qingdao 266237, China
3 National Key Laboratory of Veterinary Public Health and Safety, College of Veterinary Medicine, China Agricultural University, Beijing 100193, China
4 Key Laboratory of Animal Epidemiology of the Ministry of Agriculture, College of Veterinary Medcine, China Agricultural University, Beijing 100193, China
* Correspondence: wuchangsheng@sdu.edu.cn (C.W.); 2019194@cau.edu.cn (L.G.)
† These authors contributed equally to the work.

Abstract: Archangiumide is a unique macrolide natural product that features an endocyclic allene functionality, rendering it a prototype of a new class of secondary metabolites of microbial origin. However, its biological and/or pharmaceutical roles remain obscure. In this study, we have unveiled an antiviral potency of archangiumide that was effective against herpes simplex virus (HSV-1) replication. We found that archangiumide did not affect host cell viability, nor pathogen infectivity, but suppressed HSV-1 early replication, in terms of early replication genes, such as *ICP0*, *ICP4*, etc. Further scrutinizing the underlined master regulator, we found that HSV-1 VP16 protein expression was inhibited by archangiumide, as well as VP16 nuclear translocation. As VP16 is a coactivator of transcription, archangiumide harnessed the master regulator of HSV-1 early replication. Together, here we provide evidence that allene macrolide archangiumide possesses robust antiviral functions that may be valuable for a novel viral infection intervention, as macrolides are generally safe drugs for prolonged treatments.

Keywords: HSV-1; archangiumide; antiviral

1. Introduction

Since its first discovery in 1950s, macrolide has been one of the most commonly used families of antibiotics prescribed to treat infections of gram-positive bacteria. For example, azithromycin, clarithromycin, and erythromycin are commonly used to treat infections like pneumonia, sinusitis, pharyngitis, and tonsillitis [1]. Due to its excellent efficacy and safety record in clinical applications, as well as emerging multiple or extreme resistance, interest in exploring novel macrolides is vibrant for both natural microbial products and synthetic chemistry [2–4]. We identified archangiumide, a novel macrolide natural product produced by the myxobacterium Archangium violaceum SDU8, through a combination of genome mining and NMR-based metabolomic profiling. It is an unprecedented 19-membered allenic macrolide with a molecular formula of $C_{19}H_{26}O_6$ [5]. Its structure is characterized by the presence of an endocyclic allene, two additional E-configured alkenes, and an annulated tetrahydrofuran ring. This complex architecture not only provides a significant synthetic challenge [6] but also endows the molecule with unique chemical properties.

Under long term storage in particular conditions of illumination and room temperature, it undergoes gradual degradation. Preliminary biological activity assays demonstrated that archangiumide did not exhibit significant antitumor activity against over 20 common human tumor cells, neither antibacterial activity against Staphylococcus aureus, Bacillus subtilis, and Escherichia coli, nor antifungal activity against Candida albicans. In addition, it does not possess antioxidant (DPPH radical-scavenging) or anti-inflammatory (LPS-induced NO production in adherent cells) effects, yet its unique structure and the challenges associated with its synthesis make it a compelling candidate for further investigation in terms of therapies [5].

Although a plethora of reports has established the antibacterial functions of macrolides for decades, the long-neglected antiviral effects are now brought to spotlight, with the prominent example of azithromycin, which has been used to combat COVID-19 infection [7–9]. The exploration of macrolides as antiviral agents is further supported by the discovery of compounds like balticolid [10], a plecomacrolide with anti-HIV and anti-herpes simplex type I activity. These findings underscore the versatility of macrolide compounds in targeting different viral pathogens and suggest a broader application in antiviral strategies. Herpes simplex virus type 1 (HSV-1), a widespread human pathogen, is known to cause a spectrum of diseases ranging from mild oral and facial mucocutaneous infections to more severe and life-threatening conditions such as keratitis, encephalitis, and meningitis [11–13]. The virus's ability to establish lifelong latent infections and its increasing drug resistance highlight the urgency for novel antiviral agents [14].

In this study, we embarked on the exploration of archangiumide for antiviral therapy, specifically targeting the mechanisms by which HSV-1 infects human cells. By meticulously dissecting the infection route of HSV-1, we aimed to pinpoint archangiumide's intervention at early stages of viral infection. This approach aligns with the growing body of research that underscores the importance of early intervention in viral diseases, which is crucial for limiting viral replication and associated pathologies. As well, this research could pave the way for new treatment strategies against HSV-1 and contribute to the broader understanding of antiviral drug development.

2. Results

2.1. *Archangiumide Significantly Suppressed HSV-1 Replication*

Upon the initial discovery of archangiumide, we were not able to identify its bioactivity, as it has no antibacterial function as expected [5]. Inspired by azithromycin's effectiveness in COVID-19, here we attempt to explore the antiviral activity of archangiumide. To sail in well-charted waters, we chose HSV-1 as our model pathogen, as its infection route has been mapped in detail and HSV-1 related tests are robust as well as bio-safety accessible.

To monitor archangiumide's effect on HSV-1 replication, GFP-tagged HSV-1 viruses were infected for 24 h at MOI 0.5, with or without archangiumide. As shown in Figure 1, archanguimide suppressed HSV-1 replication in a dose-dependent manner, in terms of GFP signal intensity (Figure 1a), viral load (Figure 1b), and viral genome copy (Figure 1c). Moreover, archanguimide suppressed HSV-1 glycoprotein D (gD, Us6 mRNA) transcription at higher and lower levels of viral infection (MOI = 1 and MOI = 0.1), confirming previous data (Figure 1d). In terms of protein level, gD and VP16 proteins were also suppressed (Figure 1e,f). Last but not least, archanguimide had low toxicity to cells (Figure 1g), and these findings were not due to host cell survival issues. Inferring from the virus titer (Figure 1b), archanguimide had an IC50 of 0.039 mM against HSV-1. A 5.75 mM CC50 (50% cytotoxicity concentration) is also inferred from Figure 1f. Therefore, archanguimide has a suitable therapeutic index of 147 (CC50/IC50) for HSV-1 intervention.

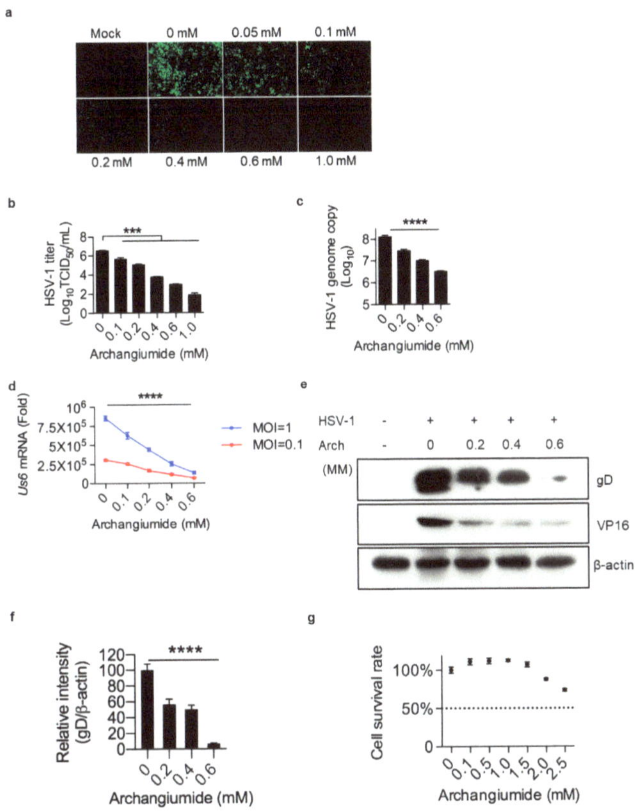

Figure 1. Archangiumide suppressed HSV-1 replication. (**a**,**b**) Archangiumide was added to HSV-1-eGFP virus-infected HeLa cells (MOI = 0.5) and cultured for 24 h. Sham infection was performed as control (Mock). GFP signal was captured (**a**). Fluorescent images were in 200× magnification. HSV-1 titer (TCID50/mL) was analyzed in culture supernatant after 24 h (**b**). Statistics were analyzed with non-parametric one-way ANOVA (Kruskl–Wallis test). ***, $p < 0.001$. n = 3. (**c**) Archangiumide was added to infected cells (MOI = 0.1). After 24 h culture, genomice DNA was extracted and the HSV-1 genome was analyzed by qPCR. Kruskl–Wallis one-way ANOVA. ****, $p < 0.0001$. n = 3. (**d**) gD transcription (Us6 mRNA) was analyzed by qPCR under high and low dose infections (MOI = 1 and MOI = 0.1). Kruskl–Wallis one-way ANOVA for each MOI group. ****, $p < 0.0001$. n = 3. (**e**) Western of gD and VP14 protein under archangiumide. β-actin as control. (**f**) Quantification of e.****, $p < 0.0001$. n = 3. (**g**) Cell survival rate under different doses of archangiumide. Dashed line represents CC50. n = 3. Data shown represent three independent experiments.

2.2. Archangiumide Was Not Able to Inactivate HSV-1 Directly

To understand the mechanism of archangiumide's antiviral effect, we first tested whether archangiumide directly inactivated the pathogen. We pre-incubated HSV-1 stock with archangiumide for 2 h, then diluted the stock to infect cells. As shown in Figure 2a–c, archangiumide pre-incubation did not affect HSV-1 infection, not even at a dose of 0.6 mM. Thus, archangiumide was not able to inactivate HSV-1 directly, and its antiviral effect had impacts on the host cells rather than the pathogen.

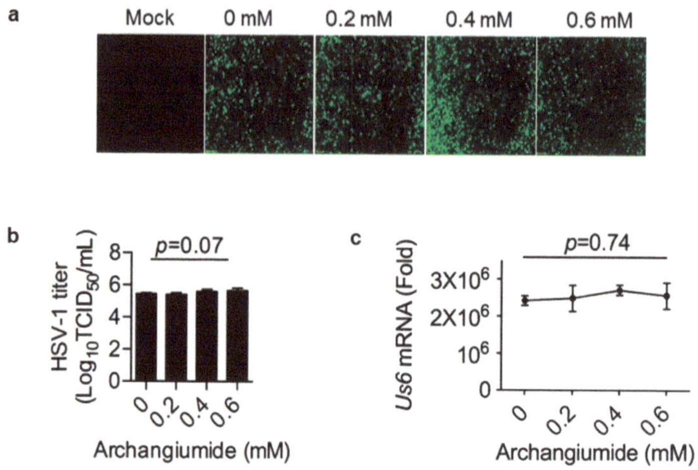

Figure 2. Archangiumide showed no effect on virus inactivation. Archangiumide was pre-incubated with HSV-1-eGFP virus stock for 2 h, then the stock was diluted into MOI = 0.1 to infect cells. (**a**) GFP signal was captured at 24 h. Fluorescent images were in 200× magnification. (**b**) HSV-1 titer (TCID50/mL) was analyzed in culture supernatant after 24 h. Kruskl–Wallis one-way ANOVA. n = 3. (**c**) gD transcription (Us6 mRNA) was analyzed by qPCR after 24 h. Kruskl–Wallis one-way ANOVA. n = 3. Data shown represent three independent experiments.

2.3. Archangiumide Did Not Block HSV-1's Attachment to the Cell, nor Penetration

HSV-1 followed a classical virus–host interaction route, in terms of attachment to the host cell, penetration into the cell, release of the genome, replication of the genome, etc. [12,13]. Therefore, step by step, we scrutinized archangiumide's intervention in HSV-1's infection route. First, we tested archangiumide's effect on viral attachment to the cell. HSV-1 was allowed to attach to pre-chilled cells for 4 h at 4 °C with or without archangiumide. Then, the infected cells were cultured in fresh medium for 24 h. As shown in Figure 3a–c, archangiumide did not suppress HSV-1 attachment to cells. Next, we tested archangiumide's effect on viral penetration into cells. After HSV-1 attachment to the pre-chilled cell, the viral medium was removed, then the attached HSV-1 was allowed to penetrate cells at 37 °C for 15 min with or without archangiumide. Unpenetrated HSV-1 was removed with an acidic PBS wash, and infected cells were cultured in fresh medium for 24 h. As shown in Figure 3d–f, archangiumide did not suppress HSV-1 penetration into cells either.

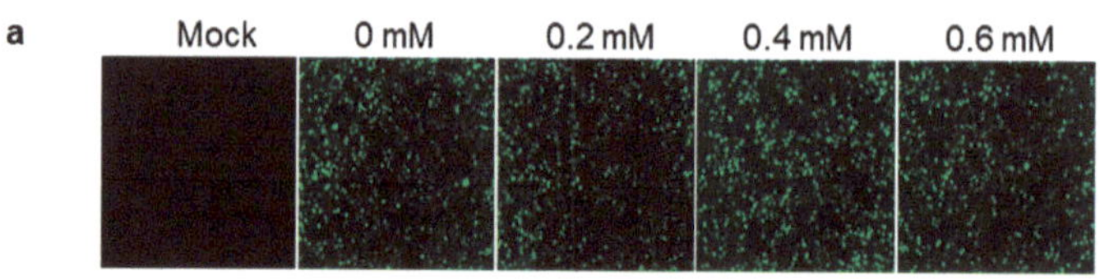

Figure 3. *Cont.*

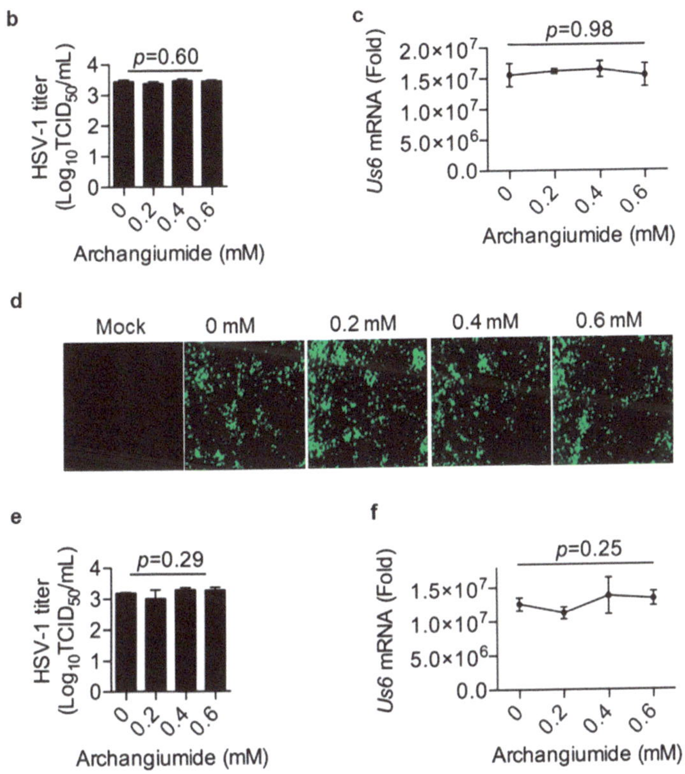

Figure 3. Archangiumide showed no effects on viral attachment and penetration. (**a**–**c**) With or without archangiumide, HSV-1 (MOI = 0.01) was loaded onto cells at 4 °C for 2 h for attachment, then removed. Infected cells were cultured in fresh medium and the GFP signal was captured at 24 h (**a**). Fluorescent images were in 200× magnification. (**b**) HSV-1 titer (TCID50/mL) was analyzed in culture supernatant after 24 h. Kruskl–Wallis one-way ANOVA. n = 3. (**c**) gD transcription (Us6 mRNA) was analyzed by qPCR after 24 h. Kruskl–Wallis one-way ANOVA. n = 3. (**d**–**f**) HSV-1 (MOI = 0.01) was loaded onto cells at 4 °C for 2 h for attachment, then removed. With or without archangiumide, attached HSV-1 was stimulated to penetrate at 37 °C for 15 min. After using an acidic PBS wash (pH = 3) to remove non-penetrated virus then neutralizing cells with a basal PBS wash (pH = 11), infected cells were cultured 24 h for analysis. (**d**) GFP signal was captured at 24 h. Fluorescent images were in 200× magnification. (**e**) HSV-1 titer (TCID50/mL) was analyzed in culture supernatant after 24 h. Kruskl–Wallis one-way ANOVA. n = 3. (**f**) gD transcription (Us6 mRNA) was analyzed by qPCR after 24 h. Kruskl–Wallis one-way ANOVA. n = 3. Data shown represent three independent experiments.

2.4. Archangiumide Suppressed Early Replication of Intracellular HSV-1

As archangiumide showed no effect on viral entry of cells, we next explored the timing of its antiviral effect on intracellular viruses after entry. At different time points post infection, 0.4 mM archangiumide was added to suppress HSV-1 replication. The data show that archangiumide was effective as early as 2 h post infection (Figure 4a,b), and an extremely strong antiviral effect (***, $p < 0.001$) was observed within 6 h. We next analyzed early replication genes (ICP0, ICP4, ICP22, ICP27, and UL29) and confirmed that archangiumide suppressed these genes' transcription as predicted (Figure 4c).

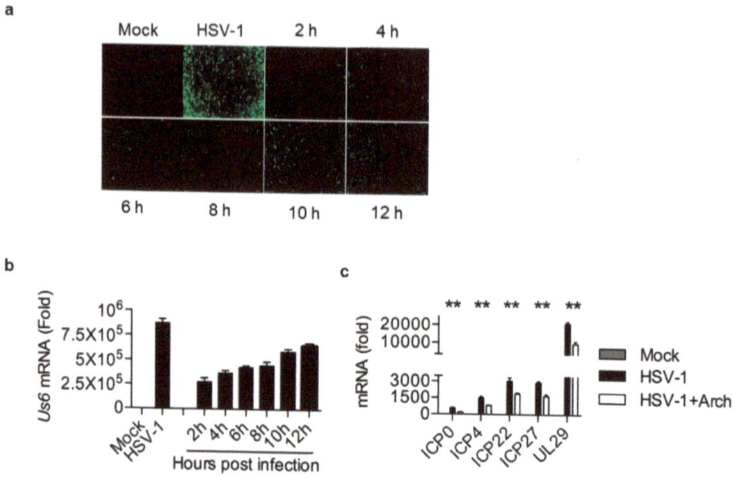

Figure 4. Archangiumide suppressed early viral replication. At different times after HSV-1 infection (MOI = 0.1), 0.4 mM archangiumide was added to culture systems to suppress replication. (**a**) GFP signal was captured at 24 h. Fluorescent images were in 200× magnification. (**b**) gD transcription (Us6 mRNA) was analyzed by qPCR after 24 h. Unpaired test, two-tailed. **, $p < 0.01$, n = 3. (**c**) Early-replication-related gene transcriptions (ICP0, ICP4, ICP22, ICP27, UL29 mRNA) were analyzed by qPCR before 6 h post infection. Kruskl–Wallis one-way ANOVA for each gene. n = 3. Data shown represent three independent experiments.

2.5. *Archangiumide Blocked VP16 Nuclear Translocation to Suppress Early Replication*

To decipher archangiumide's early intervention in HSV-1 replication, we turned to the literature for a master regulator of early replication and focused on VP16. VP16 is well known for its transactivator function, which forms a complex with various transcriptional factors to regulate HSV-1 early replication [14–17]. Therefore, we analyzed VP16 at an early stage of infection, and found that archangiumide down-regulated VP16 mRNA and protein expression instantly (Figure 5a,b). As a transactivator, VP16's nuclear translocation is a prerequisite for its function; therefore, we analyzed VP16's distribution under archangiumide. The data in Figure 5c show that VP16's nuclear translocation was blocked by archangiumide. Together, we found that archangiumide suppressed HSV-1 early replication by down-regulation of VP16's expression and nuclear translocation.

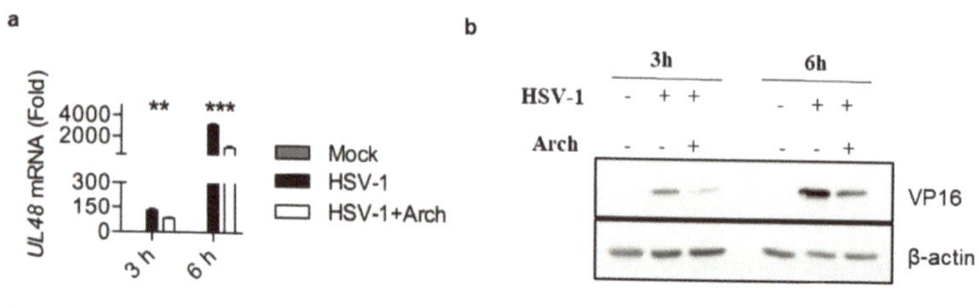

Figure 5. *Cont.*

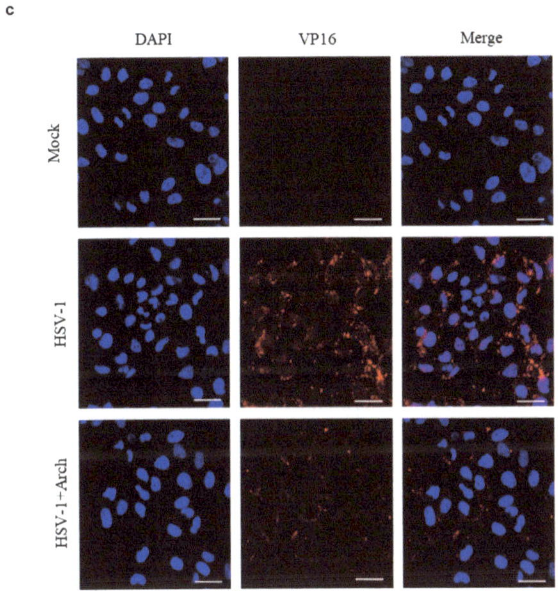

Figure 5. Archangiumide down-regulated VP16 expression and blocked its nuclear translocation. 0.4 mM archangiumide was added to HSV-1-infected cells (MOI = 4). (**a**) At 3 h and 6 h, VP16 (UL48 mRNA) was analyzed by qPCR. Kruskl–Wallis one-way ANOVA for each timepoint. ***, $p < 0.001$, **, $p < 0.01$, n = 3. (**b**) Western blot of VP16 protein under archangiumide. β-actin as control. (**c**) Intracellular distribution of VP16. DAPI as nuclear indicator. Bars indicate 10 μm. Data shown represent three independent experiments.

3. Discussion

Here, we have reported that macrolide archangiumide suppressed HSV-1 replication in the early stages of infection. Archangiumide is a novel macrolide natural product produced by the myxobacterium archangium violaceum SDU8 [5]. Firstly, we evaluated the inhibitory activity of archangiumide on HSV-1 replication by detecting the virus titer, virus genome copy, and mRNA level. The results indicated that archangiumide blocked HSV-1 replication efficiently. The therapeutic index was calculated according to IC50 and CC50, which indicated the promising prospect of archangiumide as an anti-HSV-1 inhibitor (Figure 1). Then, we attempted to discover the mechanism by which archangiumide inhibits HSV-1 replication. Antiviral agents block virus proliferation by inactivating viral particles directly or by inhibiting the virus replication cycle. Our data demonstrate that archangiumide could not inactivate HSV-1 viral particles directly (Figure 2). That means archangiumide acted on HSV-1 replication cycle. Therefore, step by step, we scrutinized archangiumide's intervention in HSV-1's replication. The data indicate that archangiumide did not block HSV-1's attachment to the cells, nor its penetration (Figure 3). We next explored the timing of its antiviral effect on intracellular viruses after entry and found that archangiumide was effective at as early as 2 h post infection, and extremely strong antiviral effect was observed within 6 h (Figure 4a,b). To decipher archangiumide's early intervention in HSV-1 replication, we turned to the literature for a master regulator of early replication and focused on VP16. VP16 is well known for its transactivator function, which forms a complex with various transcriptional factors to regulate HSV-1 early replication [18–20]. Therefore, we analyzed VP16 at early stage of infection, and found that archangiumide down-regulated VP16 mRNA and protein expression instantly (Figure 5a,b). As a transactivator, VP16's nuclear translocation is a prerequisite for its function; therefore, we analyzed

VP16's distribution under archangiumide. The data in Figure 5c show that VP16's nuclear translocation was blocked by archangiumide. Together, we found that archangiumide suppressed HSV-1 early replication by down-regulation of VP16's expression and nuclear translocation.

Although we have confidence in our conclusions, several limitations of this study should be acknowledged. First, as the first-ever allene bearing macrolide, whether archangiumide's viral suppression is directly associated with the allene group is an open question, as is the question of why it did not exhibit traditional antibacterial function. As the total synthesis of this molecule has recently been mapped out [6], it is worthwhile to test these pharmaceutical questions with structural analogs without allene group. Second, although HSV-1 is the most successful virus in terms of infectivity, additional work is planned to explore more pathological viruses, such as COVID-19 and influenza. Also, deviating from HSV-1's DNA genome nature, other RNA viruses need to be tested. Third, we presented the evidence of direct intervention of archangiumide in HSV-1 replication here. However, in vivo, more complex pharmacological and immunological effects of archangiumide are next to be explored. Last but not least, HSV-1 has a lytic-latency life cycle in vivo [11], which is the key survival technique for the pathogen. Therefore, before archangiumide's translation from bench to bed, its impact on HSV-1 lytic-latency cycle should be monitored in animal models.

In recent years, HSV-1 infection has increased and caused a great challenge to public health. Due to drug resistance, natural products have been chosen for the development of new anti-HSV-1 agents. Many such efforts have been made [14]. For example, peniterphenyl A, from a deep-sea-derived *Penicillium* sp., showed anti-HSV-1/2 activity in vitro [21], and a hydroethanolic extract of *Tanacetum parthenium* (L.) Sch.Bip. (Asteraceae) has shown in vivo anti-HSV-1 activity and is safe for oral and topical application [22].

4. Materials and Methods

4.1. Isolation of Archangiumide

The method for the fermentation and isolation of archangiumide was described previously [11]. Archangiumide stock was prepared in 1 mM DMSO, then diluted to the desired concentration in 2% FBS/DMEM before experiments.

4.2. Cells and Virus

Cell passage was cultured in DMEM containing 10% FBS. Vero was cultivated to produce HSV-1-eGFP. HeLa cells lines were cultivated to perform all infection tests. HSV-1-eGFP was kindly provided by professor Xiaojia Wang (China Agricultural University, Beijing, China). For drug efficacy tests, HSV-1-eGFP stock was diluted to a certain MOI in 2% FBS/DMEM, then was allowed to infect 90% confluent HeLa cells at 37 °C. Cytopathic effects were monitored each 6 h and infected cells were typically collected for further tests at 24 h post infection.

4.3. TCID50 and Genome Copy

For the viral titer, samples were serial diluted from 10^{-1} to 10^{-8}. The eGFP signal was captured in diluted-sample-infected cells. TCID50 was calculated using the Reed–Muench method. Eight technical repeats were performed for every diluted sample. For genome copy analysis, HSV-1 VP16 was cloned in a plasmid vector. A standard curve was generated by VP16 qPCR (ΔCT) of serial-diluted plasmid. The genome DNA of HSV-1-infected cells was extracted (Solarbio Life Sciences, Beijing, China) and then analyzed by VP16 qPCR. After fitting to the standard curve, the HSV-1 genome copy of the infected cells was calculated.

4.4. Cell Survival Rate

MTT-based cell viability was analyzed following the producer's manual (Beyotime Biotechnol, Shanghai, China). Briefly, 5000 HeLa cells were seeded in 96-well plates, then archangiumide was added to the culture system for 24 h. An amount of 10 μL MTT solution (5 mg/mL) was added and cultivated at 37 °C for 4 h. An amount of 100 μL formazan solution was added and mixed gently. After 4 h, OD570 was analyzed. Four technical repeats were performed.

4.5. Real-Time qPCR

HSV-1 mRNA transcriptions were analyzed by RT and real-time qPCR (Mei5bio, Beijing, China). Primers used: Us6 (gD), forward, 5′-CGTCCGGAAAC AACCCTACA-3′, reverse, 5′-CCCAGGTTATCCTCGCTGAC-3′. UL48 (VP16), forward, 5′-CTTCAGGTATGGCGAGTCCC-3′, reverse, 5′-GGTGTTCGTCGTCTTCGGAT-3′. ICP0, forward, 5′-GTCTGTCGCATTTGCACCTC-3′, reverse, 5′-CTTCTGTGGTGATGCCGGAG-3′. ICP4, forward, 5′-CTCCATGGTAGAGGAGGCCG-3′, reverse, 5′-CATCGTCGTCGGCT CGAAAG-3′. ICP22, forward, 5′-ATCAGCTGTTTCGGGTCCTG-3′, reverse, 5′-CCATCAG GTAACAGTCGCGT-3′. ICP27, forward, 5′-ATGTGCATCCACCACAACCT-3′, reverse, 5′-TCCTTAATGTCCGCCAGACG-3′. UL29, forward, 5′-ATGAACAGCTGCAACGGGTA-3′, reverse, 5′-GTCGTTACCGAGGGCTTCAA-3′.

4.6. Western Blot

Protein sample preparation and western blot were performed as instructed (Beyotime Biosciences, Beijing, China). Monoclonal antibodies against gD (sc-21719) were purchased from SANTA, VP16 (AB110226) was purchased from Abcam (Waltham, MA, USA), and β-actin (A5441) was commercially available (Sigma-Aldrich, St. Louis, MO, USA).

4.7. Immunofluorescence

Fixed and permeabilized cells were stained with anti-VP16 antibody (AB110226, Abcam) and then labelled with goat anti-mouse IgG (H + L)-Alexa Fluor 555 antibody (Gene Protein Link, Beijing, China). DAPI was co-stained to label nuclei. Fluorescent images were captured with an A1-SIM Laser Scanning Confocal Microscope (Nikon, Minato City, Japan).

Author Contributions: Conceptualization, L.G. and C.W.; methodology, Y.L. and J.-Q.H.; investigation, Y.L. and J.-Q.H.; data curation, Y.L. and L.G.; writing—original draft preparation, L.G.; writing—review and editing, L.G., C.W., W.-H.F., Y.L. and J.-Q.H.; supervision, L.G. and C.W.; funding acquisition, L.G. All authors have read and agreed to the published version of the manuscript.

Funding: This research was funded by the National Key Research and Development Program of China (2024YFD1800200) and the Earmarked Fund for Modern Agro-Industry Technology Research System (No. CARS-40), China.

Data Availability Statement: Data are contained within the article.

Conflicts of Interest: The authors declare no conflicts of interest.

References

1. Dinos, G.P. The macrolide antibiotic renaissance. *Br. J. Pharmacol.* **2017**, *174*, 2967–2983. [CrossRef] [PubMed]
2. Fernandes, P.; Martens, E.; Pereira, D. Nature nurtures the design of new semi-synthetic macrolide antibiotics. *J. Antibiot.* **2017**, *70*, 527–533. [CrossRef] [PubMed]
3. Myers, A.G.; Clark, R.B. Discovery of Macrolide Antibiotics Effective against Multi-Drug Resistant Gram-Negative Pathogens. *Acc. Chem. Res.* **2021**, *54*, 1635–1645. [CrossRef] [PubMed]

4. Kim, K.; Jung, S.; Kim, M.; Park, S.; Yang, H.J.; Lee, E. Global Trends in the Proportion of Macrolide-Resistant Mycoplasma pneumoniae Infections: A Systematic Review and Meta-analysis. *JAMA Netw. Open* **2022**, *5*, e2220949. [CrossRef] [PubMed]
5. Hu, J.-Q.; Wang, J.-J.; Li, Y.-L.; Zhuo, L.; Zhang, A.; Sui, H.-Y.; Li, X.-J.; Shen, T.; Yin, Y.; Wu, Z.-H.; et al. Combining NMR-Based Metabolic Profiling and Genome Mining for the Accelerated Discovery of Archangiumide, an Allenic Macrolide from the Myxobacterium Archangium violaceum SDU8. *Org. Lett.* **2021**, *23*, 2114–2119. [CrossRef]
6. Sutro, J.L.; Furstner, A. Total Synthesis of the Allenic Macrolide (+)-Archangiumide. *J. Am. Chem. Soc.* **2024**, *146*, 2345–2350. [CrossRef] [PubMed]
7. Batiha GE, S.; Zayed, M.A.; Awad, A.A.; Shaheen, H.M.; Mustapha, S.; Herrera-Calderon, O.; Pagnossa, J.P.; Algammal, A.M.; Zahoor, M.; Adhikari, A.; et al. Management of SARS-CoV-2 Infection: Key Focus in Macrolides Efficacy for COVID-19. *Front. Med.* **2021**, *8*, 642313. [CrossRef] [PubMed]
8. Pani, A.; Lauriola, M.; Romandini, A.; Scaglione, F. Macrolides and viral infections: Focus on azithromycin in COVID-19 pathology. *Int. J. Antimicrob. Agents* **2020**, *56*, 106053. [CrossRef]
9. Khoshnood, S.; Shirani, M.; Dalir, A.; Moradi, M.; Haddadi, M.H.; Sadeghifard, N.; Birjandi, F.S.; Yashmi, I.; Heidary, M. Antiviral effects of azithromycin: A narrative review. *Biomed. Pharmacother.* **2022**, *147*, 112682. [CrossRef] [PubMed]
10. Shushni, M.A.M.; Singh, R.; Mentel, R.; Lindequist, U. Balticolid: A new 12-membered macrolide with antiviral activity from an ascomycetous fungus of marine origin. *Mar. Drugs* **2011**, *9*, 844–851. [CrossRef] [PubMed]
11. Amin, I.; Vajeeha, A.; Younas, S.; Afzal, S.; Shahid, M.; Nawaz, R.; Khan, M.U.; Idrees, M. HSV-1 Infection: Role of Viral Proteins and Cellular Receptors. *Crit. Rev. Eukaryot. Gene Expr.* **2019**, *29*, 461–469. [CrossRef] [PubMed]
12. Zhu, S.; Viejo-Borbolla, A. Pathogenesis and virulence of herpes simplex virus. *Virulence* **2021**, *12*, 2670–2702. [CrossRef] [PubMed]
13. Ahmad, I.; Wilson, D.W. HSV-1 Cytoplasmic Envelopment and Egress. *Int. J. Mol. Sci.* **2020**, *21*, 5969. [CrossRef] [PubMed]
14. Treml, J.; Gazdova, M.; Smejkal, K.; Sudomova, M.; Kubatka, P.; Hassan, S.T.S. Natural Products-Derived Chemicals: Breaking Barriers to Novel Anti-HSV Drug Development. *Viruses* **2020**, *12*, 154. [CrossRef] [PubMed]
15. Yu, W.; Geng, S.; Suo, Y.; Wei, X.; Cai, Q.; Wu, B.; Zhou, X.; Shi, Y.; Wang, B. Critical Role of Regulatory T Cells in the Latency and Stress-Induced Reactivation of HSV-1. *Cell Rep.* **2018**, *25*, 2379–2389.e2373. [CrossRef] [PubMed]
16. Pietila, M.K.; Bachmann, J.J.; Ravantti, J.; Pelkmans, L.; Fraefel, C. Cellular state landscape and herpes simplex virus type 1 infection progression are connected. *Nat. Commun.* **2023**, *14*, 4515. [CrossRef]
17. Reske, A.; Pollara, G.; Krummenacher, C.; Chain, B.M.; Katz, D.R. Understanding HSV-1 entry glycoproteins. *Rev. Med. Virol.* **2007**, *17*, 205–215. [CrossRef] [PubMed]
18. Wysocka, J.; Herr, W. The herpes simplex virus VP16-induced complex: The makings of a regulatory switch. *Trends Biochem. Sci.* **2003**, *28*, 294–304. [CrossRef]
19. Herrera, F.J.; Triezenberg, S.J. VP16-dependent association of chromatin-modifying coactivators and underrepresentation of histones at immediate-early gene promoters during herpes simplex virus infection. *J. Virol.* **2004**, *78*, 9689–9696. [CrossRef] [PubMed]
20. Triezenberg, S.J.; Kingsbury, R.C.; McKnight, S.L. Functional dissection of VP16, the trans-activator of herpes simplex virus immediate early gene expression. *Genes Dev.* **1988**, *2*, 718–729. [CrossRef]
21. Chen, W.; Zhang, J.; Qi, X.; Zhao, K.; Pang, X.; Lin, X.; Liao, S.; Yang, B.; Zhou, X.; Liu, S.; et al. p-Terphenyls as Anti-HSV-1/2 Agents from a Deep-Sea-Derived Penicillium sp. *J. Nat. Prod.* **2021**, *84*, 2822–2831. [CrossRef] [PubMed]
22. Benassi-Zanqueta, E.; Marques, C.F.; Valone, L.M.; Pellegrini, B.L.; Bauermeister, A.; Ferreira, I.C.P.; Lopes, N.P.; Nakamura, C.V.; Filho, B.P.D.; Natali, M.R.M.; et al. Evaluation of anti-HSV-1 activity and toxicity of hydroethanolic extract of *Tanacetum parthenium* (L.) Sch.Bip. (Asteraceae). *Phytomedicine* **2019**, *55*, 249–254. [CrossRef]

International Journal of
Molecular Sciences

MDPI

Article

Evolutionary Patterns and Genotype-Specific Amino Acid Mutations of Tick-Borne Encephalitis Virus

Ruichen Wang [1], Anqi Gu [1,2], Fan Li [1], Qian Ma [2], Qikai Yin [1], Kai Nie [1], Shihong Fu [1], Qianqian Cui [1], Songtao Xu [1], Hao Li [2,*] and Huanyu Wang [1,*]

1 National Key Laboratory of Intelligent Tracking and Forecasting for Infectious Diseases, NHC Key Laboratory of Biosafety, National Institute for Viral Disease Control and Prevention, Chinese Center for Disease Control and Prevention, Beijing 102206, China; wangrc96@163.com (R.W.); 17852850389@163.com (A.G.); lifan@ivdc.chinacdc.cn (F.L.); yinqk@ivdc.chinacdc.cn (Q.Y.); niekai@ivdc.chinacdc.cn (K.N.); fush@ivdc.chinacdc.cn (S.F.); cuiqq@ivdc.chinacdc.cn (Q.C.); xust@ivdc.chinacdc.cn (S.X.)

2 Chinese Center for Disease Control and Prevention, Beijing 102206, China; maqian@chinacdc.cn

* Correspondence: lihao@chinacdc.cn (H.L.); wanghy@ivdc.chinacdc.cn (H.W.)

Abstract: Tick-borne encephalitis virus (TBEV) is a significant tick-borne flavivirus responsible for severe human diseases. Here, we analyzed the genetic diversity and evolutionary dynamics of TBEV using 263 genome sequences from the NCBI database and identified key amino acid mutations. TBEV sequences were classified into five genotypes—Baikalian, European, Far-Eastern, Himalaya, and Siberian—showing ORF nucleotide similarity of 81.5% to 88.0% and amino acid similarity of 93.0% to 96.4%. Extensive recombination between genotypes was not observed. Entropy analyses revealed highly variable sites distributed across the Baikalian (n = 2), European (n = 3), Far-Eastern (n = 5), and Siberian (n = 13) genotypes. Each genotype exhibited specific amino acid mutations. Positive selection analysis identified sites under selection in the full dataset (n = 2), as well as in the European (n = 6), Far-Eastern (n = 7), and Siberian (n = 4) genotypes. By integrating highly variable sites, shared genotype-specific mutations, and positively selected sites, we identified 37 key amino acid positions, primarily located on the surfaces of viral proteins. These positions may have a potential impact on protein function and pathogenicity, though further studies are required to validate and evaluate these effects comprehensively. This study provides the first comprehensive analysis of mutational landscapes across TBEV genotypes, uncovering potential critical mutations that may shape viral biology and pathogenicity, and offers valuable insights for further exploration of TBEV characteristics.

Keywords: TBEV; genetic evolution; amino acid mutations; protein structure

1. Introduction

Tick-borne encephalitis (TBE) is a natural focal infectious disease caused by bites from ticks carrying the tick-borne encephalitis virus (TBEV), characterized primarily by central nervous system symptoms and potentially leading to fatal outcomes [1]. TBE is primarily endemic in the vast regions of Europe and Asia between latitudes 39° N and 65° N, with approximately 10,000–12,000 cases reported globally each year [2,3]. The annual incidence rate of TBE varies significantly across different regions. In Europe, the reported incidence is approximately 2.19 cases per 100,000 population [4]. In Russia, the annual incidence is lower, estimated at 0.67 cases per 100,000 population [5]. In China, the incidence ranges from 0.09 to 0.44 cases per 100,000 population [6]. TBEV is primarily transmitted by ticks, with at least 24 tick species capable of spreading the virus. Among them, *Ixodes persulcatus* and *Ixodes ricinus* are the main vectors of TBEV [7]. TBEV has also been reported to spread through the consumption of unpasteurized milk and dairy products [8,9]. The host range of TBEV is broad, encompassing ticks and various vertebrate hosts such as cattle, sheep, dogs, horses, deer, and rodents [7]. Humans serve as terminal hosts for TBEV [10].

TBEV has been classified by the International Committee on Taxonomy of Viruses (ICTV) under the family *Flaviviridae* and the genus *Flavivirus* (recently renamed as *Orthoflavivirus*) [11]. TBEV is an enveloped, single-stranded positive-sense RNA virus with a genome approximately 11 kb in length. It contains a single open reading frame (ORF) that encodes three structural proteins (anchC (the precursor of the C protein), PreM, and E) and eight non-structural proteins (NS1, NS2a, NS2b, NS3, NS4a, 2K, NS4b, and NS5) [12]. Based on the genetic diversity of viral genome sequences, TBEV is classified into five main genotypes [13], including the traditionally recognized European, Siberian, and Far-Eastern genotypes, as well as the recently defined Baikalian [14] and Himalaya [15] genotypes. The virulence of different TBEV genotypes varies significantly, with case fatality rates among the three traditional genotypes ranked from highest to lowest as Far-Eastern (5.0–20.0%), Siberian (6.0–8.0%), and European (0.5–2.0%) [16]. Currently, there is no conclusive evidence that the Baikalian and Himalaya genotypes infect or cause disease in humans.

The evolutionary characteristics of TBEV have long been a focus of research. TBEV is considered the slowest-evolving tick-borne flavivirus, with divergence times among its genotypes dating back to ancient periods (95% HPD, 6373–13,208 years) [17]. However, even a few mutations in TBEV can lead to significant changes in its virological properties, including plaque size, virulence, neuroinvasiveness in mice, drug resistance, and host adaptability [18–21]. Therefore, this study conducted an in-depth investigation into the evolutionary patterns of TBEV and the mutational differences among its various genotypes. The findings provide valuable insights into the evolutionary characteristics of different TBEV genotypes. Additionally, as live TBEV research typically requires BSL-3 laboratory conditions, this study identifies key amino acid sites that can serve as references for future investigations into the virological properties of TBEV.

2. Results

2.1. The Phylogenetic Tree of TBEV

A phylogenetic tree was reconstructed based on the sequences included in this study. The analyzed TBEV sequences clearly clustered into five evolutionary branches (bootstrap support > 99%), corresponding to the European, Himalaya, Siberian, Baikalian, and Far-Eastern genotypes (Figure 1A). In terms of host origin, TBEV is predominantly found in mammals (humans and rodents) and arthropods (ticks), but has also been detected in birds. Occasionally, TBEV has been reported in mosquitoes, lice, shrews, antelopes, marmots, and pikas. Human infections have been reported for the European, Far-Eastern, and Siberian genotypes, while the Baikalian genotype (found in ticks and rodents) and the Himalaya genotype (found in marmots) have not yet been associated with human infection. TBEV strains from different hosts did not form distinct host-specific clusters, but were interspersed across the phylogenetic tree. This finding highlights the broad host range of TBEV and suggests that the virus lacks strong host specificity.

From an administrative regional perspective, TBEV is primarily distributed across Europe and Asia, including continental countries and island nations such as the UK and Japan (Figure 1A). Notably, for simplicity in statistical analysis, TBEV from Russia's Far East was grouped with Europe. When examining the administrative distribution of genotypes, it was observed that the European, Siberian, and Far-Eastern genotypes are found in more than one country, indicating cross-regional transmission for these three genotypes. The

Far-Eastern genotype is primarily distributed in Russia and China, with sporadic detections in Germany. The European genotype exhibits the widest spread, having been detected in at least 15 countries across Europe and Asia. Furthermore, European genotype sequences from different countries did not form distinct country-specific clusters, suggesting that this genotype may be associated with more frequent long-distance transmission events compared to other genotypes. The Siberian genotype is also found in multiple countries across Europe and Asia, but it is primarily concentrated in Russia. Among the sequences analyzed in this study, the Baikalian genotype was exclusively detected in Russia, while the Himalaya genotype was only found in China.

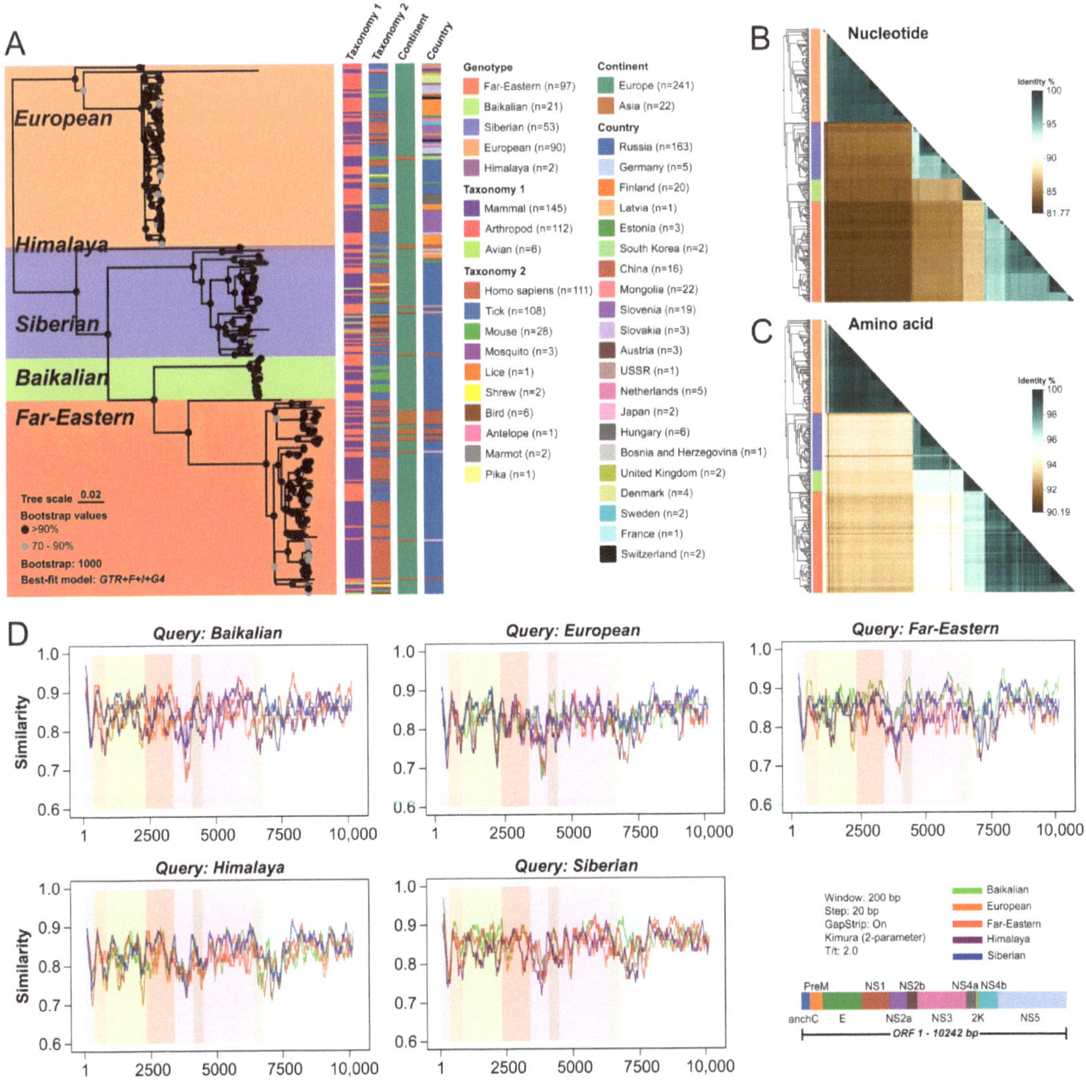

Figure 1. Phylogenetic and similarity analysis of five TBEV genotypes. (**A**) Maximum likelihood (ML) tree illustrating the genetic relationships among the five TBEV genotypes, with detailed information

on genotypes, isolates, and locations of isolation. The number of sequences for each category is shown in the legend. Heatmaps of pairwise nucleotide similarity (**B**) and amino acid similarity (**C**). The color gradient reflects the degree of similarity. (**D**) Nucleotide similarity within different genome regions across groups, detected using Simplot software (Version: 3.5.1). The background shading indicates distinct genomic regions.

Analysis based on species and administrative regions revealed that the main differences among TBEV sequences lie between genotypes. Therefore, subsequent analyses were conducted using genotypes as the grouping criterion.

2.2. Genotypic Similarity Analysis of TBEV

The ORF regions of TBEV strains from different genotypes were compared for nucleotide and amino acid similarity. Nucleotide similarity among TBEV strains ranged from 81.7% to 100.0% (Figure 1B and Table 1), while amino acid similarity ranged from 90.2% to 100.0% (Figure 1C). Between different genotypes, nucleotide similarity was 81.5–88.0%, and amino acid similarity was 93.0–96.4%. Specifically, at the nucleotide level, the European and Himalaya genotypes exhibited the greatest divergence (81.5%), while the European and Far-Eastern genotypes showed the highest similarity (88.0%) (Figure 1B). At the amino acid level, the European and Himalaya genotypes again displayed the greatest divergence (93.0%), while the Far-Eastern and Baikalian genotypes exhibited the highest similarity (96.4%) (Figure 1C). These findings underscore the significant differences among TBEV genotypes.

Table 1. The mean nucleotide and amino acid similarity between genotypes.

Number	TBEV Genotypes	TBEV Genotypes				
		1	2	3	4	5
1	European	***	81.5%	82.6%	82.1%	88.0%
2	Himalaya	93.0%	***	82.9%	82.6%	81.7%
3	Siberian	94.2%	93.9%	***	85.0%	83.6%
4	Baikalian	94.0%	93.9%	95.8%	***	86.1%
5	Far-Eastern	93.5%	93.3%	95.1%	96.4%	***

Note: '***' indicates 100%; the upper right quadrant indicates the nucleotide similarity; the lower left quadrant indicates the amino acid similarity.

A sliding window analysis was performed to examine the distribution of nucleotide similarity across the genomes of different TBEV genotypes. Nucleotide similarity in each genome window between genotypes was generally below 95%, suggesting the absence of widespread recombination events between genotypes (Figure 1D). However, recombination analyses of individual sequences revealed potential intra-genotype recombination events in the European (n = 3), Far-Eastern (n = 10), and Siberian (n = 12) genotypes (Supplementary Table S2). Overall, the NS2a region was identified as the most divergent region at the nucleotide level, while the NS5 region exhibited the highest nucleotide similarity (Figure 1D). Further analysis revealed that similarity levels between a given genotype and others were not consistent across all genome regions. For instance, the Siberian genotype displayed the greatest divergence from the Himalaya genotype in the NS2a region, while showing higher similarity with the Himalaya genotype in the early part of the NS1 region. This variation in similarity levels across genome regions was observed to varying degrees for all genotypes and was also evident in amino acid analyses (Supplementary Figure S1). These findings highlight the necessity of conducting in-depth analyses of different regions across TBEV genotypes.

2.3. Variability of Amino Acid Sites Across Different Genotypes of TBEV

Amino acid mutations can lead to differences in viral characteristics, so we further analyzed the mutational patterns among different TBEV genotypes from multiple perspectives. The mutation landscape of TBEV revealed that, while most positions showed consistent amino acid usage across genotypes, some loci exhibited variation (Figure 2A and Supplementary Data S2). Information entropy analysis identified numerous high-variability amino acid sites (Entropy > 1) (Figure 2B). These highly variable sites are often closely associated with changes in virological properties. Across the entire dataset, 102 high-variability sites were identified. These included 2, 3, 5, and 13 sites in the Baikalian, European, Far-Eastern, and Siberian genotypes, respectively. Due to the limited sequence data available for the Himalaya genotype, entropy analysis could not provide meaningful results for this group.

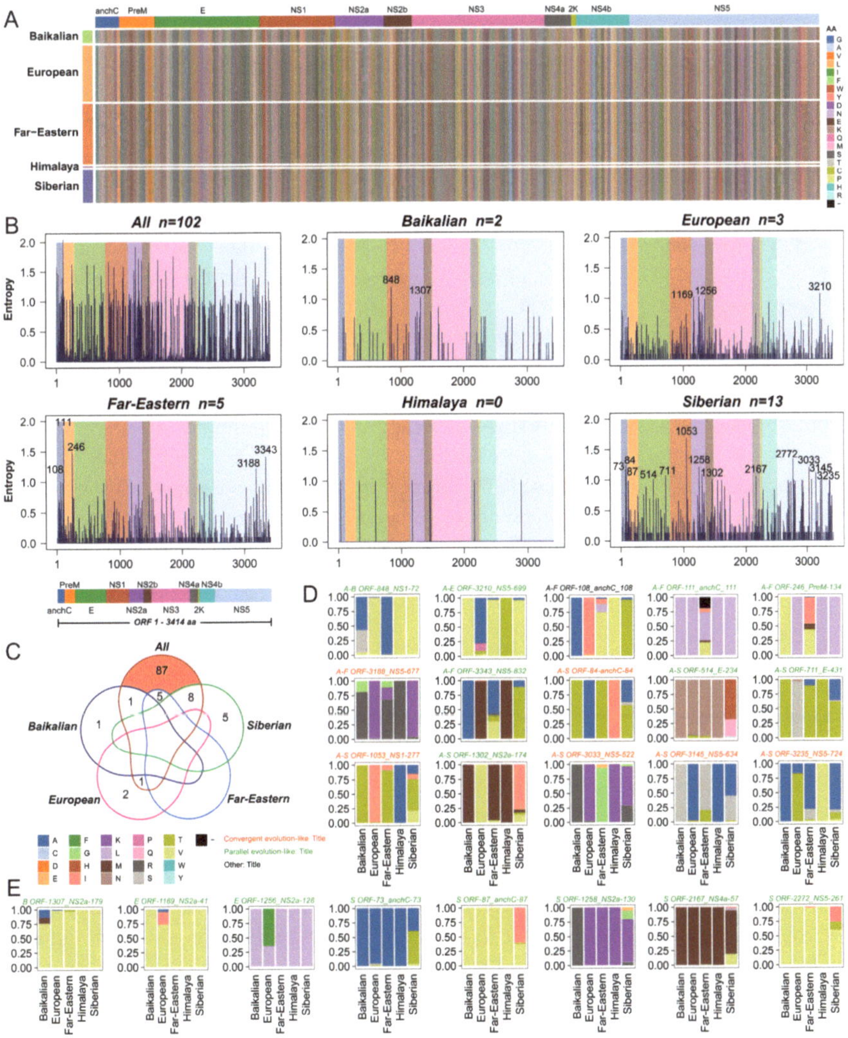

Figure 2. Mutation landscape and analysis of hypervariable sites across the five TBEV genotypes. (**A**) The mutation landscape of the analyzed sequences is displayed as a heatmap, where different

colors represent distinct nucleotides. (**B**) Shannon entropy was calculated for each amino acid position across different datasets, with entropy values greater than 1 identified as hypervariable sites, which are annotated accordingly. (**C**) The sharing of hypervariable sites across different datasets is visualized. Different lines represent different genotypes, and the intensity of the background color indicates the quantity. (**D**) Amino acid composition of shared hypervariable sites is illustrated using stacked bar charts, with colors representing different amino acids. (**E**) Amino acid composition of non-shared hypervariable sites is similarly shown as stacked bar charts. The evolutionary patterns of these sites are indicated: red denotes convergence-like evolution, green indicates parallel-like evolution, and black represents other types.

In the full dataset, highly variable sites were distributed across all protein regions except the 2K protein. However, the distribution of high-variability sites differed among individual genotypes. For the Baikalian genotype, high-variability sites were located in NS1 (n = 1) and NS2a (n = 1). For the European genotype, they were found in NS1 (n = 1), NS2a (n = 1), and NS5 (n = 1). The Far-Eastern genotype exhibited high-variability sites in anchC (n = 2), PreM (n = 1), and NS5 (n = 2). In the Siberian genotype, these sites were distributed across anchC (n = 3), E (n = 2), NS1 (n = 1), NS2a (n = 2), NS4a (n = 1), and NS5 (n = 4). Notably, high-variability sites were not shared across genotypes, although some overlap was observed between genotype-specific sites and those identified in the full dataset (Figure 2C–E).

The mutation patterns at shared sites revealed two main modes. The first is convergent evolution-like, exemplified by anchC-84 in the Siberian genotype. The primary mutations, 84T and 84A, were conserved in the Baikalian (T), Far-Eastern (T), and European (A) genotypes, while the minor mutation 84S was not observed in other genotypes. The second mode is parallel evolution-like, as observed in anchC-111 in the Far-Eastern genotype. While other genotypes exhibited strong conservation of L (>90%) at this site, the Far-Eastern genotype showed a more diverse pattern, including gap, V, I, and M mutations. A unique case was identified in the Far-Eastern genotype at anchC-108, a high-variability site where the primary amino acid V and the mutations A and I were found in other genotypes, but a unique mutation L was also present, indicating a complex mutation pattern.

Based on these observations, we identified six convergent evolution-like sites (one in the Far-Eastern genotype and five in the Siberian genotype) and sixteen parallel evolution-like sites (two in the Baikalian genotype, three in the European genotype, three in the Far-Eastern genotype, and eight in the Siberian genotype). These findings highlight diverse mutational patterns across TBEV genotypes, which may have implications for viral adaptation and evolution.

2.4. Genotype-Specific Amino Acid Mutations

This study also focused on genotype-specific amino acid mutations, which may play a critical role in the virological differences among TBEV genotypes. A total of 50, 314, 67, 450, and 322 genotype-specific mutation sites were identified in the Baikalian, European, Far-Eastern, Himalaya, and Siberian genotypes, respectively. These genotype-specific mutation sites were distributed across all 11 gene regions (Figure 3A), but the proportion of mutations within each region varied among genotypes (Figure 3B). In general, all genotypes exhibited relatively fewer mutation sites in the 2K protein region, though the proportion of mutations in this region was higher. Conversely, the NS5 protein region contained a larger number of mutation sites, but these did not represent the highest proportions. The high number of mutations complicated further analysis, as some genotype-specific mutations may occur randomly. Therefore, we focused on genotype-specific mutations with a prevalence greater than 60%, which can be considered as parallel evolution-like sites. After filtering, the number of genotype-specific mutation sites was reduced to 16, 54, 33, 63, and 22 for the Baikalian, European, Far-Eastern, Himalaya, and Siberian genotypes, respectively. Some mutation sites were shared among genotypes, such as anchC-3, NS1-285, and NS3-376 shared between the

European and Far-Eastern genotypes, NS3-37 and NS5-445 shared between the European and Himalaya genotypes, NS3-126 shared between the European and Siberian genotypes, NS2a-208 shared between the European and Siberian genotypes, and NS5-877 shared between the Baikalian and Far-Eastern genotypes (Figure 3C,D). Notably, shared mutation sites were also observed between genotypes with distant evolutionary relationships (Figures 1A and 3D), suggesting that these shared sites likely result from convergent or parallel evolution under similar selective pressures rather than direct genetic inheritance.

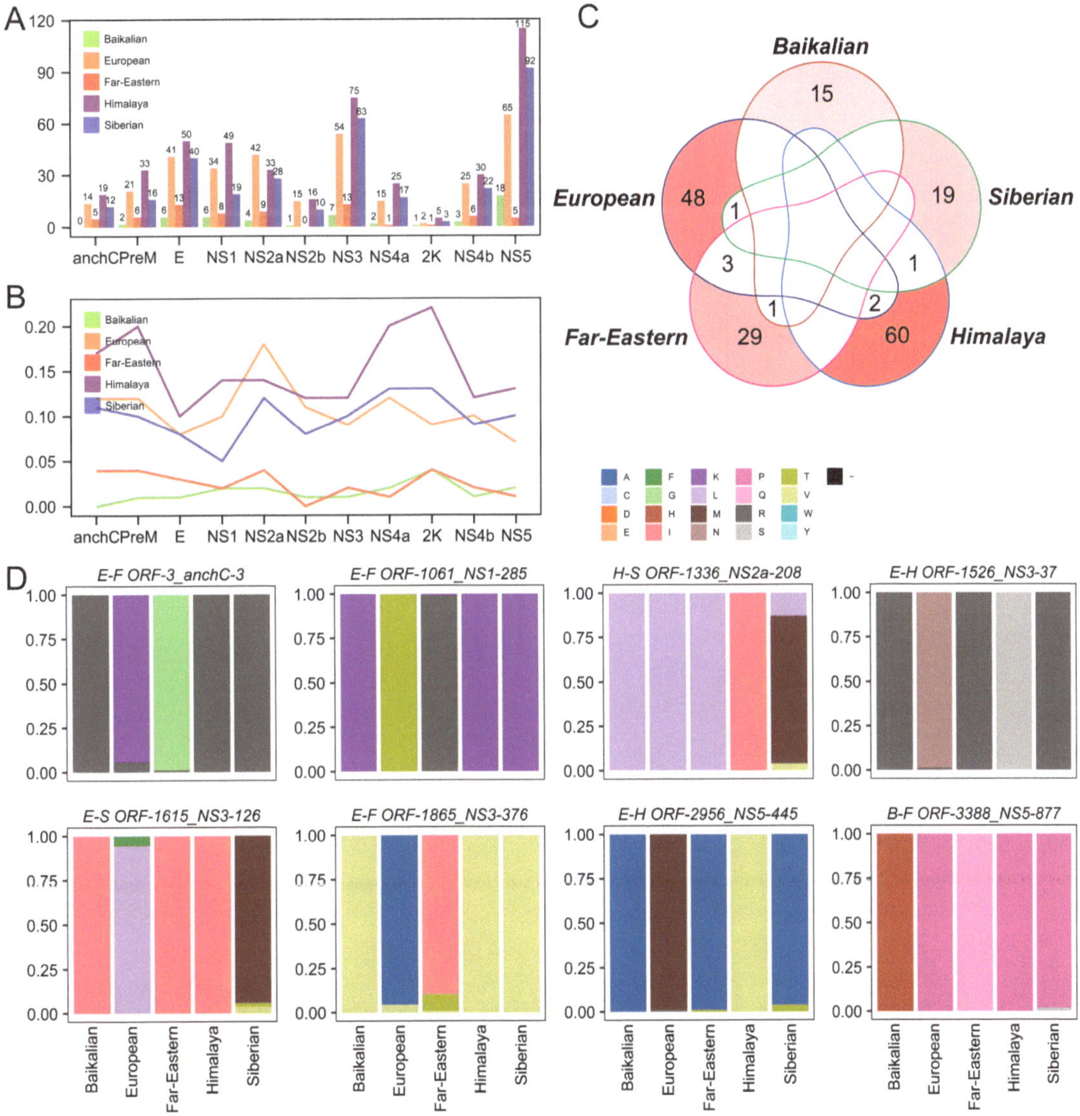

Figure 3. Analysis of genotype-specific amino acid mutations across the five TBEV genotypes. (**A**) The number of genotype-specific amino acid mutations in different protein-coding regions is displayed for each genotype. (**B**) The proportion of genotype-specific amino acid mutations relative to the length of the corresponding protein-coding region. (**C**) The sharing patterns of positions with genotype-specific amino acid mutations among the five genotypes. Different lines represent different genotypes, and

the intensity of the background color indicates the quantity. (**D**) The amino acid composition of shared mutation sites is illustrated.

2.5. Selective Pressure Analysis of Different TBEV Genotypes

Both information entropy analysis and genotype-specific amino acid mutations have identified numerous key amino acid sites, prompting further investigation into the evolutionary forces driving these mutations. Selection pressure analysis revealed the presence of positively selected sites at both the overall TBEV and genotype-specific levels, with consistent results across multiple algorithms (Figure 4A). To minimize overestimation, only sites identified as under positive selection by at least two algorithms were included in further analysis (Figure 4B). At the overall TBEV level, two positively selected sites were identified: NS1-175 and NS3-184. At the genotype level, six positively selected sites were detected in the European genotype (anchC-24, anchC-31, anchC-32, NS1-271, NS4a-55, and NS5-571), seven in the Far-Eastern genotype (anchC-111, E-463, NS2a-190, NS3-184, NS4b-24, NS5-832, and NS5-867), and four in the Siberian genotype (E-279, NS1-277, NS1-291, and NS5-522). No positively selected sites were identified in the Baikalian genotype, and due to limited sequence data, selection pressure analysis was not feasible for the Himalaya genotype.

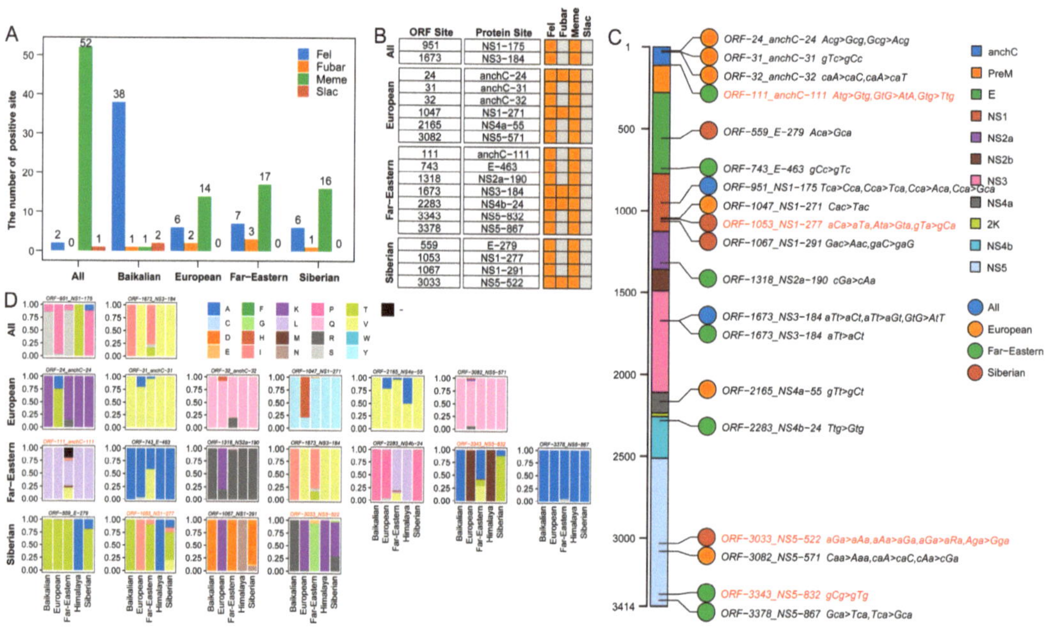

Figure 4. Selection pressure analysis of the five TBEV genotypes. (**A**) The number of positively selected sites identified across different datasets using various algorithms. (**B**) Positively selected sites confirmed by two or more algorithms. Orange indicates positive, while gray indicates negative. (**C**) Schematic representation of the positions of positively selected sites within genotypes, along with the major codon mutation patterns at these sites. (**D**) Amino acid composition of positively selected sites. The red annotations in (**C**,**D**) indicating sites that were also identified as highly variable sites.

Across the genome, positively selected sites were distributed across eight proteins: anchC (n = 4), E (n = 2), NS1 (n = 4), NS2a (n = 1), NS3 (n = 1), NS4a (n = 1), NS4b (n = 1), and NS5 (n = 1), with no positive selection detected in the PreM and 2K proteins (Figure 4C). Notably, NS3-184 was identified as a positively selected site at both the overall TBEV level and in the Far-Eastern genotype. Four sites were identified as both high-variability sites

(Figure 2B) and positively selected sites (Figure 4D): anchC-111 (ORF-111) in the Far-Eastern genotype, NS1-277 (ORF-1053) and NS5-522 (ORF-3033) in the Siberian genotype, and NS5-832 (ORF-3343) in the Far-Eastern genotype. However, none of the genotype-specific amino acid mutation sites were found to be under positive selection.

2.6. Mapping of Mutation Sites on Protein Structure

By incorporating all high-variability sites across genotypes (n = 23), shared genotype-specific amino acid mutation sites (n = 8), and all positively selected sites (n = 19), a comprehensive dataset of 37 key amino acid mutation sites associated with TBEV genotypes was compiled (with overlaps between methods). These sites are distributed across nine TBEV proteins, including anchC (n = 7), PreM (n = 1), E (n = 4), NS1 (n = 6), NS2a (n = 3), NS3 (n = 4), NS4a (n = 1), NS4b (n = 1), and NS5 (n = 10). Mapping these sites onto protein structures enabled further analysis of the potential impacts of mutations (Figure 5).

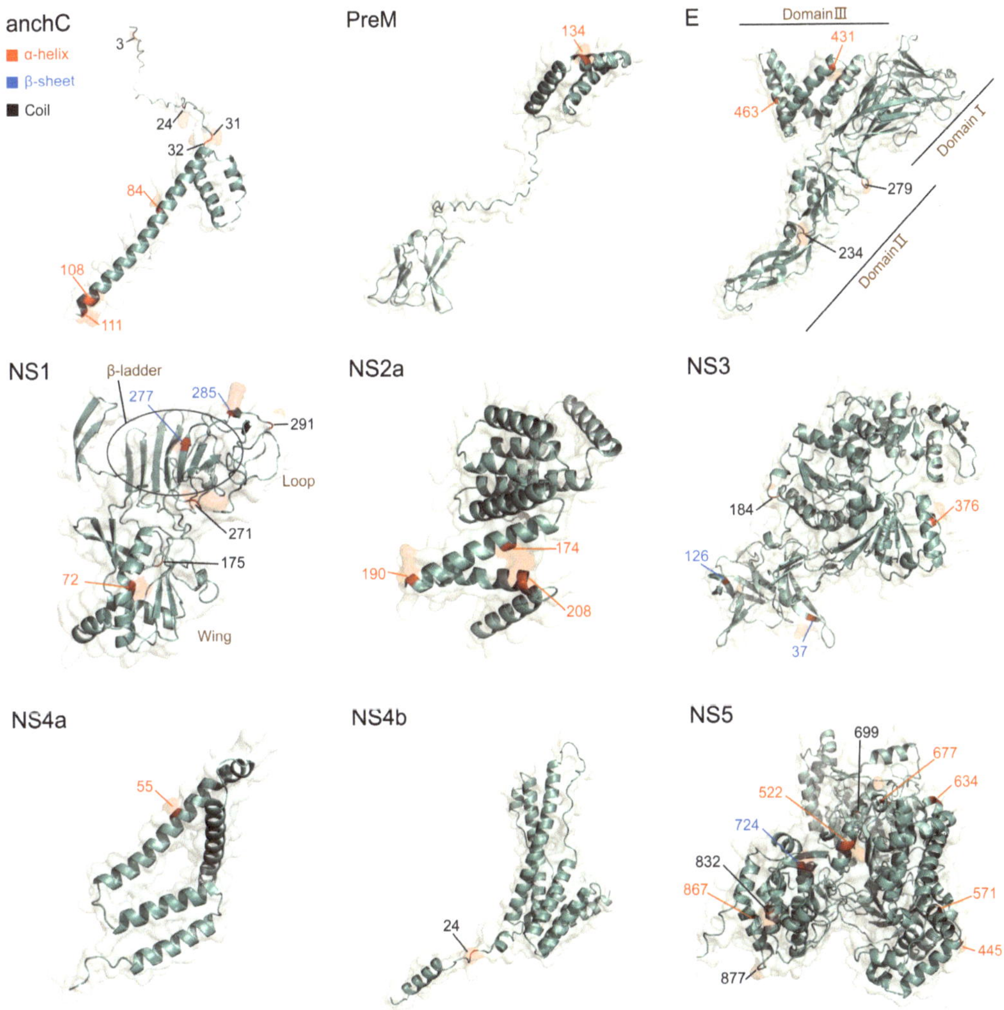

Figure 5. Structural mapping of selected key amino acid sites on the protein structure. The text labeling the key sites is annotated with secondary structure elements: red represents alpha helices,

blue indicates beta sheets, and black denotes random coils. Key amino acid sites are highlighted on the structure.

In terms of location, the key sites were predominantly found on the protein surfaces. Structurally, these sites were distributed across all three major secondary structures, with a higher number of sites located on α-helices (n = 18) and coils (n = 14), while fewer were associated with β-sheets (n = 5). In the structural proteins anchC, PreM, and E, all key sites were located on α-helices and coils, with no sites associated with β-sheets. On the E protein, two key sites (E-234 and E-279) were identified in the coil structure of Domain II, while two others (E-463 and E-431) were located on the α-helix structure of Domain III. No key sites were found in Domain I of the E protein.

Among the non-structural proteins, NS5 had the highest number of key mutation sites, which were distributed across all three major domains of this protein. In the extracellular protein NS1, the β-ladder region contained the key site NS1-277, the Wing region contained NS1-72, and the Loop region contained NS1-291. In NS3, the β-ladder region also harbored two key sites: NS3-37 and NS3-126.

Overall, the majority of the identified key sites are located within structural domains closely associated with protein function, highlighting their potential biological significance and the value of further investigation into their roles in viral evolution and pathogenesis.

3. Discussion

This study analyzed 263 TBEV genome sequences to investigate their genetic diversity and amino acid characteristics. The results revealed significant genetic and evolutionary differences among the five genotypes: Baikalian, European, Far-Eastern, Himalaya, and Siberian. High-variability amino acid sites were identified in four genotypes (excluding Himalaya), and genotype-specific mutations were observed in all five genotypes, with some sites shared between genotypes. Selection pressure analysis revealed positively selected sites at the overall TBEV level and in specific genotypes (European, Far-Eastern, and Siberian). By integrating high-variability sites, shared genotype-specific mutation sites, and positively selected sites, a total of 37 key amino acid sites were identified.

TBEV has evolved relatively independently. The divergence of TBEV genotypes occurred hundreds or even thousands of years ago, leading to the general consensus that the genotypes have evolved independently. However, a recent study suggests that the Baikalian genotype may have originated from a recombination event between the Siberian and Far-Eastern genotypes [22]. The data and methods used in this study did not identify sufficient evidence for inter-genotype recombination events (Supplementary Table S2), even under the condition that a single algorithm hit was sufficient for positive recombination detection. Additionally, the nucleotide similarity between the Baikalian genotype and the other four genotypes was below 95% across nearly all genomic regions. Therefore, while it is possible that the origin of the Baikalian genotype may have some potential historical relationship with the Siberian and Far-Eastern genotypes, Baikalian has maintained relatively independent evolution over the course of hundreds or even thousands of years, and this is even more evident for the other genotypes. However, potential recombination events occasionally occur within genotypes, particularly in the European, Far-Eastern, and Siberian genotypes, consistent with previous studies [23].

TBEV genotypes exhibit both unique and shared evolutionary characteristics. In this study, high-variability site analysis, genotype-specific amino acid mutation analysis, and selection pressure analysis were used to identify amino acid sites critical for each genotype. Some key sites showed variations specific to a single genotype, such as anchC-111 in the Far-Eastern genotype. Others were shared between certain genotypes, for instance, shared evolutionary sites were observed between the European and Far-Eastern,

European and Himalaya, European and Siberian, Himalaya and Siberian, and Baikalian and Far-Eastern genotypes during the genotype-specific amino acid mutation analysis. This indicates that while each genotype has unique mutational features, similar evolutionary events may occur at the same sites across genotypes, displaying convergent or parallel evolutionary characteristics. Convergent and parallel evolution events play a crucial role in understanding how different subtypes or evolutionary branches acquire distinct or similar virological features, which is essential for decoding a virus's adaptive strategies [24–26]. Strict convergent evolution refers to the acquisition of identical amino acid mutations at the same site across different genotypes or evolutionary branches, fitting a convergent evolution model, as previously reported in studies on SARS-CoV-2 [27]. In contrast, strict parallel evolution refers to the acquisition of different amino acid mutations at the same site across genotypes or branches, also conforming to a convergent model, as demonstrated in studies on highly pathogenic avian influenza [28]. However, the limited data availability for TBEV makes it challenging to conduct comprehensive analyses of strict convergent or parallel evolution. Therefore, this study adopted approaches resembling convergent and parallel evolution to define potential evolutionary events.

For example, anchC-111 in the Far-Eastern genotype, NS1-277 and NS5-522 in the Siberian genotype, and NS5-832 in the Far-Eastern genotype were identified as both high-variability sites and positively selected sites. Among these, NS1-277 and NS5-522 were classified as convergent evolution-like sites, while anchC-111 and NS5-832 were classified as parallel evolution-like sites. All genotype-specific amino acid mutation sites identified in this study were categorized as parallel evolution-like sites. These findings highlight that TBEV genotypes exhibit both unique and shared evolutionary characteristics, with evidence of potential convergent and parallel evolutionary processes shaping their diversity.

The key amino acid mutation sites identified in TBEV are closely associated with virological characteristics. The final dataset included 37 amino acid sites distributed across nine proteins, with most located on the protein surface and some positioned within α-helices or β-sheets, structures closely linked to protein tertiary structure. This underscores the importance of the selected sites. A literature review revealed no experimental studies directly targeting the 37 sites identified in this study. However, mutations at sites proximal to those identified here, or mutations in similar flaviviruses, have demonstrated biological significance.

For example, the D277A mutation in the E protein of TBEV has been linked to enhanced host receptor binding or cell fusion capabilities [29]. In this study, E-279, a nearby site, was identified as a key mutation, suggesting that similar experimental studies on E-279 may yield important insights. Likewise, the E-426 site in the E protein of TBEV has been associated with plaque size formation [30], potentially impacting viral virulence or growth characteristics. This study identified E-431, a nearby site, as a key mutation, further suggesting its potential biological relevance. In addition, mutations in the E protein, including D67G, T68A, E84K, F119V, E122G, A123K, N154L/Q, S158R, G159R, K171E, D181Y, E201K, D203G, D277A, D308K, T310K, K311E, G368R, Y384H, H390Y, T426I, D483E, and H496R, are also considered to influence the neuroinvasiveness and neurovirulence of TBEV [21]. Although none of our final selected sites overlap with these mutations, we have provided the amino acid composition at these sites for different TBEV genotypes in Supplementary Data S2, which may help to understand the occurrence of these mutations. Langat virus (LGTV), a tick-borne flavivirus closely related to TBEV, requires the NS5 region spanning amino acids 355–735 for its IFN inhibition function [31]. TBEV's NS5 protein has also been shown to inhibit IFN signaling [32]. Within this region, this study identified seven key amino acid sites (NS5-455, NS5-522, NS5-571, NS5-634, NS5-677, NS5-699, and NS5-724). Notably, NS5-522 was identified as a positively selected site in the Siberian genotype, while NS5-571 was a positively selected site in the European genotype. These findings suggest that these sites may be under selective pressure related to IFN inhibition, highlighting their potential importance in further experimental studies.

This study included all currently available TBEV sequences with detailed background information. However, the uneven distribution of sequences across genotypes and time points limited certain analyses, such as the identification of convergent and parallel evolution sites. Additionally, the potential key sites identified through bioinformatics approaches in this study still require experimental validation to confirm their functional relevance. Given that live TBEV research must be conducted in BSL-3 or higher biosafety laboratories, further experimental studies face certain practical challenges. Moreover, TBEV sequences derived from terminal or incidental hosts may harbor mutations associated with host-specific adaptive evolution, warranting further investigation in future studies. As Supplementary Data for this research, mutation information categorized by host, country, and temporal factors is available in Supplementary Data S2 for reference by the scientific community.

4. Materials and Methods

4.1. Dataset Collection and Preprocessing

The full name of the virus, tick-borne encephalitis virus, was used to retrieve its complete or near-complete genome sequences (greater than 90% of full length) from the NCBI database, with data up to 31 December 2023. A total of 263 sequences, which included information on collection date, geographic location, and host, were obtained and are provided in Supplementary Table S1. The ORF regions were extracted for analysis.

4.2. Phylogenetic Analysis

Multiple sequence alignment was performed using Mafft software (Version: 7.450, 23 August 2019) [33]. The most appropriate substitution model was determined using ModelFinder software (Version: 2.1.4) according to Bayesian information criterion (BIC) [34]. A maximum likelihood (ML) tree was generated using IQ-TREE software (Version: 2.1.4, 30 April 2021) with 1000 bootstrap replicates [35,36]. The resulting phylogenetic tree was visualized using Chiplot (www.chiplot.online, accessed on 1 December 2024) [37].

4.3. Recombination Analysis

Recombination detection was performed using RDP5 (Version: Beta 5.64), employing seven different detection models: RDP, GENECONV, Bootscan, Maxchi, Chimaera, SiSscan, and 3Seq [38]. A potential recombination event was considered if one model provided supporting evidence with a p-value threshold of $p < 0.01$. This loose criterion is aimed at identifying potential recombination events between genotypes as thoroughly as possible.

4.4. Shannon Entropy Calculation

The algorithm calculates the Shannon entropy for each position in a multiple sequence alignment (MSA), which quantifies the degree of variability at each nucleotide or amino acid position across all sequences. Entropy is computed by measuring the frequency distribution of characters at each position. In Equation (1), H represents the information entropy at a specific position in an amino acid sequence, quantifying the uncertainty or variability at that position. The term i denotes the index of each possible amino acid at the given position in the sequence, where i ranges from 1 to n, the total number of possible amino acids and gap. The term p_i is the probability of the i-th amino acid or gap occurring at that position, calculated based on the frequency of that amino acid or gap across multiple sequences. Higher entropy values indicate greater variability and diversity at a given position, suggesting that the site is less conserved, while lower entropy values indicate that the position is more conserved across sequences. The calculation of Shannon entropy was carried out using a custom Python (Version: 3.10) script developed for this purpose.

$$H = -\sum_{i=1}^{n} p_i log2(i) \quad (1)$$

4.5. Genotype-Specific Amino Acid Analysis

A specific genotype is first extracted from the dataset, and its amino acid sequence is aligned with those of other genotypes. The analysis focuses on identifying amino acid differences between the selected subtype and the other genotypes. A unique amino acid substitution at a given position in the selected genotype, absent in all other sequences at the same position, is considered a genotype-specific amino acid for that particular genotype.

4.6. Selective Pressure Analysis

Selection pressure on internal branches was assessed using the FEL, FUBAR, MEME, and SLAC algorithms implemented in HyPhy (Version: 2.5.48, 2 March 2023) [39]. The required tree file for the analyses was generated using IQ-TREE software (Version: 2.1.4, 30 April 2021) [35]. For the FEL, MEME, and SLAC algorithms, results were considered significant when $p < 0.1$; for the FUBAR algorithm, results were considered significant when the posterior probability was greater than 0.9. Only sites identified as positively selected by two or more algorithms were included in the final results.

4.7. Evolutionary Patterns Analysis

Due to the limited availability of TBEV sequences and their uneven distribution across the timeline, it is difficult to apply strict definitions for analyzing convergent or parallel evolution in TBEV. Therefore, a relaxed definition was adopted to identify convergent evolution-like or parallel evolution-like patterns. For mutations resembling parallel evolution, in simple terms, the mutation type at a specified highly variable site in a given genotype does not occur in other genotypes. For mutations resembling convergent evolution, in simple terms, the mutation type at a specified highly variable site in a given genotype corresponds to the predominant amino acid (>90%) at the same site in other genotypes.

4.8. Protein Structure Prediction

The consensus sequences of the TBEV dataset used in this study were generated using Geneious Prime software (Version: 2022.2.2, Biomatters), with the most frequent amino acids included in the consensus sequence. From the consensus sequences, the amino acid sequences for each protein were extracted. Subsequently, protein structures were predicted using the AlphaFold Server (https://alphafoldserver.com/, accessed on 8 December 2024). The predicted protein structures were then visualized and annotated using PyMOL software (Version: 3.0.3, 12 March 2024, Schrödinger, Inc., New York, NY, USA). The sequences used for structure prediction and the prediction results are provided in Supplementary Data S1.

4.9. Software

In the methods described above, several key software tools not previously mentioned were used for specific tasks. Sequence preprocessing, including information extraction, filtering, grouping, and mutation statistics, was performed using the SeqProcessor web application developed by our team (https://sp-app.streamlit.app/, accessed on 11 November 2024) [40]. Pairwise sequence identity was calculated using BioEdit (Version: 7.7), and the average distance between groups was computed using MEGA7 (Version: 7.0) [41]. The sliding window similarity analysis between different genotype groups was performed using SimPlot (Version: 3.5.1) [42]. For data visualization, ggplot2 v3.4.4 package in R v4.3.0 was used to create basic plots, while ComplexHeatmap v2.16.0 package [43] was utilized to generate heatmaps. ggVennDiagram v1.5.2 package was employed to create Venn diagrams, and trackViewer v1.36.2 package [44] was used to annotate important positions on the genome. The images were stitched together using Adobe Illustrator 2024 software, without any modification of the original data that could potentially affect the analysis or interpretation of the results.

5. Conclusions

This study applied three different methods to analyze mutations in 263 full-genome sequences of TBEV, constructing a mutation characteristic database for different TBEV genotypes and selecting 37 key amino acid sites. These results provide critical insights into the genetic diversity, evolutionary characteristics, and virological properties of different TBEV genotypes. They also offer important directions for future research on functional sites across TBEV genotypes.

Supplementary Materials: The supporting information can be downloaded at: https://www.mdpi.com/article/10.3390/ijms26030954/s1.

Author Contributions: Conceptualization, R.W.; Methodology, R.W.; Software, R.W.; Validation, Q.C.; Formal analysis, R.W.; Investigation, R.W. and F.L.; Resources, S.F. and H.L.; Data curation, A.G. and Q.Y.; Writing—original draft, R.W. and A.G.; Writing—review & editing, Q.M., H.L. and H.W.; Visualization, R.W.; Supervision, S.X. and H.W.; Funding acquisition, K.N., H.L. and H.W. All authors have read and agreed to the published version of the manuscript.

Funding: This research was funded by National Key R&D Program of China, grant number 2022YFC2602304 and 2022YFC2302700.

Institutional Review Board Statement: Not applicable.

Informed Consent Statement: Not applicable.

Data Availability Statement: All available data are provided in the manuscript and Supplementary Materials.

Conflicts of Interest: The authors declare no conflicts of interest.

References

1. Valarcher, J.F.; Hägglund, S.; Juremalm, M.; Blomqvist, G.; Renström, L.; Zohari, S.; Leijon, M.; Chirico, J. Tick-borne encephalitis. *Rev. Sci. et Tech. (Int. Off. Epizoot.)* **2015**, *34*, 453–466. [CrossRef] [PubMed]
2. Süss, J. Epidemiology and ecology of TBE relevant to the production of effective vaccines. *Vaccine* **2003**, *21* (Suppl. S1), S19–S35. [CrossRef]
3. Vaccines against tick-borne encephalitis: WHO position paper. *Relev. Epidemiol. Hebd.* **2011**, *86*, 241–256.
4. Wondim, M.A.; Czupryna, P.; Pancewicz, S.; Kruszewska, E.; Groth, M.; Moniuszko-Malinowska, A. Epidemiological Trends of Trans-Boundary Tick-Borne Encephalitis in Europe, 2000–2019. *Pathogens* **2022**, *11*, 704. [CrossRef]
5. Andaev, E.; Nikitin, A.; Yatsmenko, E.; Verigina, E.; Tolmacheva, M.; Ayugin, N.; Matveeva, V.; Balakhonov, S. Trends in epidemic process development of tick-borne encephalitis in the Russian Federation, laboratory diagnosis, prophylaxis and forecast for 2021. *Probl. Osob. Opasnykh Infektsii [Probl. Part. Danger. Infect.]* **2021**, *1*, 6–16. (In Russian) [CrossRef]
6. Chen, X.; Li, F.; Yin, Q.; Liu, W.; Fu, S.; He, Y.; Lei, W.; Xu, S.; Liang, G.; Wang, S.; et al. Epidemiology of tick-borne encephalitis in China, 2007–2018. *PLoS ONE* **2019**, *14*, e0226712. [CrossRef]
7. Michelitsch, A.; Wernike, K.; Klaus, C.; Dobler, G.; Beer, M. Exploring the Reservoir Hosts of Tick-Borne Encephalitis Virus. *Viruses* **2019**, *11*, 669. [CrossRef]
8. Wójcik-Fatla, A.; Krzowska-Firych, J.; Czajka, K.; Nozdryn-Płotnicka, J.; Sroka, J. The Consumption of Raw Goat Milk Resulted in TBE in Patients in Poland, 2022 "Case Report". *Pathogens* **2023**, *12*, 653. [CrossRef] [PubMed]
9. Dabas, R.; Sharma, N.; Taksande, A.B.; Prasad, R.; Munjewar, P.K.; Wanjari, M.B. Breast Milk: A Potential Route of Tick-Borne Encephalitis Virus Transmission from Mother to Infant. *Cureus* **2023**, *15*, e41590. [CrossRef]

10. Mlera, L.; Bloom, M.E. The Role of Mammalian Reservoir Hosts in Tick-Borne Flavivirus Biology. *Front. Cell Infect. Microbiol.* **2018**, *8*, 298. [CrossRef] [PubMed]
11. Simmonds, P.; Becher, P.; Bukh, J.; Gould, E.A.; Meyers, G.; Monath, T.; Muerhoff, S.; Pletnev, A.; Rico-Hesse, R.; Smith, D.B.; et al. ICTV Virus Taxonomy Profile: Flaviviridae. *J. Gen. Virol.* **2017**, *98*, 2–3. [CrossRef]
12. Pustijanac, E.; Buršić, M.; Talapko, J.; Škrlec, I.; Meštrović, T.; Lišnjić, D. Tick-Borne Encephalitis Virus: A Comprehensive Review of Transmission, Pathogenesis, Epidemiology, Clinical Manifestations, Diagnosis, and Prevention. *Microorganisms* **2023**, *11*, 1634. [CrossRef] [PubMed]
13. Ecker, M.; Allison, S.L.; Meixner, T.; Heinz, F.X. Sequence analysis and genetic classification of tick-borne encephalitis viruses from Europe and Asia. *J. Gen. Virol.* **1999**, *80 Pt 1*, 179–185. [CrossRef]
14. Kovalev, S.Y.; Mukhacheva, T.A. Reconsidering the classification of tick-borne encephalitis virus within the Siberian subtype gives new insights into its evolutionary history. *Infect. Genet. Evol. J. Mol. Epidemiol. Evol. Genet. Infect. Dis.* **2017**, *55*, 159–165. [CrossRef] [PubMed]
15. Dai, X.; Shang, G.; Lu, S.; Yang, J.; Xu, J. A new subtype of eastern tick-borne encephalitis virus discovered in Qinghai-Tibet Plateau, China. *Emerg. Microbes Infect.* **2018**, *7*, 74. [CrossRef]
16. Süss, J. Tick-borne encephalitis 2010: Epidemiology, risk areas, and virus strains in Europe and Asia-an overview. *Ticks Tick-Borne Dis.* **2011**, *2*, 2–15. [CrossRef] [PubMed]
17. Bondaryuk, A.N.; Kulakova, N.V.; Belykh, O.I.; Bukin, Y.S. Dates and Rates of Tick-Borne Encephalitis Virus-The Slowest Changing Tick-Borne Flavivirus. *Int. J. Mol. Sci.* **2023**, *24*, 2921. [CrossRef] [PubMed]
18. Helmová, R.; Hönig, V.; Tykalová, H.; Palus, M.; Bell-Sakyi, L.; Grubhoffer, L. Tick-Borne Encephalitis Virus Adaptation in Different Host Environments and Existence of Quasispecies. *Viruses* **2020**, *12*, 902. [CrossRef]
19. Růžek, D.; Gritsun, T.S.; Forrester, N.L.; Gould, E.A.; Kopecký, J.; Golovchenko, M.; Rudenko, N.; Grubhoffer, L. Mutations in the NS2B and NS3 genes affect mouse neuroinvasiveness of a Western European field strain of tick-borne encephalitis virus. *Virology* **2008**, *374*, 249–255. [CrossRef]
20. Lindqvist, R.; Rosendal, E.; Weber, E.; Asghar, N.; Schreier, S.; Lenman, A.; Johansson, M.; Dobler, G.; Bestehorn, M.; Kröger, A.; et al. The envelope protein of tick-borne encephalitis virus influences neuron entry, pathogenicity, and vaccine protection. *J. Neuroinflamm.* **2020**, *17*, 284. [CrossRef] [PubMed]
21. Kellman, E.M.; Offerdahl, D.K.; Melik, W.; Bloom, M.E. Viral Determinants of Virulence in Tick-Borne Flaviviruses. *Viruses* **2018**, *10*, 329. [CrossRef] [PubMed]
22. Sukhorukov, G.A.; Paramonov, A.I.; Lisak, O.V.; Kozlova, I.V.; Bazykin, G.A.; Neverov, A.D.; Karan, L.S. The Baikal subtype of tick-borne encephalitis virus is evident of recombination between Siberian and Far-Eastern subtypes. *PLoS Neglected Trop. Dis.* **2023**, *17*, e0011141. [CrossRef] [PubMed]
23. Norberg, P.; Roth, A.; Bergström, T. Genetic recombination of tick-borne flaviviruses among wild-type strains. *Virology* **2013**, *440*, 105–116. [CrossRef] [PubMed]
24. Meyerson, N.R.; Sawyer, S.L. Two-stepping through time: Mammals and viruses. *Trends Microbiol.* **2011**, *19*, 286–294. [CrossRef]
25. Gutierrez, B.; Escalera-Zamudio, M.; Pybus, O.G. Parallel molecular evolution and adaptation in viruses. *Curr. Opin. Virol.* **2019**, *34*, 90–96. [CrossRef] [PubMed]
26. Storz, J.F. Causes of molecular convergence and parallelism in protein evolution. *Nat. Rev. Genet.* **2016**, *17*, 239–250. [CrossRef]
27. Ito, J.; Suzuki, R.; Uriu, K.; Itakura, Y.; Zahradnik, J.; Kimura, K.T.; Deguchi, S.; Wang, L.; Lytras, S.; Tamura, T.; et al. Convergent evolution of SARS-CoV-2 Omicron subvariants leading to the emergence of BQ.1.1 variant. *Nat. Commun.* **2023**, *14*, 2671. [CrossRef] [PubMed]
28. Escalera-Zamudio, M.; Golden, M.; Gutiérrez, B.; Thézé, J.; Keown, J.R.; Carrique, L.; Bowden, T.A.; Pybus, O.G. Parallel evolution in the emergence of highly pathogenic avian influenza A viruses. *Nat. Commun.* **2020**, *11*, 5511. [CrossRef] [PubMed]
29. Khasnatinov, M.A.; Ustanikova, K.; Frolova, T.V.; Pogodina, V.V.; Bochkova, N.G.; Levina, L.S.; Slovak, M.; Kazimirova, M.; Labuda, M.; Klempa, B.; et al. Non-hemagglutinating flaviviruses: Molecular mechanisms for the emergence of new strains via adaptation to European ticks. *PLoS ONE* **2009**, *4*, e7295. [CrossRef] [PubMed]
30. Romanova, L.; Gmyl, A.P.; Dzhivanian, T.I.; Bakhmutov, D.V.; Lukashev, A.N.; Gmyl, L.V.; Rumyantsev, A.A.; Burenkova, L.A.; Lashkevich, V.A.; Karganova, G.G. Microevolution of tick-borne encephalitis virus in course of host alternation. *Virology* **2007**, *362*, 75–84. [CrossRef] [PubMed]
31. Park, G.S.; Morris, K.L.; Hallett, R.G.; Bloom, M.E.; Best, S.M. Identification of residues critical for the interferon antagonist function of Langat virus NS5 reveals a role for the RNA-dependent RNA polymerase domain. *J. Virol.* **2007**, *81*, 6936–6946. [CrossRef] [PubMed]
32. Werme, K.; Wigerius, M.; Johansson, M. Tick-borne encephalitis virus NS5 associates with membrane protein scribble and impairs interferon-stimulated JAK-STAT signalling. *Cell. Microbiol.* **2008**, *10*, 696–712. [CrossRef] [PubMed]
33. Katoh, K.; Standley, D.M. MAFFT multiple sequence alignment software version 7: Improvements in performance and usability. *Mol. Biol. Evol.* **2013**, *30*, 772–780. [CrossRef] [PubMed]

34. Kalyaanamoorthy, S.; Minh, B.Q.; Wong, T.K.F.; von Haeseler, A.; Jermiin, L.S. ModelFinder: Fast model selection for accurate phylogenetic estimates. *Nat. Methods* **2017**, *14*, 587–589. [CrossRef]
35. Nguyen, L.T.; Schmidt, H.A.; von Haeseler, A.; Minh, B.Q. IQ-TREE: A fast and effective stochastic algorithm for estimating maximum-likelihood phylogenies. *Mol. Biol. Evol.* **2015**, *32*, 268–274. [CrossRef] [PubMed]
36. Hoang, D.T.; Chernomor, O.; von Haeseler, A.; Minh, B.Q.; Vinh, L.S. UFBoot2: Improving the Ultrafast Bootstrap Approximation. *Mol. Biol. Evol.* **2018**, *35*, 518–522. [CrossRef] [PubMed]
37. Xie, J.; Chen, Y.; Cai, G.; Cai, R.; Hu, Z.; Wang, H. Tree Visualization By One Table (tvBOT): A web application for visualizing, modifying and annotating phylogenetic trees. *Nucleic Acids Res.* **2023**, *51*, W587–W592. [CrossRef]
38. Martin, D.P.; Varsani, A.; Roumagnac, P.; Botha, G.; Maslamoney, S.; Schwab, T.; Kelz, Z.; Kumar, V.; Murrell, B. RDP5: A computer program for analyzing recombination in, and removing signals of recombination from, nucleotide sequence datasets. *Virus Evol.* **2021**, *7*, veaa087. [CrossRef] [PubMed]
39. Kosakovsky Pond, S.L.; Poon, A.F.Y.; Velazquez, R.; Weaver, S.; Hepler, N.L.; Murrell, B.; Shank, S.D.; Magalis, B.R.; Bouvier, D.; Nekrutenko, A.; et al. HyPhy 2.5-A Customizable Platform for Evolutionary Hypothesis Testing Using Phylogenies. *Mol. Biol. Evol.* **2020**, *37*, 295–299. [CrossRef] [PubMed]
40. Wang, R.; Wang, H. SeqProcessor (0.1.0). Zenodo. 2024. Available online: https://zenodo.org/records/14545888 (accessed on 11 November 2024). [CrossRef]
41. Kumar, S.; Stecher, G.; Tamura, K. MEGA7: Molecular evolutionary genetics analysis version 7.0 for bigger datasets. *Mol. Biol. Evol.* **2016**, *33*, 1870–1874. [CrossRef]
42. Lole, K.S.; Bollinger, R.C.; Paranjape, R.S.; Gadkari, D.; Kulkarni, S.S.; Novak, N.G.; Ingersoll, R.; Sheppard, H.W.; Ray, S.C. Full-length human immunodeficiency virus type 1 genomes from subtype C-infected seroconverters in India, with evidence of intersubtype recombination. *J. Virol.* **1999**, *73*, 152–160. [CrossRef]
43. Gu, Z. Complex heatmap visualization. *iMeta* **2022**, *1*, e43. [CrossRef]
44. Ou, J.; Zhu, L.J. trackViewer: A Bioconductor package for interactive and integrative visualization of multi-omics data. *Nat. Methods* **2019**, *16*, 453–454. [CrossRef] [PubMed]

International Journal of
Molecular Sciences

MDPI

Article

The High Levels of Soluble Receptors for Tumor Necrosis Factor and Heart Injury in Children with the Pediatric Inflammatory Multisystem Syndrome Associated with Coronavirus Infection: Is This Just a Coincidence? A Proof-of-Concept Study

Maciej Marczak [1], Alicja Krejner-Bienias [1], Agnieszka Jasińska [1], Marek Kulus [1], Paweł Miklis [1], Katarzyna Grzela [1,*] and Tomasz Grzela [2,*]

[1] Department of Pediatric Pulmonology and Allergy, Medical University of Warsaw, 02-091 Warsaw, Poland
[2] Department of Histology and Embryology, Medical University of Warsaw, 02-004 Warsaw, Poland
* Correspondence: katarzyna.grzela@wum.edu.pl (K.G.); tomasz.grzela@wum.edu.pl (T.G.)

Abstract: (1) Pediatric inflammatory multisystem syndrome (PIMS) is a relatively rare complication of coronavirus disease (COVID-19). So far, it is unclear why COVID-19 in children has usually mild or asymptomatic courses, whereas PIMS, which develops several weeks after COVID-19, is a serious life-threatening condition. (2) In this proof-of-concept study, using the ELISA method, we compared selected clinical and immunological parameters in small groups of children with PIMS and COVID-19. Children with various inflammatory diseases were included as a control. (3) Patients with PIMS revealed significantly higher levels of pro-inflammatory molecules (C-reactive protein and IL-6) and markers of heart injury (troponin I and N-terminal prohormone of brain natriuretic peptide) as compared to other groups. Moreover, these markers correlated with increased levels of soluble receptors for tumor necrosis factor (sTNF-R1 and sTNF-R2). (4) Our observation may be a step forward to better understand the phenomenon of mild COVID-19 in children and its severe complications in PIMS. It is hypothesized that the delayed inflammation results in excessive cardiomyocyte damage and the release of sTNF-R1 and -R2. Therefore, possibly the involvement of the TNF pathway in PIMS could be explored as a potential therapeutic target. However, further studies are required to validate this approach.

Keywords: COVID-19; coronavirus disease; heart injury; NT-proBNP; PIMS; pediatric inflammatory multisystem syndrome; TNF; tumor necrosis factor; TNF-R; tumor necrosis factor-receptor

1. Introduction

Pediatric inflammatory multisystem syndrome (PIMS) associated with coronavirus infection is a systemic inflammatory disease, also known as multisystem inflammatory syndrome in children (MIS-C), which was described for the first time in May 2020 [1]. It is estimated that PIMS/MIS-C develops in 1 in 1000 children with coronavirus [2], and its development results from dysregulation of the immune system that occurs several—usually three to four and up to six—weeks after mild or asymptomatic coronavirus infection [3]. The diagnosis is based on the history of coronavirus disease (COVID-19) or virus exposure (confirmed with antibody tests), specific clinical symptoms (high fever, abdominal discomfort/pain, skin rash, conjunctivitis, fatigue), and elevated levels of inflammatory markers, similar to those reported in a "cytokine storm" in an adult's COVID-19, predominantly interleukin-6 (IL-6) [4].

IL-6 belongs to the most important regulatory factors in the immune system. It is produced mainly by monocytes and macrophages and has pleiotropic effects on different cell types. IL-6 initiates the immune response, stimulates B cells to produce immunoglobulins, and induces the differentiation of Th17 lymphocytes [5]. The results of numerous studies showed markedly elevated levels of IL-6, which is useful as an inflammatory marker in the diagnosis and assessment of disease severity [6–8]. Therefore, since IL-6 has previously been implicated in severe COVID-19, it may possibly also be involved in the inappropriate hyperactivation of immunity in PIMS/MIS-C. Of note, the biological agents that block the IL-6 pathway are currently considered as one of therapeutic options in PIMS/MIS-C [4].

There are several knowledge gaps existing in PIMS pathophysiology. First of all, it is unclear why the initial course of COVID-19 in children is rather mild, whereas its later phase, PIMS, is associated with serious life-threatening complications. The next unresolved question is the pathomechanism of the most serious PIMS complications predominantly affecting the cardiovascular system. They include acute myocarditis with left ventricular dysfunction, dilatation or aneurysm formation in coronary arteries, pericardial effusion, hypotension, and shock [9].

The aforementioned complications in PIMS may suggest the involvement of several molecules, which play a role in cardiovascular homeostasis but also reveal some immunoregulatory properties. Among possible candidates displaying such two-directional action is adenosine deaminase (ADA) [10]. ADA is an enzyme that irreversibly deaminates adenosine to inosine. Since adenosine reveals some anti-inflammatory activity, the maintenance of its concentrations and activity seems to be essential for the protection against endothelial dysfunction and vascular inflammation. Indeed, the upregulated ADA activity and decreased levels of adenosine were reported in several cardiovascular pathologies, including atherosclerosis, acute myocardial infarction, myocardial ischemia–reperfusion injury, hypertension, and thrombosis [11]. Moreover, due to the immunosuppressive action of adenosine against T cells, ADA is thought to play an important role in cell-mediated immunity by promoting the proliferation, differentiation, and maturation of lymphocytes [12]. An increase in ADA concentration and activity was observed in several inflammatory diseases, e.g., rheumatoid arthritis [10,13]. It is plausible that ADA might also be involved in COVID-19 and PIMS immunology.

Other molecules with such dual properties could be the soluble forms of receptors for tumor necrosis factor (sTNF-R1 and sTNF-R2). They both are the cleaved-off portions of surface TNF-Rs, which may act as decoys or neutralizing molecules against TNF, thus modulating the signaling pathways of this pro-inflammatory cytokine [14]. The high levels of soluble TNF receptors were reported in various conditions, including infections, autoimmune diseases, or malignancy [15]. Also, in adults with COVID-19, the levels of sTNF-R1 and sTNF-R2 correlated with disease severity [16], and the highest concentrations of both molecules were observed in patients admitted to the intensive care unit [17]. Recently it has been found that the serum levels of both soluble TNF receptors could also be used as a sensitive marker of heart injury, since they increase after myocardial infarction and, moreover, that increase correlates with infarct size and left ventricular end-diastolic volume [18]. Furthermore, high levels of sTNF-R1 and sTNF-R2 may be associated with mortality in patients with ST-segment elevation myocardial infarction (STEMI) [19]. One could hypothesize that soluble TNF receptors might also be engaged in the phenomenon of post-COVID-19 course toward PIMS in children, particularly contributing to the pathomechanism of serious cardiovascular complications in these patients. However, no such studies have been undertaken so far.

Therefore, the aim of our proof-of-concept study was to assess the serum levels of the well-known immunoregulatory cytokine IL-6 but also the aforementioned "bi-directional" molecules, among them ADA, as well as both soluble TNF receptors, and the aim was also to analyze their possible association with some clinical parameters in small group of children with COVID-19 and PIMS. A small heterogeneous group of patients with various non-COVID-19 inflammatory diseases served as a control.

2. Results

This proof-of-concept study involved group of 60 children, either with a confirmed diagnosis of COVID-19 (n = 28), PIMS (n = 14) or other non-COVID-19 inflammatory (INFLAM) diseases (n = 18). With non-equal variances and a very small size of patient groups, some of the tested data did not conform to a normal distribution. The short summary of demographic and clinical features of study groups is shown in Table 1.

Table 1. Characteristics of patient groups.

Parameter/Feature	COVID-19 (n = 28)	PIMS (n = 14)	INFLAM (n = 18)
Demographics			
Sex (female/male; n=) Age (mean/median; ±SD)	14/14 5.6/1.5; ±6.2	6/8 7.8/8.5; ±4.8	10/8 6.1/5.5; ±4.8
Age groups (n=) 0–2 years 3–7 years 8–18 years	 16 2 10	 2 4 8	 6 5 7
Clinical features			
Body temperature [°C] (mean/median; ±SD)	38.7/38.6; ±0.9	39.5/39.6; ±0.9	38.2/38.3; ±0.8
Tissue oxygenation (SpO_2) [%] (mean/median; ±SD)	93.6/96.5; ±6.5 *,#	97.0/98.0; ±3.3 *	97.8/98.0; ±3.0 #
Pulmonary injury (n=) Cardiac injury (n=) Kidney injury (n=)	7 1 * 0	3 11 *,# 0	3 1 # 0
Other symptoms/comments	gastric symptoms—5 tracheobronchitis—1 otitis—1 urin. tract infection—1	mediastinitis—1 gastric symptoms—14 skin rash—9 conjunctivitis—7	resp. tract infection—8 gastric symptoms—4 vasculitis—2 otitis—2 urin. tract infection—2 arthritis—1

*,#: differences between respective groups are statistically significant.

The patients from the COVID-19 group were relatively younger (mean age 5.6 ± 6.2 years) as compared to others (mean 7.8 ± 4.8 years for PIMS and 6.1 ± 4.8 years for INFLAM group), but this difference was non-significant. More than half of them were younger than 2 years, whereas nearly one-third of patients in this group were older than 8 years. Despite high fever (mean body temperature 38.7 ± 0.9 °C) and significantly decreased tissue oxygenation (mean saturation 93.6 ± 6.5%), all patients from the COVID-19 group presented a relatively mild course of disease. Besides mild-to-moderate gastric symptoms (mainly abdominal pain) in less than one-fifth of the patients from this group, one-fourth of them revealed mild pulmonary injury, as verified in chest computed tomography. Only one patient from this group displayed features of benign cardiomyositis with moderate arrhythmia and slightly increased fatigue.

In contrast to the aforementioned, patients from the PIMS group revealed a more severe course of disease. Apart from higher mean body temperature (39.5 ± 0.9 °C), all patients presented some gastric symptoms (abdominal pain with diarrhea, nausea, and/or vomiting) of various intensity, headache, and fatigue. Nearly one-fifth of PIMS patients displayed mild pulmonary injury confirmed in chest computed tomography. In two-thirds of the patients, some skin symptoms (mainly skin rash or erythema) were observed. Notably, more than three-fourths of the patients from this group presented moderate-to-severe symptoms of heart injury with arrhythmia, hypotension, chest pain, edema, and increased fatigue.

The third group, designated as a non-specific control, intentionally was highly heterogeneous and involved patients with various diseases of inflammatory background but with excluded coronavirus etiology. The mean body temperature in the INFLAM group (38.2 ± 0.8 °C) was lower as compared to other groups, although this difference was non-significant. Nearly half of the patients from the INFLAM group suffered from moderate-to-severe infection of the respiratory tract, mainly of viral etiology, except for three cases with moderate pulmonary injury, as verified in chest computed tomography. One-fourth of the patients from the INFLAM group revealed mild gastrointestinal symptoms, two patients presented moderate urinary tract infection, another two with acute otitis media, two with moderate cardiomyositis, and one patient with mild arthritis.

The comparison of selected parameters from routine blood tests did not show any significant differences in the mean numbers of total white blood cells (WBCs) and platelets (PLTs), as well as mean concentrations of hemoglobin (Hb) and creatinine (Table 2). The mean concentration of C-reactive protein (CRP) was significantly higher in the PIMS group compared to either COVID-19 or INFLAM patients (Table 2 and Figure 1A).

Table 2. The mean/median values (±SD; 95% CI) of selected laboratory data in study groups.

Parameter [Normal Range]	COVID-19 (n = 28)	PIMS (n = 14)	INFLAM (n = 18)
WBC [4.5–13 × $10^3/\mu L$]	9.2/9.2; ±4.2; 7.7–10.8	9.7/9.9; ±4.1; 7.5–11.8	11.1/8.6; ±6.6; 7.9–14.2
PLT [150–400 × $10^3/\mu L$]	289.9/291.5; ±82.1; 259.5–320.3	232.4/233.5; ±124.1; 167.4–297.4	277.2/236.0; ±111.3; 224.3–330.1
CRP [0–10 mg/L]	2.4/1.6; ±2.2; 1.6–3.2 *	15.9/15.5; ±10.2; 10.6–21.3 *,#	3.1/1.7; ±4.2; 1.0–5.1 #
Hb [11–14 g/dL]	12.2/12.2; ±1.9; 11.5–12.9	11.4/11.4; ±1.3; 10.7–12.1	11.8/12.4; ±2.1; 10.8–12.8
Creatinine [0.2–0.7 mg/dL]	0.4/0.3; ±0.2; 0.3–0.4	0.6/0.4; ±0.4; 0.4–0.8	0.3/0.3; ±0.1; 0.3–0.4
Troponin I [<19 µg/mL]	4.7/4.6; ±1.9; 2.8–6.6	47.5/43.3; ±38.2; 26.0–69.1 *	1.9/0.9; ±1.8; 0.3–3.5 *
NT-proBNP [<125 pg/mL]	98/39; ±95; 15–181 *	8307/4571; ±9721; 2807–13807 *,#	207/44; ±425; −132–549 #
D-dimer [0–550 ng/mL]	907/730; ±566; 453–1360 *	2572/2549; ±1777; 1606–3538 *	1431/1108; ±1262; 557–2305

Abbreviations: SD—standard deviation; 95% CI—confidence interval; WBC—white blood cell; PLT—platelet; CRP—C-reactive protein; Hb—hemoglobin; NT-proBNP—N-terminal prohormone of brain natriuretic peptide; *, #: differences between respective groups are statistically significant.

The assessment of laboratory markers of cardiovascular injury revealed significantly higher mean concentrations of troponin I (Figure 1B), N-terminal prohormone of brain natriuretic peptide—NT-proBNP (Figure 1C), and the product of fibrin degradation, D-dimer (Figure 1D), in the PIMS group as compared to COVID-19 or INFLAM patients. Interestingly, no significant differences in the mean levels of the abovementioned markers were observed between COVID-19 and INFLAM groups.

The analysis of selected immunological factors was shortly summarized in Table 3.

Table 3. The mean/median values (±SD; 95% CI) of selected immunological factors in study groups.

Parameter [Concentration]	COVID-19 (n = 28)	PIMS (n = 14)	INFLAM (n = 18)
ADA [pg/mL]	6.16/3.34; ±6.50; 3.76–8.56	8.07/9.02; ±5.98; 4.94–11.20	5.62/3.53; ±5.60; 3.05–8.19
IL-6 [pg/mL]	9.78/3.68; ±12.26; 5.24–14.32 *	20.12/14.65; ±16.63; 11.41–28.83 *	14.43/6.64; ±14.04; 7.94–20.92
TNF-R1 [ng/mL]	4.51/4.09; ±1.64; 3.90–5.12 *	8.37/7.69; ±5.47; 5.51–11.23 *	5.84/4.75; ±4.01; 3.99–7.69
TNF-R2 [ng/mL]	14.89/13.78; ±7.09; 12.26–17.52 *	33.46/29.19; ±21.78; 22.05–44.87 *	20.16/15.44; ±12.91; 14.20–26.12

Abbreviations: SD—standard deviation; 95% CI—confidence interval; ADA—adenosine deaminase; IL-6—interleukin-6; TNF-R—tumor necrosis factor receptor; *: differences between respective groups are statistically significant.

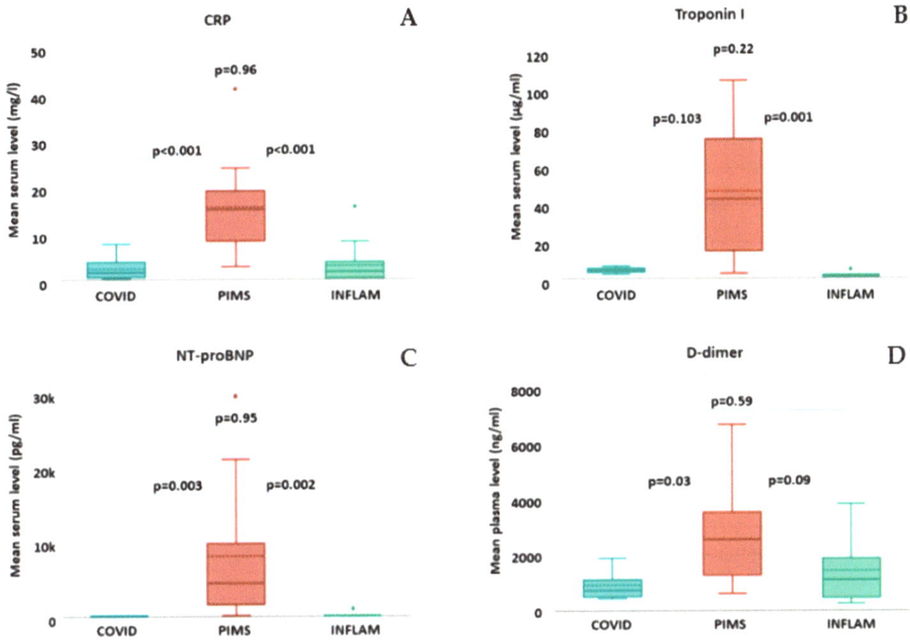

Figure 1. The concentrations of selected laboratory markers in study groups of COVID-19 (n = 28), PIMS (n = 14), and INFLAM (n = 18). The box plots represent the standard data structure; the boxes correspond to the interquartile range (with quartile Q1: lower edge; Q2 or median: solid line within the box; dashed line: mean; and Q3: upper edge,) and whiskers correspond to "non-outlier" minimum and maximum values. Each panel shows results for respective markers. (**A**) C-reactive protein; (**B**) troponin I; (**C**) N-terminal prohormone of brain natriuretic peptide; and (**D**) D-dimer. The differences between groups with $p < 0.05$ are considered statistically significant.

It was shown that patients from the PIMS group presented the highest mean and median values among all study groups for all tested molecules, although in the case of ADA, this difference did not reach statistical significance (Table 3, Figure 2A). The assessment of serum concentrations of IL-6 revealed that in all tested groups, the mean levels of this cytokine were above the normal limit (>1.9 pg/mL); In the PIMS group, the levels were the highest, in INFLAM, they were intermediate, whereas in COVID-19, they were the lowest. However, the difference appeared statistically significant only between the COVID-19 and PIMS groups (Table 3, Figure 2B).

The assessment of serum levels of both soluble TNF receptors revealed that in all groups, their mean levels were above the normal range (>0.5 ng/mL for sTNF-R1 and >2.5 ng/mL for sTNF-R2) [19]. Notably, in the PIMS group, the mean values were highest and significantly different from those in the COVID-19 group. However, although higher than the INFLAM group, they did not reach statistical significance in a comparison of PIMS vs. INFLAM. Also, the mean levels of sTNF-R1 and -R2 did not differ between the COVID-19 and INFLAM groups (Table 3, Figure 3).

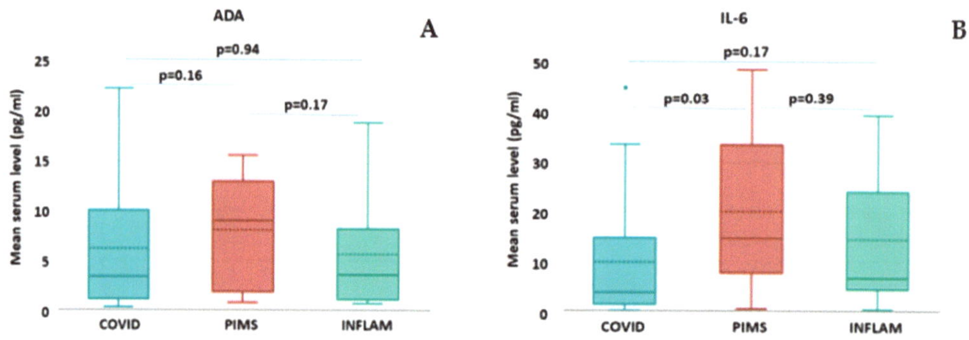

Figure 2. The concentrations of selected immunomodulators in study groups of COVID-19 (n = 28), PIMS (n = 14), and INFLAM (n = 18). The box plots represent the standard data structure; the boxes correspond to the interquartile range (with quartile Q1: lower edge; Q2 or median: solid line within the box; dashed line: mean; and Q3: upper edge,) and whiskers correspond to "non-outlier" minimum and maximum values. Each panel shows results for respective molecules, namely (**A**) adenosine deaminase; (**B**) IL-6. The differences between groups with $p < 0.05$ are considered statistically significant.

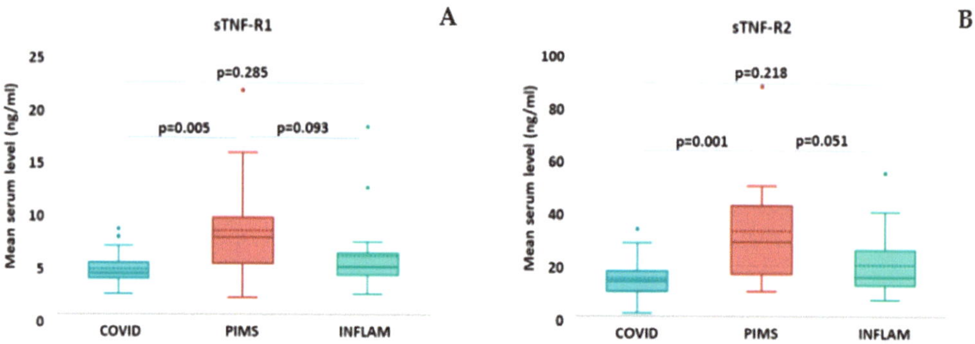

Figure 3. The concentrations of soluble TNF receptors in study groups of COVID-19 (n = 28), PIMS (n = 14), and INFLAM (n = 18). The box plots represent the standard data structure; the boxes correspond to the interquartile range (with quartile Q1: lower edge; Q2 or median: solid line within the box; dashed line: mean; and Q3: upper edge,) and whiskers correspond to "non-outlier" minimum and maximum values. Each panel shows results for respective molecules, namely (**A**) sTNF-R1; (**B**) sTNF-R2. The differences between groups with $p < 0.05$ are considered statistically significant.

The associations between the individual levels of selected immunomodulators were assessed in all patients. Very strong correlation (r = 0.9, at $p < 0.01$) was observed between serum concentrations of both soluble TNF receptors, as well as between IL-6 and either sTNF-R1 or sTNF-R2 (Table 4).

Table 4. The correlation between serum levels of selected immunoregulatory molecules.

Immunoregulator	ADA	IL-6	sTNF-R1	sTNF-R2
ADA	-	r = 0.11; p = 0.56	r = −0.01; p = 0.97	r = 0.02; p = 0.92
IL-6	r = 0.11; p = 0.56	-	r = 0.64; p < 0.01 *	r = 0.61; p < 0.01 *
sTNF-R1	r = −0.01; p = 0.97	r = 0.64; p < 0.01 *	-	r = 0.90; p < 0.01 *
sTNF-R2	r = 0.02; p = 0.92	r = 0.61; p < 0.01 *	r = 0.90; p < 0.01 *	-

*: correlation is statistically significant.

Furthermore, the individual serum levels of ADA, IL-6, sTNF-R1, and -R2 in all patients were also tested in regard to their possible association with molecular markers of cardiovascular injury—troponin I, NT-proBNP, and D-dimer. It was found that serum levels of troponin I revealed a statistically significant correlation with concentrations of both sTNF-R1 and sTNF-R2, whereas NT-proBNP was found to correlate only with levels of sTNF-R2. No other significant associations were observed either in the case of D-dimer, ADA, or IL-6 (Table 5).

Table 5. The correlation between molecular markers of cardiovascular injury and selected immunoregulatory molecules.

Cardiovascular Marker vs. Immunoregulator	Troponin I	NT-proBNP	D-dimer
ADA	r = 0.11; p = 0.62	r = 0.34; p = 0.17	r = 0.33; p = 0.09
IL-6	r = 0.16; p = 0.48	r = 0.20; p = 0.36	r = 0.27; p = 0.18
sTNF-R1	r = 0.45; p = 0.04 *	r = 0.29; p = 0.18	r = 0.22; p = 0.26
sTNF-R2	r = 0.49; p = 0.02 *	r = 0.40; p = 0.05 *	r = 0.17; p = 0.41

*: correlation is statistically significant.

3. Discussion

Our study has shown for the first time the significant increase in soluble receptors for TNF in patients with PIMS and that this increase significantly correlates with cardiovascular complications in these patients. However, it is unclear why these complications appear after several weeks ofCOVID-19, especially since the latter has usually a mild or even asymptomatic course. One can speculate that this is the consequence of an immature immune system in children, where the cytokine storm, typical for severe COVID-19 in adults, is relatively mild and develops rather as the prolonged inflammatory response. This assumption could be supported by the observations of increased levels of pro-inflammatory cytokines, mainly IL-6, but also TNF [6,20], which although lower when compared to those in adults' COVID-19, are still elevated above the normal levels. Together with the outstanding vascular tropism of coronavirus via angiotensin-converting enzyme (ACE)-2 and, possibly, dipeptydyl peptidase (DPP)-4 [21,22], they may promote chronic inflammation in the vascular system and, finally, lead to the exacerbation and development of severe cardiovascular complications. One cannot exclude that the increased levels of ADA, which at least in our study were similar in the COVID-19 and PIMS groups, could also be involved in that process.

The most intriguing finding is the increased secretion of soluble receptors for TNF. The first possible explanation is based on canonical cellular response to the stimulus. The cell exposure to TNF may lead to increased expression of membrane receptors for this cytokine, which in turn is followed by the increased shedding of their soluble forms [23]. The latter are considered as the natural modulators of the response to TNF.

Also, it is plausible that the extensive damage of cardiomyocytes contributes to increased levels of soluble TNF receptors in circulation. Presumably, this scenario may result from either the direct action of various cardiotoxic agents or indirect mechanisms, including coronavirus-mediated endothelial injury of coronary arteries with subsequent myocardial ischemia [22]. Therefore, it may even be independent from high levels of TNF. However, further studies are needed to confirm the aforementioned hypotheses.

It is noteworthy that the exact role of TNF and its receptors in various cardiovascular complications remains unclear. Surprisingly, several studies have shown that in some patients with systemic inflammatory conditions, the therapeutic inhibition of TNF by a monoclonal antibody (e.g., infliximab) led to the severe worsening of heart function [23,24]. The observed phenomenon may be explained by the dual action of the TNF molecule that results from the substantially different function of its both receptors.

Accordingly, TNF-R1 is constitutively expressed on the surface of most cells. It belongs to the family of death receptors and its triggering by TNF leads to the formation of a death-inducing signaling complex, the activation of the caspase pathway, and the induction of programmed cell death. However, in the presence of some specific adaptor proteins, including TNF receptor associated factor 2 (TRAF2), the inhibitor of apoptosis (IAP), and receptor-interacting protein (RIP), the alternative pathways involving c-Jun N-terminal kinase (JNK) and nuclear factor (NF)-κB could be activated, leading to cell survival and the production of pro-inflammatory cytokines [23].

In contrast to the aforementioned, the expression of TNF-R2 is restricted mainly to immune cells but also endothelial cells and cardiomyocytes, and it is induced upon stimulation with pro-inflammatory factors. Binding of TNF to TNF-R2 activates the canonical and non-canonical signaling pathways with NF-κB and phosphatidylinositol-3-kinase (PI3K), which promote cell proliferation and survival [23]. Furthermore, TNF-R2, also in its soluble form, was identified as the molecule involved in a unique mechanism called reverse signaling. In this mechanism, TNF-R2 binds to the membrane-bound TNF and indirectly induces mitogen-activated protein kinase (MAPK) pathways, thus resulting in an immunosuppressive effect on TNF-expressing inflammatory cells, mainly T lymphocytes and monocytes [23,25].

Such complexity of TNF signaling may suggest that the molecular targeting of the TNF-dependent pathways could offer some therapeutic potential for patients with PIMS. Notably, the TNF-focused strategies may selectively block the pathogenic mechanisms only [23]; however, future research should clarify the safety and specificity of such approaches.

There are several limitations of our study. The first and main limitation is its proof-of-concept nature with the single time point of analysis, which certainly will require further research. The next issue is the very small number of patients involved. This was due to the low occurrence of symptomatic COVID-19 in children and outstanding rarity of PIMS (0.1% of children with coronavirus) [2]. Obviously, the small study groups significantly reduced the observation sensitivity and allowed for the detection of only strong effects or differences. Hence, although no significant differences were observed in some comparisons, the small sample size may limit the ability to detect such subtle variations.

Also, the collected data did not reveal a normal distribution, thus increasing the risk of result distortion and unintended bias in their interpretation. Nevertheless, even despite aforementioned restrictions, we still were able to observe statistically significant differences between study groups in some parameters.

Another weak point of this report is the size and composition of the INFLAM group. It was very small and highly heterogeneous, actually involving patients sharing two common features, namely (1) the presence of inflammatory disease, either infectious or non-infectious, and (2) excluded COVID-19 etiology of inflammation. Hence, the INFLAM group was intended to be a kind of non-COVID-19 control that could serve as a comparator to the COVID-19 and PIMS groups. Nevertheless, one has to keep in mind that with the small size and high heterogeneity, the real scientific value of that group is rather anecdotal and should be considered as a form of screening to indicate some directions for future research. Accordingly, when considering our findings concerning soluble TNF receptors and heart injury in PIMS, the most interesting selection for further more detailed comparative studies may be the group of patients with some cardiovascular diseases, e.g., Kawasaki disease.

4. Materials and Methods

Thisstudy involved three small groups of children, (1) one with active coronavirus disease (COVID-19), (2) with post-COVID-19 pediatric inflammatory multisystem syndrome (PIMS), and (3) with various non-COVID-19 inflammatory (INFLAM) diseases. All study participants were recruited among children referred to the Department of Pediatrics, Allergology and Pulmonology at the University Hospital of the Medical University of Warsaw between December 2020 and December 2021 during the second-to-fourth wave of theCOVID-19 pandemic.

All procedures involving patients were performed according to the Declaration of Helsinki. Whenever possible, due to patients' age, children and/or all their parents or legal guardians gave written informed consent to participate in the study, which was approved by the Ethics Committee at the Medical University of Warsaw (approval no. KB/212/2020, dated 14 December 2020).

The patients were qualified for the COVID-19 group based on the suspected or already diagnosed infection with coronavirus (SARS-CoV-2, variants alpha or delta). The presence of typical symptoms and the COVID-19 diagnosis were verified by polymerase chain reaction and specific IgM/IgG antibody testing.

The children recruited to the PIMS group conformed to specified inclusion criteria as follows: COVID-19 or asymptomatic infection (confirmed by specific antibody testing) within the last 6 weeks, continuous fever above 38.5 °C for at least 3 days, and the occurrence of clinical symptoms involving at least two systems, namely gastrointestinal tract (usually with abdominal pain, diarrhea, nausea and/or vomiting), respiratory system (with cough, chest pain, and/or dyspnea), cardiovascular system (chest pain, arrhythmia, hypotension, faint, edema), or skin (erythema, rash, itching) issues.

The non-COVID-19 inflammatory group involved children with various diseases of either infectious (bacterial or viral, except those caused by coronavirus) or non-infectious origin. The main inclusion criteria concerned typical symptoms of inflammation, e.g., increased body temperature, pain, edema, and some other disease-specific symptoms, as well as elevated laboratory markers of inflammation.

All children were subjected to standard clinical examination. Apart from a routine laboratory assessment, including blood morphology, basic biochemistry, and coagulology, additional blood tests were focused on the measurement of serum levels of inflammation markerC-reactive protein, D-dimer, and two molecular indicators of heart injury (troponin I and N-terminal prohormone of brain natriuretic peptide–NT-proBNP). Furthermore, 2 mL of serum sample was kept frozen at −80 °C until being used for immunological tests.

The concentrations of IL-6, ADA, and both soluble TNF receptors—sTNF-R1 and -R2—in serum samples were measured in duplicate, using respective enzyme-linked immunosorbent assay (ELISA) tests according to detailed protocols provided by the manufacturer. Human ADA ELISA Kit was purchased from Invitrogen/Life Technologies Co. (Carlsbad, CA, USA). Human IL-6 High Sensitivity ELISA was from Invitrogen/Bender MedSystems GmbH (Vienna, Austria), whereas both Human sTNF-RI and sTNF-RII ELISA Kits were purchased from Invitrogen/Life Technologies Co. (Frederic, MD, USA).

The statistical analysis was performed using DATA-tab software (2024) (https://datatab.net; DATAtab e.U., Graz, Austria) and MedCalc Software 18.11v (https://medcalc.org; MedCalc Software Ltd., Ostend, Belgium), for both URL accessed between October–November 2024.

The Shapiro–Wilk test was used to assess the distribution of analyzed variables. The basic demographic and clinical features of patients within study groups, namely the age, body temperature, tissue oxygenation, and selected laboratory parameters, were analyzed using descriptive statistics involving the calculation of the arithmetic mean, median, standard deviations, confidence intervals, and quartiles. Then, they were compared between groups using the non-parametric Kruskal–Wallis test. The occurrence of main symptoms, due to the small sample size, was compared between groups using Fisher's exact test. The observed differences were considered statistically significant at $p < 0.05$.

The correlations between selected parameters, such as heart injury markers (D-dimer, troponin I, and NT-proBNP) and immunoregulatory molecules (ADA, IL-6, and sTNF-R1 and -R2), as well as cross-talk between the latter, were analyzed using Pearson's correlation test with a significance level $p < 0.05$.

5. Conclusions

Our proof-of-concept observation may be a step forward to better understand the phenomenon of mild COVID-19 in children and the pathomechanism of its severe cardiovascular complications in PIMS. We hypothesize that, possibly, the delayed immune response to coronavirus infection may result in the activation of the TNF pathway in the heart, the overexpression of TNF receptors, and the shedding of their soluble forms (sTNF-R1 and -R2) into circulation. Alternatively, inflammatory reactions or cardiotoxic effects of the coronavirus itself may result in excessive damage of cardiomyocytes and the release of soluble TNF receptors. However, both scenarios require validation in future research.

As shown in previous studies, the increased levels of sTNF-R1 and -R2 are indicators of poor prognostics; nevertheless, their exact role and the mechanism of that increase remain unclear. Our findings suggest that the TNF pathway may be explored as a potential therapeutic target to prevent PIMS complications in children and, presumably, in severe COVID-19 in adults. However, due to high complexity of TNF-related signaling, the molecular targeting should be highly selective and block the pathogenic mechanism only. Therefore, this issue still requires further studies.

Author Contributions: Conceptualization, M.M., K.G., and T.G.; methodology, K.G., T.G. and A.J.; validation, A.K.-B., A.J. and T.G.; formal analysis, T.G.; investigation, M.M. and K.G.; resources, M.M. and P.M.; data curation, M.M., A.J., A.K.-B., and P.M.; writing—original draft preparation, M.M., A.K.-B., K.G. and T.G.; writing—review and editing, T.G.; visualization, T.G.; supervision, K.G., T.G. and M.K.; project administration, K.G.; funding acquisition, M.K. All authors have read and agreed to the published version of the manuscript.

Funding: This research received no external funding.

Institutional Review Board Statement: The study was conducted in accordance with the Declaration of Helsinki and was approved by the Ethics Committee at the Medical University of Warsaw (protocol code KB/212/2020, approval date 14 December 2020).

Informed Consent Statement: Whenever possible due to patients' age, the written informed consent was obtained from children involved in the study and from all their parents or legal guardians.

Data Availability Statement: Due to law restrictions aimed to protect patient confidentiality, the original data are publicly not available; however, the data presented in this report are available from the corresponding author upon request.

Conflicts of Interest: The authors declare no conflicts of interest.

References

1. Royal College of Paediatrics and Child Health. Guidance: Paediatric Multisystem Inflammatory Syndrome Temporally Associated with COVID-19. 2020. Available online: https://www.rcpch.ac.uk/sites/default/files/2020-05/COVID-19-Paediatric-multisystem-%20inflammatory%20syndrome-20200501.pdf (accessed on 18 November 2024).
2. Godfred-Cato, S.; Bryant, B.; Leung, J.; Oster, M.E.; Conklin, L.; Abrams, J.; Roguski, K.; Wallace, B.; Prezzato, E.; Koumans, E.H.; et al. COVID-19-Associated Multisystem Inflammatory Syndrome in Children—United States, March–July 2020. *MMWR Morb. Mortal. Wkly. Rep.* **2020**, *69*, 1074–1080, Erratum in *MMWR Morb. Mortal. Wkly. Rep.* **2020**, *69*, 1229. [CrossRef]
3. Schvartz, A.; Belot, A.; Kone-Paut, I. Pediatric Inflammatory Multisystem Syndrome and Rheumatic Diseases DuringSARS-CoV-2 Pandemic. *Front. Pediatr.* **2020**, *8*, 605807. [CrossRef]

4. Henderson, L.A.; Canna, S.W.; Friedman, K.G.; Gorelik, M.; Lapidus, S.K.; Bassiri, H.; Behrens, E.M.; Kernan, K.F.; Schulert, G.S.; Seo, P.; et al. American College of Rheumatology Clinical Guidance for Multisystem Inflammatory Syndrome in Children Associated With SARS-CoV-2 and Hyperinflammation in Pediatric COVID-19: Version 3. *Arthritis Rheumatol.* **2022**, *74*, e1–e20. [CrossRef]
5. Kang, S.; Narazaki, M.; Metwally, H.; Kishimoto, T. Historical overview of the interleukin-6 family cytokine. *J. Exp. Med.* **2020**, *217*, e20190347, Erratum in *J. Exp. Med.* **2020**, *217*, jem.2019034704212020c. [CrossRef] [PubMed]
6. Lee, P.Y.; Day-Lewis, M.; Henderson, L.A.; Friedman, K.G.; Lo, J.; Roberts, J.E.; Lo, M.S.; Platt, C.D.; Chou, J.; Hoyt, K.J.; et al. Distinct clinical and immunological features of SARS-CoV-2-induced multisystem inflammatory syndrome in children. *J. Clin. Investig.* **2020**, *130*, 5942–5950. [CrossRef]
7. Gruber, C.N.; Patel, R.S.; Trachtman, R.; Lepow, L.; Amanat, F.; Krammer, F.; Wilson, K.M.; Onel, K.; Geanon, D.; Tuballes, K.; et al. Mapping systemic inflammation and antibody responses in Multisystem Inflammatory Syndrome in Children (MIS-C). *Cell* **2020**, *183*, 982–995. [CrossRef]
8. Diorio, C.; Henrickson, S.E.; Vella, L.A.; McNerney, K.O.; Chase, J.; Burudpakdee, C.; Lee, J.H.; Jasen, C.; Balamuth, F.; Barrett, D.M.; et al. Multisystem inflammatory syndrome in children and COVID-19 are distinct presentations of SARS-CoV-2. *J. Clin. Investig.* **2020**, *130*, 5967–5975. [CrossRef] [PubMed]
9. Whittaker, E.; Bamford, A.; Kenny, J.; Kaforou, M.; Jones, C.E.; Shah, P.; Ramnarayan, P.; Fraisse, A.; Miller, O.; Davies, P.; et al. PIMS-TS Study Group and EUCLIDS and PERFORM Consortia. Clinical Characteristics of 58 Children With a Pediatric Inflammatory Multisystem Syndrome Temporally Associated with SARS-CoV-2. *JAMA* **2020**, *324*, 259–269. [CrossRef]
10. Antonioli, L.; Colucci, R.; La Motta, C.; Tuccori, M.; Awwad, O.; Da Settimo, F.; Blandizzi, C.; Fornai, M. Adenosine deaminase in the modulation of immune system and its potential as a novel target for treatment of inflammatory disorders. *Curr. Drug Targets* **2012**, *13*, 842–862. [CrossRef] [PubMed]
11. Kutryb-Zajac, B.; Mierzejewska, P.; Slominska, E.M.; Smolenski, R.T. Therapeutic Perspectives of Adenosine Deaminase Inhibition in Cardiovascular Diseases. *Molecules* **2020**, *25*, 4652. [CrossRef]
12. Cortés, A.; Gracia, E.; Moreno, E.; Mallol, J.; Lluís, C.; Canela, E.I.; Casadó, V. Moonlighting adenosine deaminase: A target protein for drug development. *Med. Res. Rev.* **2015**, *35*, 85–125. [CrossRef] [PubMed]
13. Vinapamula, K.S.; Pemmaraju, S.V.; Bhattaram, S.K.; Bitla, A.R.; Manohar, S.M. Serum Adenosine Deaminase as Inflammatory Marker in Rheumatoid Arthritis. *J. Clin. Diagn. Res.* **2015**, *9*, BC08–BC10. [CrossRef]
14. Zoccali, C.; Tripepi, G.; Stel, V.; Fu, E.L.; Mallamaci, F.; Dekker, F.; Jager, K.J. Decoy receptors as biomarkers for exploring aetiology and designing new therapies. *Clin. Kidney J.* **2024**, *17*, sfae222. [CrossRef]
15. Diez-Ruiz, A.; Tilz, G.P.; Zangerle, R.; Baier-Bitterlich, G.; Wachter, H.; Fuchs, D. Soluble receptors for tumour necrosis factor in clinical laboratory diagnosis. *Eur. J. Haematol.* **1995**, *54*, 1–8. [CrossRef]
16. Aciole, M.R.; Gonçales, J.P.; Neves, P.A.F.; Soares, C.R.P.; de Oliveira, M.I.; de Melo, H.R.L.; Neto, R.G.d.L.; Moura, L.C.R.V.; Araújo, P.S.R.; de Lorena, V.M.B. Levels of soluble TNF receptors (sTNFR1 and sTNFR2) increase with clinical worsening of patients and are related to COVID-19 mortality. *Immunobiology* **2024**, *229*, 152748. [CrossRef]
17. Sancho Ferrando, E.; Hanslin, K.; Hultström, M.; Larsson, A.; Frithiof, R.; Lipcsey, M. Uppsala Intensive Care COVID-19 Research Group. Soluble TNF receptors predict acute kidney injury and mortality in critically ill COVID-19 patients: A prospective observational study. *Cytokine* **2022**, *149*, 155727. [CrossRef] [PubMed]
18. Paccalet, A.; Crola Da Silva, C.; Mechtouff, L.; Amaz, C.; Varillon, Y.; de Bourguignon, C.; Cartier, R.; Prieur, C.; Tomasevic, D.; Genot, N.; et al. Serum Soluble Tumor Necrosis Factor Receptors 1 and 2 Are Early Prognosis Markers After ST-Segment Elevation Myocardial Infarction. *Front. Pharmacol.* **2021**, *12*, 656928. [CrossRef]
19. Befekadu, R.; Grenegård, M.; Larsson, A.; Christensen, K.; Ramström, S. Levels of soluble tumor necrosis factor receptor 1 and 2 are associated with survival after ST segment elevation myocardial infarction. *Sci. Rep.* **2022**, *12*, 14762. [CrossRef]
20. Calcaterra, V.; Loretelli, C.; Biganzoli, D.; Abdelsalam, A.; Marano, G.; Carelli, S.; Fiori, L.; Mannarino, S.; D'auria, E.; Verduci, E.; et al. Long-term cytokine profile in multisystem inflammatory disease among children. *Cytokine* **2024**, *183*, 156744. [CrossRef] [PubMed]
21. Krejner-Bienias, A.; Grzela, K.; Grzela, T. DPP4 Inhibitors and COVID-19—Holy Grail or Another Dead End? *Arch. Immunol. Ther. Exp.* **2021**, *69*, 1. [CrossRef]
22. Lui, K.O.; Ma, Z.; Dimmeler, S. SARS-CoV-2 induced vascular endothelial dysfunction: Direct or indirect effects? *Cardiovasc. Res.* **2024**, *120*, 34–43. [CrossRef]
23. Rolski, F.; Błyszczuk, P. Complexity of TNF-α Signaling in Heart Disease. *J. Clin. Med.* **2020**, *9*, 3267. [CrossRef]

24. Grillo, T.G.; Silveira, C.F.d.S.M.P.; Quaglio, A.E.V.; Dutra, R.M.; Baima, J.P.; Bazan, S.G.Z.; Sassaki, L.Y. Acute heart failure as an adverse event of tumor necrosis factor inhibitor therapy in inflammatory bowel disease: A review of the literature. *World J. Cardiol.* **2023**, *15*, 217–228. [CrossRef]
25. Zhang, M.; Wang, J.; Jia, L.; Huang, J.; He, C.; Hu, F.; Yuan, L.; Wang, G.; Yu, M.; Li, Z. Transmembrane TNF promotes activation-induced cell death by forward and reverse signaling. *Oncotarget* **2017**, *8*, 63799–63812. [CrossRef] [PubMed]

International Journal of
Molecular Sciences

MDPI

Article

Sustained Microglial Activation Promotes Synaptic Loss and Neuronal Dysfunction after Recovery from ZIKV Infection

Nahyun Kim [1], Hanul Choi [2], Uijin Kim [1], Suyeon Kim [1], Young Bong Kim [1,2,*] and Ha Youn Shin [1,*]

[1] Department of Biomedical Science & Engineering, Konkuk University, Seoul 05029, Republic of Korea; knh64@naver.com (N.K.); rladmlwls135@naver.com (U.K.); 0924tndus123@naver.com (S.K.)
[2] Department of Bio-Industrial Technologies, Konkuk University, Seoul 05029, Republic of Korea; chlgksmf9977@hanmail.net
* Correspondence: kimera@konkuk.ac.kr (Y.B.K.); hayounshin@konkuk.ac.kr (H.Y.S.)

Abstract: Zika virus (ZIKV), transmitted by Aedes mosquitoes, has been a global health concern since 2007. It primarily causes fetal microcephaly and neuronal defects through maternal transmission and induces neurological complications in adults. Recent studies report elevated proinflammatory cytokines and persistent neurological alterations post recovery, but the in vivo mechanisms remain unclear. In our study, viral RNA loads in the brains of mice infected with ZIKV peaked at 7 days post infection and returned to baseline by day 21, indicating recovery. RNA sequencing of the cerebral cortex at 7 and 21 days revealed upregulated genes related to neuroinflammation and microglial activation. Histological analyses indicated neuronal cell death and altered neurite morphology owing to severe neuroinflammation. Additionally, sustained microglial activation was associated with increased phospho-Tau levels, constituting a marker of neurodegeneration. These findings highlight how persistent microglial activation leads to neuronal dysfunction post ZIKV recovery, providing insights into the molecular pathogenesis of ZIKV-induced brain abnormalities.

Keywords: ZIKV; long-term effects; microglial activation; neuroinflammation; synaptic loss

1. Introduction

Zika virus (ZIKV), a member of the Flaviviridae family within the genus Flavivirus, was first identified in the Zika forest of Uganda in 1947 [1]. Since the discovery of the first human infection in 1950, ZIKV has spread from Africa to Asia and subsequently to the Americas, causing significant epidemics until 2016, with sporadic cases still reported worldwide. Primarily transmitted through the bites of Aedes mosquitoes, ZIKV can also spread through sexual contact, blood transfusion, or mother-to-child transmission during pregnancy. Although ZIKV infection typically results in mild symptoms, such as fever, rash, headache, joint pain, and red eyes lasting several days to a week, it poses significant risks during pregnancy. Infection during pregnancy can lead to severe birth defects including microcephaly and neurological conditions like myelitis and encephalitis [2,3]. These neurological complications are not limited to congenital cases but also affect adults following ZIKV infection [4,5]. Studies have demonstrated elevated levels of proinflammatory cytokines, including interleukin (IL)-1β, interferon (IFN)-γ, IL-6, and IL-8, in ZIKV-infected adult patients compared with healthy individuals [6]. Additionally, ZIKV infection has been linked to the induction of Guillain–Barre syndrome, an autoimmune neurological disease [7]. Furthermore, ZIKV has been detected in adult human brain tissue [8]. ZIKV infection has also been increasingly associated with CNS pathology in adults, including acute myelitis, encephalitis, meningoencephalitis, and encephalomyelitis [4,5,9,10]. Notably, a clinical study found that ~40% of ZIKV-infected patients still exhibited neurological changes 1 year after symptom onset, with some patients showing incomplete recovery even after 2 years [11]. Collectively, these findings underscore the detrimental effects of

ZIKV infection on the mature nervous system and suggest potential long-term neurological consequences, yet the precise underlying mechanisms remain to be fully elucidated.

One of the primary cell types targeted by ZIKV within the brain comprises glial cells [12–14]. Among glial cells, microglia—the predominant innate immune cells in the central nervous system (CNS)—play a crucial role in neurodevelopment and CNS homeostasis [15]. Responsible for maintaining normal brain function and tissue integrity, activated microglia serve as the initial defense against invading pathogens, and their activation leads to the increased production of proinflammatory cytokines, chemokines, and reactive oxygen species (ROS) [16,17]. Although microglial activation is typically considered protective for the brain, prolonged activation can cause chronic inflammation, leading to irreversible CNS damage. Such persistent neuroinflammation not only impairs neuronal plasticity and memory but also contributes significantly to tissue damage in various neurodegenerative disorders [16,18]. In this context, microglial activation has been linked to the increased phosphorylation of the Tau protein, exacerbating Tau pathology—a hallmark of neurodegenerative diseases such as Alzheimer's disease [19]. In addition to the pathological accumulation of neurofibrillary tangles (NFTs) containing the hyperphosphorylated Tau protein, Alzheimer's disease is characterized by the accumulation of amyloid-β (Aβ) plaques and neuronal loss [20,21]. The accumulation of Aβ triggers microglial responses, promoting Tau protein hyperphosphorylation and NFT formation, ultimately leading to neurodegeneration and cognitive decline [22].

While several studies have reported neuroinflammation and microglial activation as consequences of ZIKV infection [23,24], the detailed mechanisms underlying brain impairments in recovering patients remain elusive. This may be attributable to the lack of a suitable animal model representing the recovery state post ZIKV infection. Evidence suggests that ZIKV can persist latently in mature neural cells [25], and its association with widespread changes in gene expression in the brain has been demonstrated [4–6]. Thus, it is crucial to move beyond acute phase symptoms to develop an understanding of the long-term neurological sequelae of ZIKV infection, including their potential to lead to neurodegenerative disorders. To this end, we established an in vivo recovery model of ZIKV infection using ZIKV-susceptible A129 mice. Through comprehensive RNA-sequencing (RNA-seq)-based transcriptome analysis, we explored gene expression profiles in the brains of recovered mice post ZIKV infection. Gene Ontology (GO) analysis revealed the significant upregulation of the biological process categories "neuroinflammatory response" and "microglial cell activation" in ZIKV-infected mice. Our findings indicate severe inflammatory pathology, neuroinflammation-induced neuronal cell death, and alterations in neurite morphology in mice recovered from ZIKV infection. Furthermore, increased levels of phospho-Tau (p-Tau) were observed together with microglial activation in recovered mouse brains. These findings shed light on the mechanism by which ZIKV affects the brain and underscore the role of this virus-induced microglial activation mechanism in neurodegeneration. They also have implications for understanding the molecular pathogenesis of ZIKV-induced abnormalities in the brain and may guide the development of effective therapeutic strategies against ZIKV infection.

2. Results

2.1. *Immunodeficient A129 Mice Recover from Infection with the ZIKV-FLR Strain after 21 Days*

To investigate the long-term effects of neuroinvasive ZIKV infection, we established a mouse model for studying the recovery phase following ZIKV infection. Type I interferon (α/β)-receptor-deficient A129 mice, which are highly susceptible to ZIKV infection, were intraperitoneally injected with 1×10^4 PFU of either the ZIKV-FLR (American and Asian lineage) or the ZIKV-MR766 strain (African lineage). Mice were monitored daily for signs of illness including changes in body weight and mortality. All mice infected with the ZIKV-MR766 strain lost body weight, quickly succumbed to infection, and reached clinical endpoints within 7 days (Figure 1a). The pattern following ZIKV-FLR infection was different. Approximately 40% of infected mice succumbed to infection within 7 days and exhibited neurological signs such as hind limb paralysis whereas the remaining 60% of ZIKV-FLR-infected mice gradually regained body weight and recovered from hind limb paralysis within 21 days (Figure 1b). To examine the impacts of ZIKV-MR766 and ZIKV-FLR strains on the mouse brain, we obtained cerebral cortices from mice in each group and measured viral load by quantitative reverse-transcription polymerase chain reaction (RT-qPCR). Mice infected with ZIKV-MR766 exhibited approximately 100-fold higher levels of ZIKV RNA in the brain relative to mice infected with ZIKV-FLR, and viral replication remained active until death (Figure 1c). In contrast, viral RNA levels in ZIKV-FLR-infected mice gradually increased, peaking at 7 days post infection (dpi), and then declined to undetectable levels, indicating that the mice were free of viral replication by 21 dpi (Figure 1d). To further investigate the presence of any remaining viral proteins in cells, we performed a Western blot analysis of the ZIKV envelope (E) protein, a critical structural component of the virus that plays a central role in viral entry into host cells. The expression of the ZIKV E protein indicates an active viral presence, with peak levels typically occurring during the acute phase of infection [26,27]. Our results showed that ZIKV E protein levels increased at 7 dpi, declined to undetectable levels by 21 dpi, and remained undetectable at 28 dpi (Figure S1). On the basis of these results, we selected the relatively mild ZIKV-FLR strain to establish a mouse model of a post-ZIKV-infection recovery state.

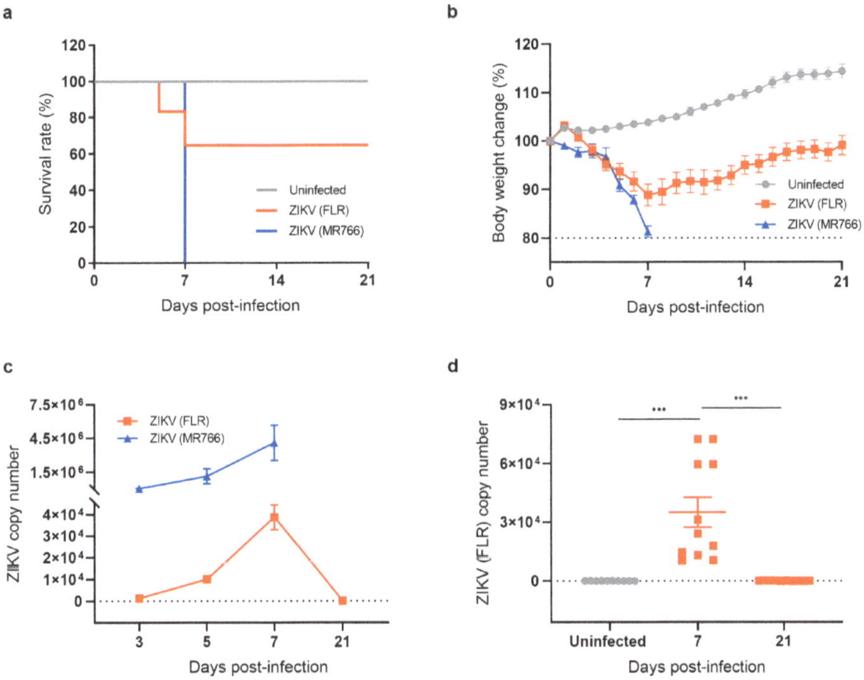

Figure 1. Generation of a mouse model of recovery from ZIKV infection. (**a**) Survival rates of mice used as uninfected controls and mice infected with ZIKV strains FLR (American and Asian strain) or MR766 (African strain). Survival of mice was monitored for 21 days. All mice infected with the ZIKV-MR766 strain succumbed within 7 dpi whereas fewer than 40% of mice infected with the ZIKV-FLR strain died, with over 60% surviving beyond 21 days. (**b**) Body weight change (percent) after ZIKV infection. (**c**) Comparison of ZIKV copy number, measured by RT-qPCR. (**d**) Quantification of ZIKV (FLR) RNA levels in the mouse brain using RT-qPCR analysis. All data represent means ± SEMs from at least 7 mice (*** $p < 0.001$). The results for the uninfected control are shown in gray, the ZIKV MR766 strain in blue, and the ZIKV FLR strain in red. The dashed line indicates the basal level.

2.2. The Transcriptional Profile of the Mouse Brain Is Altered after Recovery from ZIKV Infection

Using the ZIKV-FLR strain, we first examined whether transcriptional changes occurred in the brain tissue in ZIKV-infected mice and whether these changes persisted after recovery. The brain, particularly the cerebral cortex, is well known as a major target organ of ZIKV [8,28]. To compare transcriptome profiles between acute infection and the recovery stage, we performed RNA-seq on the cerebral cortices of ZIKV-FLR-infected mice, obtained at 7 and 21 days post infection (dpi) (Figure 2a). The cerebral cortices of A129 mice obtained at 7 dpi, exhibiting the highest viral load, were used to investigate the transcriptional profile of acute infection whereas the cerebral cortices at 21 dpi, exhibiting viral clearance, were used to examine changes in gene expression during the recovery phase. Differentially expressed genes (DEGs) in the cerebral cortex at 7 dpi and 21 dpi in ZIKV-infected mice were identified based on their normalized expression relative to that in uninfected controls. At 7 dpi, 2378 genes were upregulated and 725 genes were downregulated compared with uninfected controls; in contrast, 1063 genes were upregulated and 279 genes were downregulated at 21 dpi (Figure 2b).

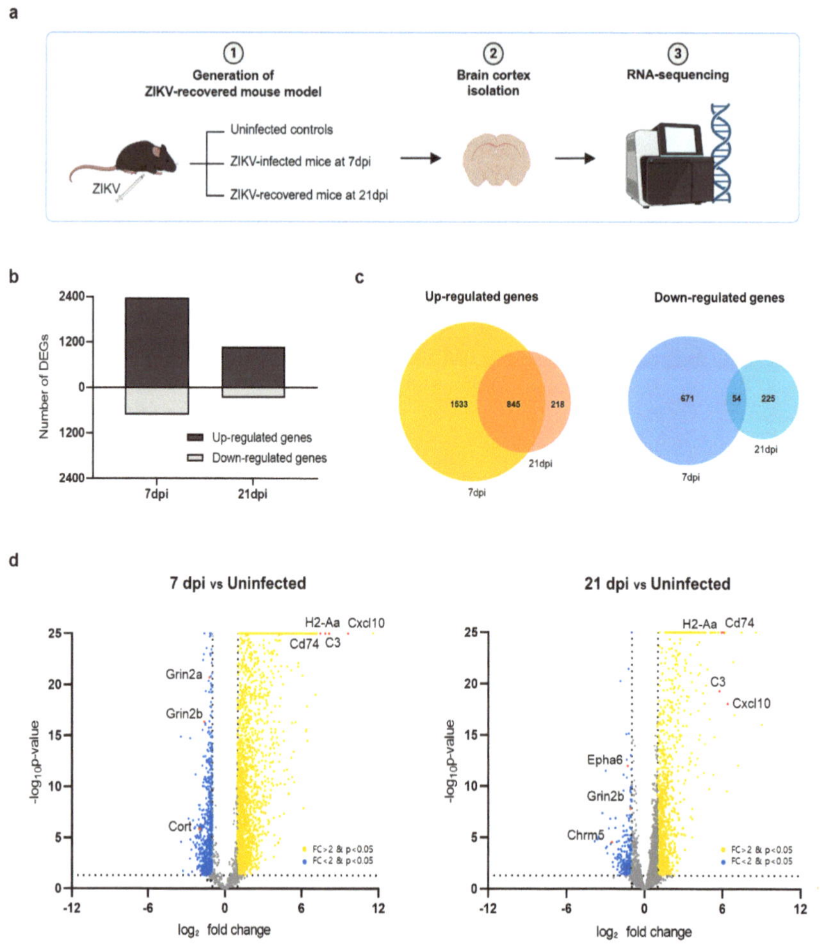

Figure 2. Transcriptome profiles of the cerebral cortex in A129 mice after ZIKV infection. (**a**) Schematic showing the experimental strategy for analyzing changes in gene expression in the mouse brain following ZIKV infection. (**b**) Bar chart displaying the total number of upregulated and downregulated

DEGs in mouse groups harvested at 7 dpi and 21 dpi (uninfected, n = 3; 7 dpi, n = 3; 21 dpi, n = 3). Gene expression levels in both groups were normalized to those in uninfected controls. (**c**) Venn diagrams showing the overlap of upregulated genes (left) and downregulated genes (right) in mice brains harvested at 7 dpi or 21 dpi, identifying genes commonly regulated at both time points. (**d**) Volcano plots depicting gene transcripts in ZIKV-infected mouse brains harvested at 7 dpi and 21 dpi compared with uninfected controls. Yellow dots: genes upregulated upon ZIKV infection; blue dots: genes downregulated upon ZIKV infection. Selected DEGs have been highlighted in red. Gray dots indicate genes with no significant change in expression. The dashed lines represent the significance thresholds for both *p*-value and log2-fold change.

We then investigated genes that were significantly up- or downregulated in common in both acute (7 dpi) and recovery (21 dpi) phases. Utilizing a Venn diagram analysis, we found that the predominant pattern was upregulation, with upregulated 845 genes and only 54 downregulated genes at both time points (Figure 2c). Among the top 20 significantly upregulated genes in both conditions were *Cxcl10*, *H2-Aa*, *C3*, and *CD74*, as illustrated in volcano plots (Figure 2d). These genes, which are known to be associated with neuroinflammation, play critical roles in regulating the immune system and inflammatory responses. Conversely, the most highly downregulated genes in common were *Grin2b*, *Cort*, and *Chrm5*, which function in neurotransmission. Overall, of the 3103 genes whose transcriptional expression levels changed upon ZIKV infection, more than half recovered to levels comparable to those in uninfected condition by 21 dpi whereas the remainder remained up- or downregulated even after recovery. Notably, genes related to neuroinflammation were activated whereas those related to neurotransmission were repressed.

2.3. ZIKV-Induced Microglial Cell Activation Results in Upregulation of Neuroinflammatory Responses and Downregulation of Neurotransmission

To better understand the relationship between functional pathways and DEG datasets obtained at 7 and 21 dpi, we conducted an unbiased enrichment analysis using Gene Ontology (GO) and the Kyoto Encyclopedia of Genes and Genomes (KEGG). The GO analysis of biological process (BP) terms revealed consistent enrichment of genes related to neuroinflammatory responses and microglial cell activation in brain tissues obtained at both 7 dpi and 21 dpi (Figure 3a). Significantly correlated BP terms included 'phagocytosis', 'neuroinflammatory response', 'microglia cell activation', 'glial cell proliferation', 'synapse pruning', and 'regulation of neuron apoptotic process'. DEG datasets correlated with molecular function (MF) and cellular component (CC) terms also indicated a close relation to neuroinflammatory processes and the activation of microglia. DEG datasets obtained at 7 and 21 dpi were also significantly correlated with specific KEGG pathways subsets, including 'cytokine-cytokine receptor interaction', 'phagosome', 'complement and coagulation cascades', 'neuroactive ligand-receptor interaction', 'NF-κB signaling pathway', and 'synapse' (Figure 3b). Accordingly, the majority of enriched terms found in both GO and KEGG analyses were related to neuroinflammatory response, microglia cell activation, and neurotransmission. Consistent with previous reports, these results demonstrate that ZIKV infection has the potential to activate microglia and induce neurological damage [23,24]. Importantly, these results also suggest that microglial activation and neuronal damage persist after recovery from ZIKV infection.

Microglia are resident macrophages that account for approximately 10–15% of cells in the mouse brain and function as a first-line defense system in the CNS [29]. Several studies have shown that microglial cells are permissive for ZIKV infection and that microglial cell activation results in neuroinflammation and is detrimental to neurotransmission [24,30–32]. To further clarify whether ZIKV-induced aberrant microglial activation correlates with neurological complications, we investigated DEGs associated with microglial activation, neuroinflammation, and neurotransmission at both 7 dpi and 21 dpi. A heatmap analysis revealed that genes associated with microglial activation and neuroinflammation, including cytokine, chemokine, complement-system, and phagocytosis genes, were significantly upregulated at 7 dpi, exhibiting up to a 10-log2-fold change (Figure 3c and Figure S2). The expression of these genes was reduced ~2-fold at 21 dpi but did not fully return to the levels of uninfected controls. On the other hand, genes associated with neurotransmission were significantly downregulated at 7 dpi and remained at similarly reduced levels at 21 dpi. These results suggest that abnormal microglial activation can persist even after recovery from ZIKV infection. This prolonged activation may contribute to ongoing neuroinflammation and potentially lead to lasting damage to neurotransmission, even in the absence of a detectable viral presence.

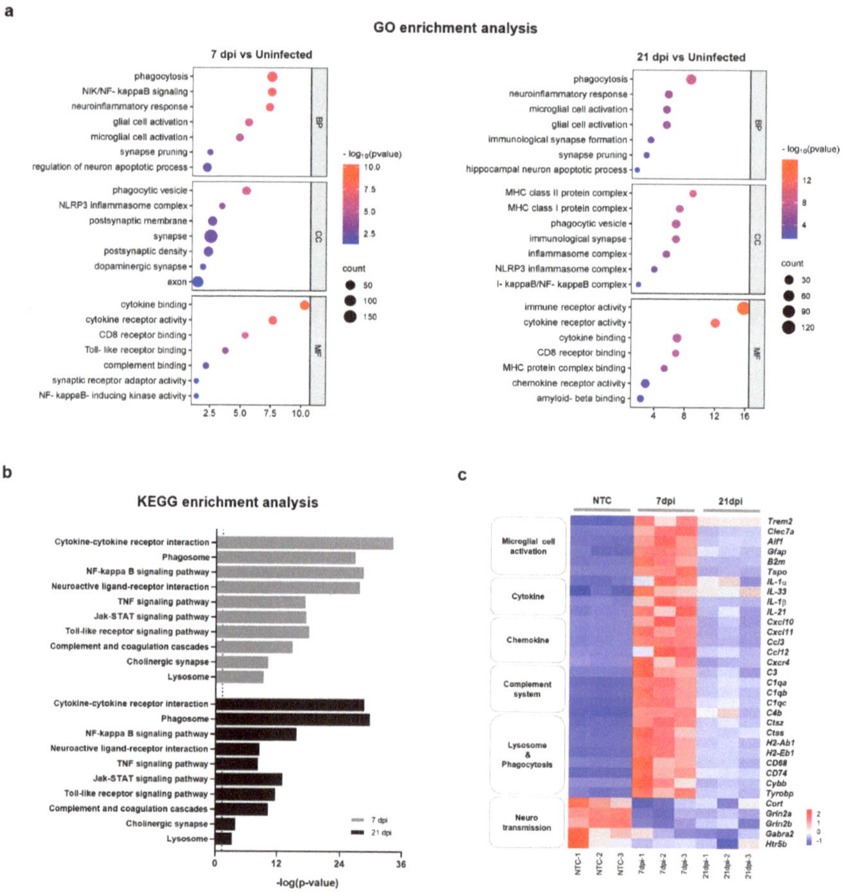

Figure 3. Functional categorization of DEGs in the brains of mice infected with and recovered from ZIKV. (**a**) Selected GO functional classifications for significant DEGs in ZIKV-infected mice brains obtained at 7 and 21 dpi. Biological processes associated with significant DEGs mainly included 'microglial cell activation', 'neuroinflammatory responses', 'synapse pruning', and 'neuronal apoptotic process'. GO: Gene Ontology; BP: biological process; CC: cellular component; MF: molecular function. (**b**) Selected KEGG pathways associated with significant DEGs in ZIKV-infected mice brains obtained at 7 and 21 dpi. Significant pathways of common DEGs included 'cytokine-cytokine receptor interaction', 'phagosome', 'complement and coagulation cascade', and 'neuroactive ligand-receptor interaction'. The vertical line represents the threshold for a p-value of 0.05. (**c**) Heatmap displaying the relative expression levels of significantly changed genes related to neuroinflammation (microglial cell activation, cytokine, chemokine, complement system, and lysosome and phagocytosis) and neurotransmission.

2.4. ZIKV-Induced Microglial Cell Activation Enhances Synaptic Pruning through the Complement System

Next, we sought to determine how ZIKV infection alters gene expression patterns and induces neurotoxic inflammation through microglial activation and potentially leads to neurodegeneration. Transcriptome profiling in ZIKV-infected and recovered mice revealed the sustained upregulation of the microglia cell activation markers, Clec7a, Aif1, Gfap, and Trem2 (Figure S2). As shown in Supplementary Figure S2, ZIKV significantly increased the expression of complement-related genes (C3, C1q, C4b, and C3ar1), together with genes related to microglial-mediated phagocytosis (Ctss, Ctsc, Cybb, CD74,CD68, H2-Ab1, and H2-Aa), in the brains of ZIKV-infected and recovered mice. Recent studies suggest that microglia-mediated synapse removal can be triggered by aging or disease conditions. In the CNS, microglia rely on classical complement cascades to mediate phagocytic signaling for the removal of excess synapses [18]. In Alzheimer's mouse models, activated microglia engage in synaptic phagocytosis, a process dependent on complement factors like C3, C1q, and CR [33]. This microglial pruning of synaptic terminals is a key mechanism underlying synapse loss in neurodegenerative disorders and viral infections. To explore this further, we investigated whether mice recovering from ZIKV infection exhibited impaired synaptic function. As seen in Figure 3a, 'synaptic pruning' was among the most significantly expressed categories in GO analyses following ZIKV infection in mice. We found that ZIKV significantly downregulated the expression of genes encoding neuropeptides or neurotransmission receptors, such as Cort, Gabra2, Grin2b, and Htr5, in the mouse brain cortex. Genes encoding nicotinic acetylcholine receptor subunits (Chrnb4 and Chrna7), adrenoceptors (Adra2b and Adra1d), muscarinic acetylcholine receptors (Chrm1 and Chrm3), and two serotonin receptors (Htr2a and Htr4), which typically participate in the regulation of cognitive function, were downregulated at both 7 dpi and 21 dpi compared with uninfected mice brains (Figure S2). Most of these genes have been shown to modulate synaptic plasticity [34–38] and CNS development [39,40].

To further validate RNA-seq data, we also performed RT-qPCR analyses on mouse cerebral cortices obtained at 7 and 21 dpi for representative genes involved in microglial cell activation, cytokine and chemokine responses, complement system activation, phagocytosis, and neurotransmission. We found that the expression of a panel of genes known to be specifically induced by microglial activation, including Clec7a, Aif1, Gfap, and Trem2, was significantly increased upon ZIKV infection (Figure 4a). Additionally, genes encoding cytokines and chemokines, including IL-1α, Il-1β, Cxcr4, and Cxcl10, were significantly upregulated in response to ZIKV infection, suggesting the presence of neuroinflammation (Figure S3). Together with microglial-activation-associated genes, genes related to complement activation, lysosomal function, and phagocytosis were significantly upregulated (Figure 4b,c). Importantly, all of these genes exhibited persistent upregulation at 21 dpi (i.e., recovery phase), and these changes were statistically significant. Notably, the C1q and C3 genes showed more than an 8-fold change in expression in ZIKV-recovered mice compared with uninfected controls. These results were consistent with the RNA-seq data (Figure S2) and suggest that ZIKV-induced microglial activation contributes to the transcriptional upregulation of genes related to complement activation, lysosomal function, and phagocytosis in the brains of ZIKV-infected and recovered mice. We also confirmed the downregulation in common in the neurotransmission-related genes, Cort, Grin2b, Gabra2, and Htr5 (Figure 4d), validating a potential contribution of microglial phagocytosis to synaptic pruning.

Furthermore, we investigated the long-term effects of ZIKV infection on the mouse brain. At 28 dpi, the ZIKV viral protein (E) was undetectable in the brain (Figure S1), indicating viral clearance. However, RT-qPCR analyses of cerebral cortices collected at 28 dpi revealed that microglial activation remained elevated (Figure S4a). Along with persistent microglial activation, we observed the significant upregulation of genes associated with the complement system, lysosomal function, and phagocytosis (Figure S4b,c). Importantly, these genes continued to show upregulation at 28 dpi. Conversely, neurotransmission-related genes were notably downregulated (Figure S4d).

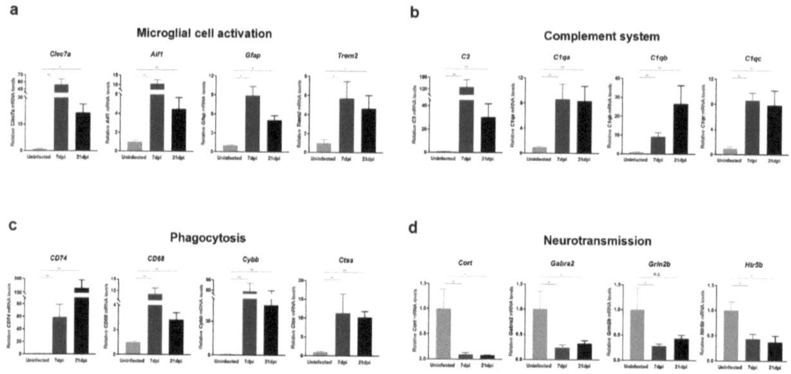

Figure 4. Transcriptional analysis of genes related to microglial activation, complement system, phagocytosis, and neurotransmission in the brains of ZIKV-infected and recovered mice. RT-qPCR analyses of genes associated with (**a**) microglial cell activation, (**b**) the complement system, (**c**) phagocytosis, and (**d**) neurotransmission are shown. mRNA levels of target genes were normalized to *Gapdh* mRNA levels. All data represent the relative expressions of target genes normalized to those of uninfected controls. Results are shown as means ± SEMs of six independent replicates (* $p < 0.05$; ** $p < 0.01$; n.s.: not significant).

Taken together, our data suggest that the aberrant activation of microglia and complement pathways persists in both ZIKV-infected and ZIKV-recovered mice, ultimately leading to synaptic loss. These findings support our hypothesis that ZIKV-induced microglial cell activation contributes to increased synapse pruning and may explain the sustained synaptic impairment observed following long-term recovery from ZIKV infection.

2.5. Sustained Microglial Cell Activation after Recovery from ZIKV Infection Induces Neurotoxicity and Neuronal Cell Death

We next examined whether sustained microglial activation following recovery from ZIKV infection results in exacerbated neuronal loss. To address this question, we evaluated histological alterations to determine the extent of pathology in mouse brain tissues. Hematoxylin and eosin (H&E) staining revealed dead neurons as eosinophilic cells with shrunken, darkly stained nuclei surrounded by a void space. In the cortex of uninfected control groups, neurons exhibited a round shape with clearly visible nuclei. In contrast, at 7 dpi, distinct neuronal damage and immune cell infiltration were observed [41], with neuronal cell death continuing through 21 dpi (Figure S5a). Additionally, Nissl staining revealed a significant decrease in both the density and number of intact neurons in the cortex at 7 dpi and 21 dpi compared to the uninfected control group (Figure S5b). These findings underscore the critical role of inflammatory responses and neuronal damage in the pathology of ZIKV infection. The inflammatory environment, including monocyte infiltration, likely contributes to the sustained activation of microglia, which, in turn, may lead to increased ROS production and further exacerbate neuronal damage. Our analysis of ROS levels in the cortex of ZIKV-infected mice supports this, revealing a significant increase in ROS compared with uninfected controls (Figure 5a). Specifically, ROS levels were approximately 6-fold higher at 7 dpi and 5-fold higher at 21 dpi compared to baseline uninfected control levels (Figure 5b). TUNEL assays conducted to assess neuronal cell death in the cortex of mice following recovery from ZIKV infection revealed a number of TUNEL-positive cells in the cortex of both ZIKV-infected and ZIKV-recovered mice (Figure 5c). A quantitative analysis showed significant differences in the number of TUNEL-labeled cells between uninfected mice and ZIKV-infected mice at both the

7 dpi and 21 dpi time points (Figure 5d). A comparison of the number of apoptotic cells in ZIKV-infected brains to uninfected controls showed a remarkable 72-fold increase in 7 dpi mice and a 14.5-fold increase in 21 dpi mice. Although the direct link between ROS generation and neuronal death should be further validated through additional studies, such as by evaluating the effects of inhibiting ROS generation, several studies have already highlighted the connection between microglial activation, ROS production, and neuronal cell death [42,43].

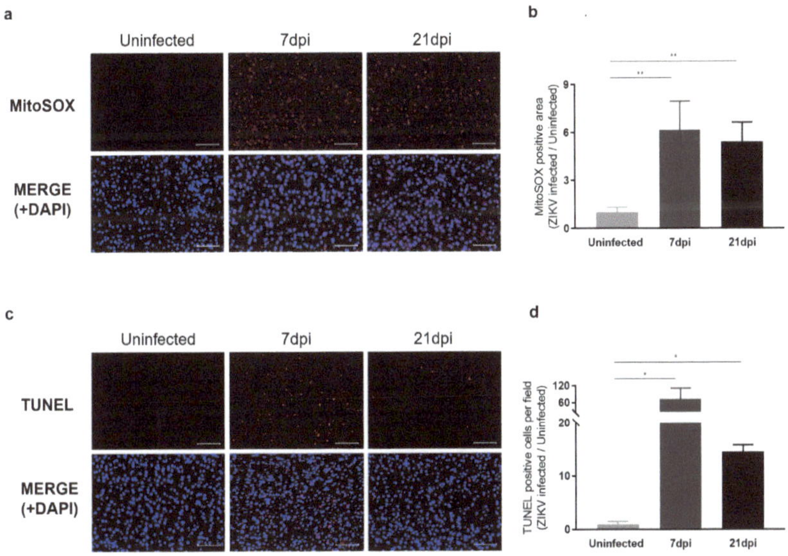

Figure 5. Characterization of neurological damage caused by persistent microglial activation after ZIKV Infection. (**a**) Representative images of mitochondrial ROS, stained with MitoSOX (Red), in the mouse cerebral cortex. Cell nuclei were counterstained with DAPI (blue). Images were visualized by fluorescence microscopy (200×). Scale bar: 20 µm. (**b**) Quantification of the relative MitoSOX-positive area. Results are shown as means ± SEMs of five independent replicates (** $p < 0.01$). (**c**) Representative images of DNA fragmentation (red) in the mouse cerebral cortex, detected by TUNEL assay. Cell nuclei were counterstained with DAPI (blue). Images were visualized by fluorescence microscopy (200×). Scale bar: 20 µm. (**d**) Quantification of TUNEL-positive cells in mouse cerebral cortices from five independent replicates (* $p < 0.05$).

Consistent with transcriptome profiling results indicating that ZIKV-induced aberrant microglial activation results in neurotoxicity, histological analyses and the quantification of neuronal apoptosis revealed that ZIKV infection was associated with neurological damage in the mouse brain. Our results further indicate that even after levels of viral RNA return to near-uninfected control levels, mice recovered from ZIKV infection display neurological impairment and the activation of neuronal cell death pathways owing to the persistence of microglial activity.

2.6. Sustained Microglial Cell Activation Following ZIKV Infection Results in Abnormal Neural Morphology and Accumulation of Phosphorylated Tau

Our RNA-seq results indicated that ZIKV infection leads to a decrease in the expression of genes associated with axon guidance and synaptic transmission. To assess whether these transcriptomic changes manifest as neuronal phenotypes, we examined the dendritic architecture of the cerebral cortex obtained at 7 and 21 dpi by immunostaining for the neurite marker MAP2 (microtubule-associated protein 2), a neuron-specific cytoskeletal protein enriched in dendrites that plays a role in determining and stabilizing neuronal morphology during development [44]. We observed a significant reduction in the number of intact dendritic segments and branches in ZIKV-infected mice (Figure 6a), indicating a detrimental impact of ZIKV on neurite morphology. A comparison of the number of neurons in ZIKV-infected brains to uninfected baseline levels showed a 40% decrease in mice at 7 dpi and a 20% decrease at 21 dpi (Figure 6b). These findings suggest that ZIKV infection triggers the activation of microglia and the complement system, leading to excessive synaptic pruning and subsequent synaptic impairment in mice.

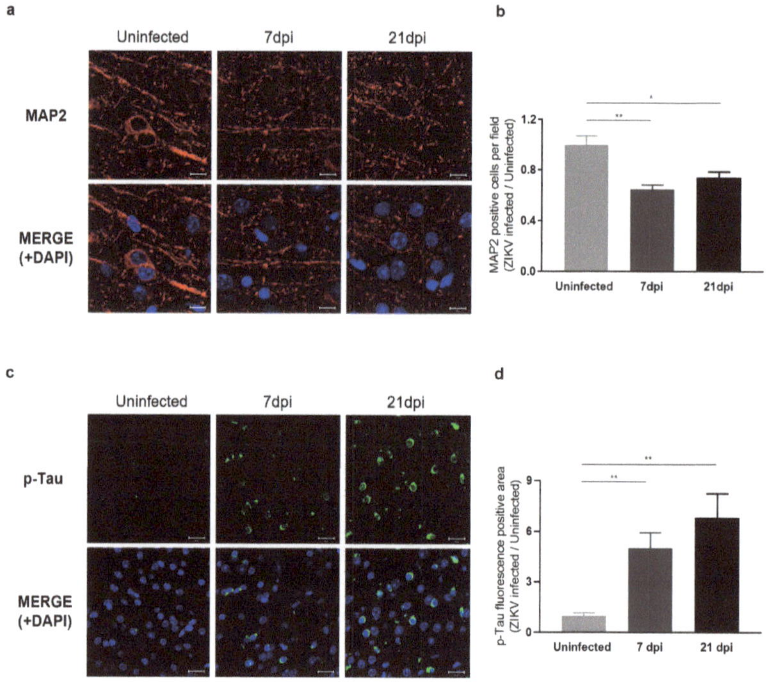

Figure 6. Characterization of neuronal phenotypes and Tau pathology produced by sustained microglial activation following ZIKV infection. (**a**) Representative images of MAP2 staining (red) in the mouse cerebral cortex. Cell nuclei were counterstained with DAPI (blue) and all images were visualized by fluorescence microscopy (800×). Scale bar: 10 μm. (**b**) Quantification of MAP2-positive cells. Results are shown as means ± SEMs of five independent replicates (* $p < 0.05$; ** $p < 0.01$). (**c**) Representative fluorescent images of p-Tau (PHF13, Ser396) in the mouse cerebral cortex. Cell nuclei were counterstained with DAPI (blue) and all images were visualized by fluorescence microscopy (400×). Scale bar: 20 μm. (**d**) Quantification of p-Tau fluorescence-positive area in mouse cerebral cortices from five independent replicates (** $p < 0.01$).

To gain further insights into the effects of ZIKV infection on the progression of neurodegenerative disorders, we examined Tau pathology in ZIKV-recovered mice. Under normal conditions, Tau protein stabilizes cellular microtubules, but under pathological conditions, the intracellular accumulation of hyperphosphorylated Tau disrupts microtubules and damages the cytoskeleton. This process has been linked to neurodegeneration and cognitive impairment [19,22,45]. Several studies have additionally reported that the accumulation of hyperphosphorylated Tau is associated with microglial activation [46–48]. Given the significance of p-Tau as a risk factor for neurodegenerative diseases, particularly Alzheimer's disease, we conducted further evaluations of Tau protein levels using immunohistochemistry (IHC). As shown in Figure 6c, p-Tau staining was significantly increased in the cortex of ZIKV-infected mice compared with uninfected controls and further increased in ZIKV-recovered mice. These data suggest that ZIKV infection leads to sustained microglial activation and increased p-Tau levels in ZIKV-recovered mice, which may contribute to neurodegeneration (Figure 6d).

3. Discussion

Using a mouse model that recapitulates the recovery phase of ZIKV infection, we have elucidated the detailed mechanisms underlying the long-term sequelae of neuroinvasive ZIKV infection and provided critical evidence of a potential association with neurodegeneration. Indeed, we observed robust viral replication in the brains of these mice, peaking at 7 dpi and clearing by 21 dpi. Based on these findings, we designated 7 dpi as the acute phase and 21 dpi as the recovery phase of ZIKV infection. We then performed an integrative analysis of genome-wide gene expression changes and histological data on the ZIKV-infected mouse cerebral cortex obtained from the acute (7 dpi) and recovery (21 dpi) phases of infection. RNA-seq results revealed that sustained microglial activation plays a central role in the persistence of neuroinflammation and the downregulation of neurotransmission during the recovery phase. Persistent microglial activation is also associated with the activation of the complement system and resultant microglial phagocytosis. In line with the transcriptomic data, our histological examinations show that persistent microglial activation is associated with significant neuronal cell death and increased ROS production. Subsequent neural-cell-death-induced excessive synaptic loss and microglial-activation-mediated p-Tau accumulation imply a potential impact on neurodegeneration.

The recent coronavirus-disease-19 (COVID-19) pandemic has highlighted the importance of investigating the long-term consequences of viral infections [49,50]. Similar to the long-term effects caused by SARS-CoV-2 and other flavivirus family members [49,51], evidence suggests that ZIKV can also cause neurological damage even after recovery from infection [8,52]. Therefore, further studies should be performed in aged mice to investigate the long-term effects of ZIKV infection and its progression towards neurodegenerative disorders. Specifically, extending the observation period to 3–6 months post infection, beyond the current 21 dpi, will be crucial for understanding the prolonged impact of ZIKV infection and its potential to induce chronic neurodegenerative changes.

Our histological data revealed that synaptic elimination was not fully restored, and p-Tau levels were even higher in the recovery phase than in the acute phase. This observation suggests that despite the apparent resolution of the acute phase of ZIKV infection, there may be ongoing pathological changes related to the accumulation of protein aggregates, such as NFTs, due to elevated p-Tau levels. Autophagy is a lysosomal degradative process that eliminates damaged organelles and protein aggregates [53]. Most neurodegenerative diseases are characterized by pathological protein aggregates that lead to the formation of NFTs due to hyperphosphorylated Tau. Impairments in autophagy are known to play a significant role in the progression of tauopathies and efficient autophagy is essential for the degradation of phosphorylated Tau, helping maintain its levels at a low threshold [54]. To further assess autophagic activity after ZIKV recovery, we performed RT-qPCR analysis on autophagy-related genes (ATGs) and autophagosome-related LC genes using ZIKV-infected brain tissues obtained at 28 dpi (Figure S6). Our results showed that autophagy-related Atg7 and Atg12 genes, as well as autophagosome-related LC3A and LC3B genes, were downregulated at 28 dpi. These genes are crucial for the elongation and closure steps of autophagosome formation, suggesting a potential impairment in the autophagic process during the recovery phase. This impairment could contribute to the accumulation of damaged proteins and cellular debris. Further studies investigating the impact of autophagy and protein aggregation on long-term recovery in mouse models of ZIKV infection could provide more insights into the relationship between autophagic defects and the pathogenesis of neurodegenerative diseases.

Several studies have demonstrated that ZIKV does not replicate efficiently in immunocompetent mice [55,56]. To more accurately model the conditions of ZIKV infection in humans, we used immunocompromised A129 mice, which are highly susceptible to ZIKV and provide valuable insights into viral replication and neuroinflammation. While A129 mice are advantageous for studying viral mechanisms, they do not fully represent how the immune system interacts with ZIKV in humans. Immunocompetent models, such as those with C57BL/6 mice, are still essential for understanding the interaction between ZIKV and a functional immune system, thereby enhancing our understanding of ZIKV pathogenesis in humans. Initially, we considered using C57BL/6 mice, a common strain of immunocompetent mice, for our experiments. These mice were intraperitoneally injected with the ZIKV-FLR strain, similar to the conditions used for A129 mice. However, we observed minimal ZIKV replication in C57BL/6 mice, which limited our ability to effectively study viral effects and pathogenesis (Figure S7). Due to these limitations, we decided to use immunocompromised A129 mice, which are highly susceptible to ZIKV. This decision allowed us to achieve more consistent and measurable viral replication, providing valuable insights into the disease. Recent research indicates that specific ZIKV strains and particular inoculation conditions can indeed induce infection in immunocompetent models [31,57]. Moreover, employing an immunocompetent mouse model with the temporal blockading of type I interferon using specific antibodies offers a promising alternative [41]. This approach allows immune responses to be elicited in immunocompetent mice while enhancing infection by temporarily blocking type I interferon. Therefore, adjusting the ZIKV strain, inoculation method, dose, and infection conditions might make it feasible to study ZIKV infection in immunocompetent mice. Additionally, there is a growing trend to use both immunocompetent and immunocompromised models to gain a comprehensive understanding of immune responses and viral pathogenesis. This approach could complement our findings and enhance the understanding of ZIKV pathogenesis in conditions that more closely resemble human infection. Therefore, incorporating both immunocompetent and immunocompromised mouse models in subsequent studies will be crucial for a comprehensive understanding of ZIKV pathogenesis, immune responses, and long-term effects.

Moreover, although we did not observe notable behavioral abnormalities in recovered mice, incorporating specific behavioral tests for cognitive impairment and motor dysfunction might be useful in investigating neurodegenerative clinical signs in these models. Further studies to investigate the effects of diet and exercise during recovery phase would also provide valuable insights for designing new strategies to enhance recovery from ZIKV infection.

In conclusion, we have demonstrated the sequelae of consequences of ZIKV infection in adult brain tissue using a ZIKV-susceptible recovery mouse model. Our findings provide a fundamental understanding of the in vivo mechanisms underlying brain impairments associated with ZIKV infection post recovery. Sustained microglial activation, even after recovery, may lead to persistent neuroinflammation and synaptic loss, resulting in long-term defects in neuronal function. Therefore, it is essential to develop effective antiviral vaccines to prevent ZIKV infection and to design potent therapeutics for treating ZIKV infection and related neuropathogenic flavivirus family members.

4. Materials and Methods

4.1. ZIKV Propagation

ZIKV-MR766 strain (African lineage, accession number AY632535) and FLR strain (American and Asian lineage, accession number KU820897) were obtained from American Type Culture Collection (ATCC, Manassas, VA, USA). ZIKV was propagated in Vero cells. Briefly, Vero cells were cultured in Dulbecco's Modified Eagle's Medium (DMEM; Capricorn Scientific, Ebsdorfergrund, Germany) supplemented with 10% fetal bovine serum (FBS; Gibco BRL, Carlsbad, CA, USA) and 1% penicillin/streptomycin (Capricorn Scientific, Ebsdorfergrund, Germany). Cells were infected with ZIKV at a multiplicity of infection (MOI) of 0.1 for 4 days at 37 °C in a humidified chamber under 5% CO_2 conditions. Following infection, the culture medium was harvested and centrifuged at $1000\times g$ and cell supernatants were stored as virus stocks at −80 °C until needed. Infectivity was assessed by determining virus titers using a standard plaque assay in Vero cells treated with serial dilutions of virus stocks.

4.2. Mice

Interferon alpha/beta (IFN-α/β) receptor-deficient A129 mice were obtained from the Korea Research Institute of Chemical Technology (KRICT, Daejeon, Republic of Korea). Female A129 mice, 10–15 weeks old, were used to investigate ZIKV-induced neurological deficits. Mice in the experimental groups were intraperitoneally injected with either the ZIKV-FLR or ZIKV-MR766 strain. Specifically, we administered a challenge dose of 1×10^4 PFU of the virus in 200 μL to ensure consistency and reproducibility in our infection experiments. The uninfected control group was intraperitoneally injected with an equivalent volume of phosphate-buffered saline (PBS). Mice were monitored daily for signs of illness including changes in body weight and mortality. All animal experiments were performed in a Biosafety Level 2 animal facility according to the relevant guidelines and laboratory procedures were approved by the Konkuk University Institutional Animal Care and Use Committee (IACUC).

4.3. RNA Isolation

Total RNA was isolated from the cerebral cortices of mouse brains using the PureLink RNA Mini Kit (Invitrogen, Carlsbad, CA, USA) according to the manufacturer's instructions. The quantity and quality of the isolated RNA were analyzed using UV spectrophotometry (NanoPhotometer N60/N50; Implen, Munchen, Germany)

4.4. Quantitative Reverse Transcription-Polymerase Chain Reaction (RT-qPCR)

cDNA was synthesized from total RNA using SuperScript III First-Strand Synthesis Supermix (Invitrogen, Carlsbad, CA, USA) and qPCR was performed on the LightCycler 96 system (Roche, Mannheim, Germany) using SsoAdvanced Universal SYBR Green Supermix (Bio-Rad, Hercules, CA, USA). Gene-specific primers used for qPCR are listed in Table S1. mRNA levels of genes were normalized to those of glyceraldehyde-3-phosphate dehydrogenase (*Gapdh*).

4.5. RNA Sequencing (RNA-Seq)

Total RNA was isolated from the cerebral cortex of mouse brains and quantified with a Victor3 fluorometer (Perkin Elmer, Norwalk, CT, USA) using the PicoGreen method (Invitrogen, Carlsbad, CA, USA). The integrity of total RNA was analyzed using a 2100 Bioanalyzer (Agilent Technologies, Palo Alto, CA, USA). RNA samples with RNA Integrity Number (RIN) values ≥ 7 were used to generate RNA-seq libraries. Isolated RNA was treated with DNase to remove DNA contamination and rRNA was removed using the Ribo-Zero rRNA removal kit (Illumina, San Diego, CA, USA). Purified total RNA was randomly fragmented, synthesized into cDNA, and then used to prepare RNA-seq libraries using the TruSeq Stranded mRNA LT Sample Prep Kit (Illumina, San Diego, CA, USA) according to the manufacturer's instructions. The quantity and quality of libraries were assessed and paired-end sequencing with an average length of 100 reads was performed using the NovaSeq 6000 system (Illumina, San Diego, CA, USA). Three replicates were used for sequencing. Complete raw and normalized RNA-seq data have been deposited in Gene Expression Omnibus (GEO) under accession number GSE265861.

4.6. RNA-Seq Data Analysis

Low-quality reads were filtered out by trimming RNA-seq data using Trimmomatic (version 0.38, USA). The remaining reads were then aligned to the UCSC mouse GRCm38/mm10 reference genome using the Bowtie2 aligner (version 2.3.4.1, Johns Hopkins University, Baltimore, MD, USA). To compare expression levels across all samples, we calculated fragments per kilobase of transcript per million mapped reads (FPKM), fold change, and variance using edge R. Transcripts with FPKM > 1 were selected for further analysis. Transcripts exhibiting log2 fold change ≥ 2 and significant pairwise variance ($p < 0.05$) were categorized as differentially expressed genes. Gene set enrichment analysis (GSEA) was conducted utilizing the KEGG database and GO terms.

4.7. H&E and Nissl Staining

Mouse whole-brain tissues were fixed in 4% paraformaldehyde, dehydrated through a series of ethanol solutions and xylene, and embedded in paraffin according to standard protocols. The paraffin blocks were sectioned into 3–5 μm thick slices, which were then cut into coronal sections to ensure comprehensive analysis of ZIKV-induced neurological defects across all brain regions. After dewaxing and rehydration, the paraffin sections were subjected to hematoxylin and eosin (H&E) staining and Nissl staining. Neurons and Nissl bodies in the cortical region were then observed using an inverted microscope (TS2-S-SM; Nikon, Tokyo, Japan).

4.8. Western Blotting

Each mouse brain cortex was lysed with cold RIPA buffer (150 mM NaCl, 50 mM Tris pH 7.5, 0.1% SDS, 1% Triton X-100, 0.5% deoxycholate, protease inhibitor cocktail (Roche, Basel, Switzerland), 1 mM orthovanadate, 10 mM NaF, 100 mM PMSF) with gentle sonication. Protein samples were denatured in sodium dodecyl sulfate (SDS) sample buffer and separated by SDS-PAGE on 10% gels. The proteins were then transferred electrophoretically to a nitrocellulose membrane. Membranes were blocked with 5% non-fat milk in Tris-buffered saline with Tween 20 (TBS-T) for 1 h at 20 °C, followed by incubation with primary antibody (Table S2) overnight at 4 °C. After washing, membranes were incubated with secondary antibody (Table S2) at room temperature for 1 h. Target protein bands were visualized using enhanced chemiluminescence (Thermo Fisher Scientific Inc., Waltham, MA, USA), and signals were detected with the ChemiDoc Imaging System (Bio-Rad, Hercules, CA, USA).

4.9. Immunohistochemistry

For immunohistochemistry, the same paraffin-embedded brain tissues as described in Section 4.7 were used. Sections (3–5 μm) were immersed in xylene for 15 min and then sequentially rehydrated in absolute ethanol and a 95% ethanol solution in distilled water. Antigens were reactivated by treatment with 0.01 M citrate buffer for 50 min at 95 °C. Slides were washed with PBS and then incubated with primary antibodies (Table S2) overnight at 4 °C. After washing with PBS, slides were incubated with secondary antibodies (Table S2) at 25 °C for 1 h and washed twice with PBS. Slides were then covered with ProLong Gold Antifade Mountant with DNA Stain DAPI (Invitrogen, Carlsbad, CA, USA) for 1 min. Slides were imaged using an inverted microscope (Eclipse Ts2R; Nikon, Tokyo, Japan) or a confocal microscope (Zeiss LSM 800, Oberkochen, Germany) at magnifications of 400× and 600×.

4.10. Measurement of ROS

ROS production in brain tissue was measured by MitoSOX Red (Invitrogen, Carlsbad, CA, USA) staining according to the manufacturer's protocol. Whole-brain tissues were rapidly frozen by immersion in dry ice and then stored at −80 °C. The frozen brain tissues were sectioned into 5–7 μm thick coronal slices. Briefly, these sections were incubated with MitoSOX Red for 15 min at 37 °C in the dark. Slides were mounted for 1 min at 25 °C in the dark using ProLong Gold Antifade Mountant with DNA Stain DAPI (Invitrogen, Carlsbad, CA, USA). ROS signals were visualized by fluorescence microscopy (Eclipse Ts2R; Nikon, Tokyo, Japan) and MitoSOX-positive areas were quantified using ImageJ software 1.54j (Media Cybernetics Inc., Rockville, MD, USA).

4.11. Measurement of Cell Apoptosis by TUNEL Assay

DNA fragmentation of apoptotic cells in brain sections was visualized by terminal deoxynucleotidyl transferase-mediated dUTP nick end labeling (TUNEL) staining using an in situ BrdU-Red DNA Fragmentation assay kit (Abcam, Cambridge, MA, USA) according to the manufacturer's instructions. Slides were then counterstained with DAPI for 1 min at 25 °C in the dark. Red fluorescent signals corresponding to TUNEL-positive cells were analyzed using a fluorescence microscope (Nikon Eclipse Ts2R, Tokyo, Japan) and TUNEL-positive cells per field were quantified using ImageJ software1.54j (Media Cybernetics Inc., Rockville, MD, USA).

4.12. Statistical Analyses

Results are shown as mean values ± standard errors of the mean (SEMs). Two-tailed unpaired *t*-tests were applied for statistical analysis.

Supplementary Materials: The supporting information can be downloaded at: https://www.mdpi.com/article/10.3390/ijms25179451/s1.

Author Contributions: Conceptualization, N.K., H.Y.S. and Y.B.K.; methodology, N.K., H.C., U.K. and S.K.; software, N.K.; validation, N.K.; formal analysis, N.K.; investigation, N.K.; resources, H.Y.S. and Y.B.K.; data curation, N.K.; writing—original draft preparation, N.K.; writing—review and editing, H.Y.S.; visualization, N.K.; supervision, H.Y.S. and Y.B.K.; project administration, H.Y.S. and Y.B.K.; funding acquisition, H.Y.S. and Y.B.K. All authors have read and agreed to the published version of the manuscript.

Funding: This research was funded by the Ministry of Health and Welfare of the Republic of Korea (grant number HI18C2177), the Konkuk University Researcher Fund in 2023, the Ministry of Food and Drug Safety (22183MFDS443), and the Korean Health Technology R&D Project (No. HV23C1857 and HV22C0263).

Institutional Review Board Statement: The animal study protocol was approved by the Institutional Animal Care and Use Committee (IACUC) of Konkuk University (KU19213; 4 December 2019).

Informed Consent Statement: Not applicable.

Data Availability Statement: The data presented in this study are available from the corresponding author upon reasonable request.

Acknowledgments: We acknowledge Macrogen for providing the RNA sequencing service and Biorender.com for providing the image sources used in the schematic figure in Figure 2a.

Conflicts of Interest: The authors declare no competing interests.

References

1. Dick, G.W.; Kitchen, S.F.; Haddow, A.J. Zika virus. I. Isolations and serological specificity. *Trans. R. Soc. Trop. Med. Hyg.* **1952**, *46*, 509–520. [CrossRef]
2. de Araujo, T.V.B.; Rodrigues, L.C.; de Alencar Ximenes, R.A.; de Barros Miranda-Filho, D.; Montarroyos, U.R.; de Melo, A.P.L.; Valongueiro, S.; de Albuquerque, M.; Souza, W.V.; Braga, C.; et al. Association between Zika virus infection and microcephaly in Brazil, January to May, 2016: Preliminary report of a case-control study. *Lancet Infect. Dis.* **2016**, *16*, 1356–1363. [CrossRef]
3. Sher, A.A.; Glover, K.K.M.; Coombs, K.M. Zika Virus Infection Disrupts Astrocytic Proteins Involved in Synapse Control and Axon Guidance. *Front. Microbiol.* **2019**, *10*, 596. [CrossRef] [PubMed]
4. da Silva, I.R.F.; Frontera, J.A.; Bispo de Filippis, A.M.; Nascimento, O.; Group, R.-G.-Z.R. Neurologic Complications Associated With the Zika Virus in Brazilian Adults. *JAMA Neurol.* **2017**, *74*, 1190–1198. [CrossRef] [PubMed]
5. Alves-Leon, S.V.; Lima, M.D.R.; Nunes, P.C.G.; Chimelli, L.M.C.; Rabelo, K.; Nogueira, R.M.R.; de Bruycker-Nogueira, F.; de Azeredo, E.L.; Bahia, P.R.; Rueda Lopes, F.C.; et al. Zika virus found in brain tissue of a multiple sclerosis patient undergoing an acute disseminated encephalomyelitis-like episode. *Mult. Scler. J.* **2019**, *25*, 427–430. [CrossRef] [PubMed]
6. Tappe, D.; Perez-Giron, J.V.; Zammarchi, L.; Rissland, J.; Ferreira, D.F.; Jaenisch, T.; Gomez-Medina, S.; Gunther, S.; Bartoloni, A.; Munoz-Fontela, C.; et al. Cytokine kinetics of Zika virus-infected patients from acute to reconvalescent phase. *Med. Microbiol. Immunol.* **2016**, *205*, 269–273. [CrossRef]
7. Pinto-Díaz, C.A.; Rodríguez, Y.; Monsalve, D.M.; Acosta-Ampudia, Y.; Molano-González, N.; Anaya, J.-M.; Ramírez-Santana, C. Autoimmunity in Guillain-Barré syndrome associated with Zika virus infection and beyond. *Autoimmun. Rev.* **2017**, *16*, 327–334. [CrossRef] [PubMed]

8. Figueiredo, C.P.; Barros-Aragão, F.G.; Neris, R.L.; Frost, P.S.; Soares, C.; Souza, I.N.; Zeidler, J.D.; Zamberlan, D.C.; de Sousa, V.L.; Souza, A.S. Zika virus replicates in adult human brain tissue and impairs synapses and memory in mice. *Nat. Commun.* **2019**, *10*, 3890. [CrossRef] [PubMed]
9. Bido-Medina, R.; Wirsich, J.; Rodriguez, M.; Oviedo, J.; Miches, I.; Bido, P.; Tusen, L.; Stoeter, P.; Sadaghiani, S. Impact of Zika Virus on adult human brain structure and functional organization. *Ann. Clin. Transl. Neurol.* **2018**, *5*, 752–762. [CrossRef]
10. Mecharles, S.; Herrmann, C.; Poullain, P.; Tran, T.H.; Deschamps, N.; Mathon, G.; Landais, A.; Breurec, S.; Lannuzel, A. Acute myelitis due to Zika virus infection. *Lancet* **2016**, *387*, 1481. [CrossRef]
11. Chang, A.Y.; Lynch, R.; Martins, K.; Encinales, L.; Cadena Bonfanti, A.A.; Pacheco, N.; Reid, S.P.; Lara Sarabia, O.E.; Gonzalez Torres, H.J.; Mejia Castillo, S.; et al. Long-term clinical outcomes of Zika-associated Guillain-Barre syndrome. *Emerg. Microbes Infect.* **2018**, *7*, 148. [CrossRef]
12. Miner, J.J.; Diamond, M.S. Understanding How Zika Virus Enters and Infects Neural Target Cells. *Cell Stem Cell* **2016**, *18*, 559–560. [CrossRef] [PubMed]
13. van den Pol, A.N.; Mao, G.; Yang, Y.; Ornaghi, S.; Davis, J.N. Zika Virus Targeting in the Developing Brain. *J. Neurosci.* **2017**, *37*, 2161–2175. [CrossRef] [PubMed]
14. Muffat, J.; Li, Y.; Omer, A.; Durbin, A.; Bosch, I.; Bakiasi, G.; Richards, E.; Meyer, A.; Gehrke, L.; Jaenisch, R. Human induced pluripotent stem cell-derived glial cells and neural progenitors display divergent responses to Zika and dengue infections. *Proc. Natl. Acad. Sci. USA* **2018**, *115*, 7117–7122. [CrossRef]
15. Davalos, D.; Grutzendler, J.; Yang, G.; Kim, J.V.; Zuo, Y.; Jung, S.; Littman, D.R.; Dustin, M.L.; Gan, W.B. ATP mediates rapid microglial response to local brain injury in vivo. *Nat. Neurosci.* **2005**, *8*, 752–758. [CrossRef]
16. Liu, B.; Hong, J.S. Role of microglia in inflammation-mediated neurodegenerative diseases: Mechanisms and strategies for therapeutic intervention. *J. Pharmacol. Exp. Ther.* **2003**, *304*, 1–7. [CrossRef]
17. Marinelli, C.; Di Liddo, R.; Facci, L.; Bertalot, T.; Conconi, M.T.; Zusso, M.; Skaper, S.D.; Giusti, P. Ligand engagement of Toll-like receptors regulates their expression in cortical microglia and astrocytes. *J. Neuroinflamm.* **2015**, *12*, 244. [CrossRef]
18. Rajendran, L.; Paolicelli, R.C. Microglia-Mediated Synapse Loss in Alzheimer's Disease. *J. Neurosci.* **2018**, *38*, 2911–2919. [CrossRef] [PubMed]
19. Lee, D.C.; Rizer, J.; Selenica, M.L.; Reid, P.; Kraft, C.; Johnson, A.; Blair, L.; Gordon, M.N.; Dickey, C.A.; Morgan, D. LPS-induced inflammation exacerbates phospho-tau pathology in rTg4510 mice. *J. Neuroinflamm.* **2010**, *7*, 56. [CrossRef]
20. Johnson, G.V.; Stoothoff, W.H. Tau phosphorylation in neuronal cell function and dysfunction. *J. Cell Sci.* **2004**, *117*, 5721–5729. [CrossRef]
21. Bachiller, S.; Jimenez-Ferrer, I.; Paulus, A.; Yang, Y.; Swanberg, M.; Deierborg, T.; Boza-Serrano, A. Microglia in Neurological Diseases: A Road Map to Brain-Disease Dependent-Inflammatory Response. *Front. Cell Neurosci.* **2018**, *12*, 488. [CrossRef] [PubMed]
22. Hickman, S.; Izzy, S.; Sen, P.; Morsett, L.; El Khoury, J. Microglia in neurodegeneration. *Nat. Neurosci.* **2018**, *21*, 1359–1369. [CrossRef] [PubMed]
23. He, Z.; Chen, J.; Zhu, X.; An, S.; Dong, X.; Yu, J.; Zhang, S.; Wu, Y.; Li, G.; Zhang, Y. NLRP3 inflammasome activation mediates Zika virus–associated inflammation. *J. Infect. Dis.* **2018**, *217*, 1942–1951. [CrossRef] [PubMed]
24. Diop, F.; Vial, T.; Ferraris, P.; Wichit, S.; Bengue, M.; Hamel, R.; Talignani, L.; Liegeois, F.; Pompon, J.; Yssel, H.; et al. Zika virus infection modulates the metabolomic profile of microglial cells. *PLoS ONE* **2018**, *13*, e0206093. [CrossRef] [PubMed]
25. Russo, F.B.; Jungmann, P.; Beltrao-Braga, P.C.B. Zika infection and the development of neurological defects. *Cell Microbiol.* **2017**, *19*, e12744. [CrossRef] [PubMed]
26. Bos, S.; Viranaicken, W.; Frumence, E.; Li, G.; Despres, P.; Zhao, R.Y.; Gadea, G. The Envelope Residues E152/156/158 of Zika Virus Influence the Early Stages of Virus Infection in Human Cells. *Cells* **2019**, *8*, 1444. [CrossRef]
27. Lunardelli, V.A.S.; de Souza Apostolico, J.; Souza, H.F.S.; Coirada, F.C.; Martinho, J.A.; Astray, R.M.; Boscardin, S.B.; Rosa, D.S. ZIKV-envelope proteins induce specific humoral and cellular immunity in distinct mice strains. *Sci. Rep.* **2022**, *12*, 15733. [CrossRef]
28. Li, C.; Xu, D.; Ye, Q.; Hong, S.; Jiang, Y.; Liu, X.; Zhang, N.; Shi, L.; Qin, C.F.; Xu, Z. Zika Virus Disrupts Neural Progenitor Development and Leads to Microcephaly in Mice. *Cell Stem Cell* **2016**, *19*, 120–126. [CrossRef]
29. Picard, K.; St-Pierre, M.-K.; Vecchiarelli, H.A.; Bordeleau, M.; Tremblay, M.-È. Neuroendocrine, neuroinflammatory and pathological outcomes of chronic stress: A story of microglial remodeling. *Neurochem. Int.* **2021**, *145*, 104987. [CrossRef] [PubMed]
30. Di Castro, M.A.; Trettel, F.; Milior, G.; Maggi, L.; Ragozzino, D.; Limatola, C. The chemokine CXCL16 modulates neurotransmitter release in hippocampal CA1 area. *Sci. Rep.* **2016**, *6*, 34633. [CrossRef]
31. Manet, C.; Mansuroglu, Z.; Conquet, L.; Bortolin, V.; Comptdaer, T.; Segrt, H.; Bourdon, M.; Menidjel, R.; Stadler, N.; Tian, G. Zika virus infection of mature neurons from immunocompetent mice generates a disease-associated microglia and a tauopathy-like phenotype in link with a delayed interferon beta response. *J. Neuroinflamm.* **2022**, *19*, 307. [CrossRef] [PubMed]
32. Chun, H.; Marriott, I.; Lee, C.J.; Cho, H. Elucidating the interactive roles of glia in Alzheimer's disease using established and newly developed experimental models. *Front. Neurol.* **2018**, *9*, 389660. [CrossRef] [PubMed]
33. Hong, S.; Beja-Glasser, V.F.; Nfonoyim, B.M.; Frouin, A.; Li, S.; Ramakrishnan, S.; Merry, K.M.; Shi, Q.; Rosenthal, A.; Barres, B.A. Complement and microglia mediate early synapse loss in Alzheimer mouse models. *Science* **2016**, *352*, 712–716. [CrossRef] [PubMed]

34. Riedel, G.; Platt, B.; Micheau, J. Glutamate receptor function in learning and memory. *Behav. Brain Res.* **2003**, *140*, 1–47. [CrossRef]
35. Endele, S.; Rosenberger, G.; Geider, K.; Popp, B.; Tamer, C.; Stefanova, I.; Milh, M.; Kortüm, F.; Fritsch, A.; Pientka, F.K. Mutations in GRIN2A and GRIN2B encoding regulatory subunits of NMDA receptors cause variable neurodevelopmental phenotypes. *Nat. Genet.* **2010**, *42*, 1021–1026. [CrossRef]
36. Close, J.L.; Yao, Z.; Levi, B.P.; Miller, J.A.; Bakken, T.E.; Menon, V.; Ting, J.T.; Wall, A.; Krostag, A.-R.; Thomsen, E.R.J.N. Single-cell profiling of an in vitro model of human interneuron development reveals temporal dynamics of cell type production and maturation. *Neuron* **2017**, *93*, 1035–1048.e5. [CrossRef]
37. Tang, G.-B.; Zeng, Y.-Q.; Liu, P.-P.; Mi, T.-W.; Zhang, S.-F.; Dai, S.-K.; Tang, Q.-Y.; Yang, L.; Xu, Y.-J.; Yan, H.-L. The histone H3K27 demethylase UTX regulates synaptic plasticity and cognitive behaviors in mice. *Front. Mol. Neurosci.* **2017**, *10*, 267. [CrossRef]
38. Govindpani, K.; Turner, C.; Waldvogel, H.J.; Faull, R.L.; Kwakowsky, A. Impaired expression of GABA signaling components in the Alzheimer's disease middle temporal gyrus. *Int. J. Mol. Sci.* **2020**, *21*, 8704. [CrossRef] [PubMed]
39. Tang, G.-B.; Mi, T.-W.; Sun, M.-L.; Xu, Y.-J.; Yang, S.-G.; Du, H.-Z.; Teng, Z.-Q.; Gao, J.; Liu, C.-M. Overexpression of serotonin receptor 5b expression rescues neuronal and behavioral deficits in a mouse model of Kabuki syndrome. *IBRO Rep.* **2020**, *9*, 138–146. [CrossRef]
40. Gonzalez-Nunez, V. Role of gabra2, GABA(A) receptor alpha-2 subunit, in CNS development. *Biochem. Biophys. Rep.* **2015**, *3*, 190–201. [CrossRef]
41. Smith, D.R.; Hollidge, B.; Daye, S.; Zeng, X.; Blancett, C.; Kuszpit, K.; Bocan, T.; Koehler, J.W.; Coyne, S.; Minogue, T.; et al. Neuropathogenesis of Zika Virus in a Highly Susceptible Immunocompetent Mouse Model after Antibody Blockade of Type I Interferon. *PLoS Negl. Trop. Dis.* **2017**, *11*, e0005296. [CrossRef]
42. Lull, M.E.; Block, M.L. Microglial activation and chronic neurodegeneration. *Neurotherapeutics* **2010**, *7*, 354–365. [CrossRef] [PubMed]
43. Correale, J. The role of microglial activation in disease progression. *Mult. Scler. J.* **2014**, *20*, 1288–1295. [CrossRef]
44. Dehmelt, L.; Halpain, S. The MAP2/Tau family of microtubule-associated proteins. *Genome Biol.* **2005**, *6*, 204.
45. Bejanin, A.; Schonhaut, D.R.; La Joie, R.; Kramer, J.H.; Baker, S.L.; Sosa, N.; Ayakta, N.; Cantwell, A.; Janabi, M.; Lauriola, M. Tau pathology and neurodegeneration contribute to cognitive impairment in Alzheimer's disease. *Brain* **2017**, *140*, 3286–3300. [CrossRef]
46. Heneka, M.T.; Kummer, M.P.; Stutz, A.; Delekate, A.; Schwartz, S.; Vieira-Saecker, A.; Griep, A.; Axt, D.; Remus, A.; Tzeng, T.-C. NLRP3 is activated in Alzheimer's disease and contributes to pathology in APP/PS1 mice. *Nature* **2013**, *493*, 674–678. [CrossRef]
47. Maphis, N.; Xu, G.; Kokiko-Cochran, O.N.; Jiang, S.; Cardona, A.; Ransohoff, R.M.; Lamb, B.T.; Bhaskar, K. Reactive microglia drive tau pathology and contribute to the spreading of pathological tau in the brain. *Brain* **2015**, *138*, 1738–1755. [CrossRef] [PubMed]
48. Jiang, S.; Maphis, N.M.; Binder, J.; Chisholm, D.; Weston, L.; Duran, W.; Peterson, C.; Zimmerman, A.; Mandell, M.A.; Jett, S.D. Proteopathic tau primes and activates interleukin-1β via myeloid-cell-specific MyD88-and NLRP3-ASC-inflammasome pathway. *Cell Rep.* **2021**, *36*, 109720. [CrossRef] [PubMed]
49. Granholm, A.-C. Long-term effects of SARS-CoV-2 in the brain: Clinical consequences and molecular mechanisms. *J. Clin. Med.* **2023**, *12*, 3190. [CrossRef]
50. Lenz, C.; Slack, M.P.; Shea, K.M.; Reinert, R.R.; Taysi, B.N.; Swerdlow, D.L. Long-Term effects of COVID-19: A review of current perspectives and mechanistic insights. *Crit. Rev. Microbiol.* **2024**, *50*, 315–328. [CrossRef]
51. Blackhurst, B.M.; Funk, K.E. Molecular and Cellular Mechanisms Underlying Neurologic Manifestations of Mosquito-Borne Flavivirus Infections. *Viruses* **2023**, *15*, 2200. [CrossRef]
52. Garber, C.; Soung, A.; Vollmer, L.L.; Kanmogne, M.; Last, A.; Brown, J.; Klein, R.S. T cells promote microglia-mediated synaptic elimination and cognitive dysfunction during recovery from neuropathogenic flaviviruses. *Nat. Neurosci.* **2019**, *22*, 1276–1288. [CrossRef] [PubMed]
53. Nixon, R.A. The role of autophagy in neurodegenerative disease. *Nat. Med.* **2013**, *19*, 983–997. [CrossRef] [PubMed]
54. Kruger, U.; Wang, Y.; Kumar, S.; Mandelkow, E.M. Autophagic degradation of tau in primary neurons and its enhancement by trehalose. *Neurobiol. Aging* **2012**, *33*, 2291–2305. [CrossRef]
55. Rossi, S.L.; Tesh, R.B.; Azar, S.R.; Muruato, A.E.; Hanley, K.A.; Auguste, A.J.; Langsjoen, R.M.; Paessler, S.; Vasilakis, N.; Weaver, S.C. Characterization of a novel murine model to study Zika virus. *Am. J. Trop. Med.* **2016**, *94*, 1362. [CrossRef] [PubMed]
56. Nazerai, L.; Pravsgaard Christensen, J.; Randrup Thomsen, A. A 'furry-tale'of Zika virus infection: What have we learned from animal models? *Viruses* **2019**, *11*, 29. [CrossRef]
57. Gorman, M.J.; Caine, E.A.; Zaitsev, K.; Begley, M.C.; Weger-Lucarelli, J.; Uccellini, M.B.; Tripathi, S.; Morrison, J.; Yount, B.L.; Dinnon, K.H., 3rd; et al. An Immunocompetent Mouse Model of Zika Virus Infection. *Cell Host Microbe* **2018**, *23*, 672–685.e6. [CrossRef]

International Journal of
Molecular Sciences

MDPI

Review

Applications of Biological Therapy for Latent Infections: Benefits and Risks

Yuan Zong, Koju Kamoi *, Miki Miyagaki, Jing Zhang, Mingming Yang, Yaru Zou and Kyoko Ohno-Matsui

Department of Ophthalmology & Visual Science, Graduate School of Medical and Dental Sciences, Tokyo Medical and Dental University, Tokyo 113-8510, Japan; zongyuan666@gmail.com (Y.Z.); miyagakimk@gmail.com (M.M.); zhangj.c@foxmail.com (J.Z.); yangmm-12@outlook.com (M.Y.); alicezouyaru519@gmail.com (Y.Z.); k.ohno.oph@tmd.ac.jp (K.O.-M.)

* Correspondence: koju.oph@tmd.ac.jp; Tel.: +81-3-5803-5302

Abstract: Biological therapies have revolutionized medical treatment by targeting the key mediators or receptors involved in inflammatory responses, thereby effectively suppressing inflammation and achieving beneficial outcomes. They are more advanced than conventional therapies using corticosteroids and immunosuppressants, offering effective solutions for autoimmune diseases, cancer, transplant rejection, and various infectious diseases, including coronavirus disease 2019. Although they exert low immunosuppressive effects, biological therapies can reactivate specific biological targets associated with infections. This review summarizes the currently available biological therapies and discusses their immunosuppressive mechanisms and clinical applications, highlighting the variations in the types and frequencies of infection recurrence induced by different biological agents. Additionally, this review describes the risk factors associated with various biological agents, thus aiding clinicians in selecting the most appropriate biological therapy.

Keywords: biological therapy; latent infection; pathogen reactivation; risk management

1. Introduction

Biological therapies primarily target key mediators or receptors involved in inflammatory responses and effectively inhibit the occurrence and progression of inflammation to achieve therapeutic goals. These therapies have revolutionized the field of medicine, offering effective treatments for autoimmune diseases, tumors, transplant rejection, and various infectious diseases, including coronavirus disease 2019 (COVID-19). They are more advanced than conventional therapies involving corticosteroids and immunosuppressants [1–4]. The currently available biological therapies, their immunosuppressive mechanisms, and clinical applications are summarized in Table 1. Despite exhibiting lower immunosuppressive effects than those of the other treatment modalities, biological therapies may induce specific biological targets related to infection [5]. For example, tumor necrosis factor (TNF)-α inhibitors may lead to tuberculosis reactivation, and anti-CD20 targeted therapy may lead to hepatitis B virus (HBV) reactivation [6–9]. Different biological agents exhibit clinical variations in the types and frequencies of latent infectious pathogen reactivation. This review discusses the risk factors and diseases associated with various biological agents to help clinicians in selecting the most appropriate biological therapy for patients. Specifically, this article focuses on the risks of latent infectious pathogen reactivation associated with different biological agents and suggests preventive measures against various infections, such as tuberculosis (TB) and chronic viral infection reactivation.

Table 1. List of biological therapies, their immunosuppressive mechanisms, and clinical applications.

Category	Representative Drugs	Primary Target	Immunosuppressive Mechanism	Clinical Applications	
				Autoimmunity Diseases	Cancers
TNF-α Inhibitors	Infliximab (Remicade), Adalimumab (Humira), Etanercept (Enbrel)	TNF-α	Inhibits TNF-α, reducing inflammation	Rheumatoid arthritis, Crohn's disease, Psoriasis	-
B-Cell Targeting Agents	Rituximab (Rituxan), Obinutuzumab (Gazyva), Ofatumumab (Arzerra)	CD20	Depletes B cells, impairing humoral immune response	Rheumatoid arthritis, Systemic lupus erythematosus, Vasculitis	Non-Hodgkin lymphoma, Chronic lymphocytic leukemia
Checkpoint Inhibitors	Pembrolizumab (Keytruda), Nivolumab (Opdivo), Ipilimumab (Yervoy)	PD-1, PD-L1, CTLA-4	Activates the immune system, potentially causing immune dysregulation	-	Melanoma, Lung cancer, Renal cell carcinoma, Hodgkin's lymphoma, Head and neck squamous cell carcinoma
IL-6 Inhibitors	Tocilizumab (Actemra), Sarilumab (Kevzara)	IL-6	Inhibits IL-6, affecting acute phase response and systemic immune reactions	Rheumatoid arthritis, Giant cell arteritis, Systemic juvenile idiopathic arthritis	-
IL-17 Inhibitors	Secukinumab (Cosentyx), Ixekizumab (Taltz)	IL-17	Inhibits IL-17, affecting neutrophil recruitment and function	Psoriasis, Psoriatic arthritis, Ankylosing spondylitis	-
JAK Inhibitors	Tofacitinib (Xeljanz), Baricitinib (Olumiant), Upadacitinib (Rinvoq)	JAK-STAT signaling pathway	Inhibits multiple cytokines signaling pathways, broadly suppressing the immune system	Rheumatoid arthritis, Psoriatic arthritis, Ulcerative colitis	-
CCR4 Inhibitors	Mogamulizumab (Poteligeo)	CCR4	Depletes Tregs and Th2 cells, impairing immune regulatory functions	-	Adult T-cell leukemia/lymphoma, Cutaneous T-cell lymphoma
IL-23 Inhibitors	Guselkumab (Tremfya), Risankizumab (Skyrizi), Tildrakizumab (Ilumya)	IL-23	Inhibits IL-23, affecting Th17 cell functions	Psoriasis, Crohn's disease	-
CD38 Inhibitors	Daratumumab (Darzalex), Isatuximab (Sarclisa)	CD38	Broadly depletes CD38-positive cells, impairing immune surveillance	-	Multiple myeloma
T-Cell Costimulation Blockers	Abatacept (Orencia) Belatacept (Nulojix)	CD80/CD86	Inhibits T-cell costimulatory signals, weakening T-cell-mediated immune responses	Rheumatoid arthritis, Organ transplantation, Psoriasis, Psoriatic arthritis	-
IL-1 Inhibitors	Anakinra (Kineret), Canakinumab (Ilaris)	IL-1	Inhibits IL-1, reducing inflammation	Rheumatoid arthritis, Neutrophilic dermatoses	Multiple myeloma
IL-12/IL-23 Inhibitors	Ustekinumab (Stelara)	IL-12/IL-23 p40 subunit	Inhibits IL-12 and IL-23, reducing Th1 and Th17 responses	Psoriasis, Crohn's disease	-
IL-4/IL-13 Inhibitors	Dupilumab (Dupixent)	IL-4Rα	Inhibits IL-4 and IL-13 signaling, reducing inflammation	Atopic dermatitis, Eosinophilic esophagitis, Asthma	-
IL-5 Inhibitors	Mepolizumab (Nucala), Reslizumab (Cinqair), Benralizumab (Fasenra)	IL-5	Inhibits IL-5, reducing eosinophil activity	Eosinophilic asthma	-
Complement Inhibitors	Eculizumab (Soliris)	C5	Inhibits complement component C5, preventing complement-mediated damage	Paroxysmal nocturnal hemoglobinuria, Atypical hemolytic uremic syndrome	-
Integrin Inhibitors	Natalizumab (Tysabri), Vedolizumab (Entyvio)	α4-integrin, α4β7-integrin	Inhibits integrin-mediated cell adhesion, reducing immune cell infiltration	Multiple sclerosis, Crohn's disease, Ulcerative colitis	-
CD52 Inhibitors	Alemtuzumab (Lemtrada)	CD52	Depletes CD52-positive cells, broadly suppressing the immune system	Multiple sclerosis	Chronic lymphocytic leukemia
GM-CSF Inhibitors	Sargramostim (Leukine)	GM-CSF	Modulates GM-CSF signaling, affecting immune cell activation	Bone marrow transplantation	Chronic myelogenous leukemia
IL-2 Receptor Antagonists	Basiliximab (Simulect), Daclizumab (Zinbryta)	IL-2Rα (CD25)	Blocks IL-2 signaling, reducing T-cell proliferation	Prevention of organ transplant rejection, Multiple sclerosis	-

Table 1. *Cont.*

Category	Representative Drugs	Primary Target	Immunosuppressive Mechanism	Clinical Applications	
				Autoimmunity Diseases	Cancers
RANKL Inhibitors	Denosumab (Prolia, Xgeva)	RANKL	Inhibits RANKL, reducing osteoclast activity	Osteoporosis	Bone metastases
Anti-VEGF Antibodies	Bevacizumab (Avastin), Ranibizumab (Lucentis)	VEGF	Inhibits VEGF, reducing angiogenesis	Age-related macular degeneration	Metastatic colorectal cancer, Non-small cell lung cancer, Glioblastoma, Ovarian cancer, Renal cell carcinoma

Abbreviations: TNF-α: Tumor Necrosis Factor-alpha; CD20: Cluster of Differentiation 20; PD-1: Programmed Death-1; PD-L1: Programmed Death-Ligand 1; CTLA-4: Cytotoxic T-Lymphocyte-associated protein 4; IL-6: Interleukin-6; IL-17: Interleukin-17; JAK-STAT signaling pathway: Janus Kinase-Signal Transducer and Activator of Transcription signaling pathway; CCR4: C-C Motif Chemokine Receptor 4; IL-23: Interleukin-23; CD38: Cluster of Differentiation 38; CD80/CD86: Cluster of Differentiation 80/86; IL-1: Interleukin-1; IL-12/IL-23 p40 subunit: Interleukin-12/Interleukin-23 p40 subunit; IL-4Rα: Interleukin-4 Receptor alpha; IL-5: Interleukin-5; C5: Complement Component 5; α4-integrin, α4β7-integrin: alpha-4 Integrin, alpha-4 beta-7 Integrin; CD52: Cluster of Differentiation 52; GM-CSF: Granulocyte-Macrophage Colony-Stimulating Factor; IL-2Rα (CD25): Interleukin-2 Receptor alpha (Cluster of Differentiation 25); RANKL: Receptor Activator of Nuclear Factor kappa-B Ligand; VEGF: Vascular Endothelial Growth Factor.

2. Mycobacterium Tuberculosis (MTB)

2.1. Pathogenesis and Latency

One third of the global population harbors latent TB infection (LTBI) caused by MTB. In individuals with a competent immune system, MTB is engulfed by the alveolar macrophages, leading to the formation of granulomas via the activation of macrophages and T cells. This containment strategy keeps bacteria dormant without overt clinical manifestations, leading to LTBI [10,11]. However, in cases of immunosuppression, the ability of the immune system to maintain the integrity of granulomas diminishes, allowing latent MTB to escape and proliferate, leading to active TB infection [12,13].

2.2. Biologics and Reactivation

TNF-α inhibitors can reactivate LTBI by inhibiting the activity of TNF-α, a critical cytokine involved in the formation and maintenance of granulomas encapsulating the bacteria and activation of macrophages responsible for the phagocytosing and killing of MTB, thereby weakening the ability of the immune system to control latent bacteria, causing the breakdown of granulomas, reducing the efficacy of macrophages, and ultimately allowing the dormant bacteria to escape and proliferate, resulting in active TB infection [14,15]. TNF inhibitors significantly increase the risk of TB [14–20]. Therefore, comprehensive screening, including tuberculin skin tests, is necessary to rule out MTB infections in patients undergoing treatment with biological agents.

Antigen-presenting cells (APCs) induce Th17 cell differentiation and promote the production of interleukin (IL)-17 and IL-22. IL-23 and IL-17 serve as critical targets for autoimmune diseases, such as psoriasis, and form an integral part of immune prophylaxis against TB [21]. To date, no clinical trials and real-world studies have reported TB reactivation in patients with treated or untreated LTBI exposed to IL-23 inhibitors [11,22–25]. Blauvelt et al. documented a case of bone TB in a 58-year-old Asian male patient following the administration of tildrakizumab (200 mg), ultimately necessitating treatment discontinuation; the patient initially tested negative for TB in the single-step Mantoux test [23]. Sporadic cases of active TB have been reported following the introduction of the IL-12/IL-23 dual inhibitor, ustekinumab, in clinical therapy, regardless of the administration of other anti-tuberculosis prophylaxes [10,26–28]. In a Phase III trial of ustekinumab, involving five clinical studies conducted in North America, Europe, and Asia, 3177 patients with psoriasis were evaluated. Among these patients, 167 with LTBI did not develop active TB after receiving isoniazid prophylaxis [29].

Anti-IL-17 therapies, including secukinumab and ixekizumab, exhibit a low risk of MTB reactivation. In a retrospective, multicenter, and multinational study, the tolerability and safety of IL-17 and IL-23 inhibitors were evaluated in patients with psoriasis with newly diagnosed LTBI who received no treatment and in those who received chemoprophylaxis for TB. In this study, only one patient who had not undergone TB screening and did not receive TB chemoprophylaxis was diagnosed with intestinal TB after 14 months of treatment with the IL-17 inhibitor, ixekizumab [11]. To date, no cases of MTB reactivation have been reported in other clinical trials and real-world studies [11,23,30].

Similar to TNF-, IL-6 is a multifunctional pro-inflammatory cytokine involved in a wide range of inflammatory responses and immune regulation, including immune reactions, synovial inflammation, hematopoiesis, and the transition from acute to chronic inflammation [31]. Anti-IL-6 receptor therapy can alleviate various symptoms related to the immune and inflammatory systems owing to its diverse roles in immune and inflammatory system dysfunction. Following a post-marketing research evaluation of all rheumatoid arthritis (RA) cases, the incidence rate of TB after exposure to the IL-6 receptor antibody tocilizumab (TCZ) was found to be 0.05% [32]. Clinical reports have been documented predominantly in countries with a high risk of TB. In a post-marketing surveillance study conducted in Japan in 2008, 3881 RA patients receiving TCZ treatment at a dose of 8 mg/kg every 4 weeks were monitored for 28 weeks, during which four new cases of active TB were observed [33]. Schiff et al. analyzed the cumulative safety data from eight clinical trials of TCZ; among the 4009 patients who received at least one dose of TCZ, seven trials reported eight cases of MTB infection, all of which were newly diagnosed and originated from countries with a high risk of TB [34]. In a retrospective investigation by Itagaki et al., patients treated with biologics and with a history of anti-mycobacterial therapy at Keio University Hospital in Japan from 1 January 2012 to 31 August 2020 were reviewed. This study included 44 individuals who had an established risk of developing TB before initiating TCZ treatment and commenced LTBI therapy as a preventive measure. Among them, two patients developed TB after starting LTBI treatment, both testing positive for interferon-gamma release assays (IGRAs) in TB screening tests [35].

During the COVID-19 pandemic, TCZ was widely used to manage patients with severe COVID-19 infections exhibiting elevated systemic inflammatory marker levels. NOH et al. documented a case of active TB emerging after TCZ administration for COVID-19 [36]. Overall, the risk of TB reactivation with TCZ is relatively low; however, proper screening before use, the implementation of preventive therapy measures, and close monitoring during treatment are essential for risk management.

Owing to its immunosuppressive effects, similar to those of anti-TNF therapy, the Janus kinase (JAK) inhibitor tofacitinib (CP-690,550) also increases the risk of infection, including with TB [37]. Experimental evidence indicates that tofacitinib diminishes the ability of the host to control TB in animal models while enhancing the reactivation of LTBI [38]. Winthrop et al. reviewed the phase II, phase III, and long-term extension clinical trial data of tofacitinib for RA before April 2013. They found that among 5671 patients treated with tofacitinib, 26 cases of active TB were reported, making it the most common opportunistic infection; most TB cases occurred in patients receiving higher doses of the medication. Of the 26 cases, 21 (81%) occurred in countries with a high TB burden [37]. Before initiating tofacitinib therapy, it is essential to conduct screening and treatment for LTBI, particularly in regions with a high prevalence of TB, similar to the approach used with anti-TNF therapies.

Abatacept (CTLA4-Ig), classified as a T-cell stimulation blocker, binds to the co-stimulatory molecules CD80/86 on APCs and CD28 on T cells, thereby inhibiting T-cell activation [39]. Recent studies have revealed the crucial role of the CD28 signaling pathway in combating MTB infections [9,40]. However, current clinical reports indicate that active TB in patients treated with abatacept is a rare event, with a significantly lower incidence than that in patients with RA treated with TNF inhibitors [9,41].

In conclusion, the risk of LTBI reactivation with biological therapies is evident, particularly in patients residing in regions with high TB prevalence or in those with additional TB risk factors. The most significant and well-documented risk is associated with anti-TNF therapy, whereas the risk associated with other non-anti-TNF biologics is relatively low. Nevertheless, before initiating biological therapy, physicians should assess the TB risk of the patient, and, if necessary, conduct TB screening and prophylactic treatment to ensure patient safety and well-being.

3. HBV

3.1. Pathogenesis and Latency

HBV belongs to a family of hepatotropic DNA viruses and is highly prevalent in Asia and Africa. HBV can result in acute and chronic hepatitis, which may progress to liver cirrhosis or hepatocellular carcinoma [42]. Globally, approximately 300 million people are chronically infected with hepatitis B, with more than 500,000 deaths annually from HBV-related diseases [43]. In adults, >95% of HBV infections are self-limiting; however, replication-competent HBV DNA can persist in the liver for several years or even decades. HBV reactivation (HBVr) can occur in patients with evident chronic infection (hepatitis B surface antigen [HBsAg]-positive) and in patients with past markers of HBV infection, as evidenced by the presence of antibodies to the hepatitis B core antigen (anti-HBc) with or without anti-HBs, in the absence of circulating HBsAg [44]. Immunosuppressive therapies, including biologics, can induce immunosuppression, allowing the reactivation of a previously dormant or low-replicating HBV, leading to HBVr [42]. The HBV infection status is closely linked to the risk of HBVr. The risk of hepatitis B reactivation is significantly increased by 5–8 fold in patients who are positive for HBsAg compared with those who are negative for HBsAg but positive for anti-HBc [45,46].

3.2. Biologics and Reactivation

Table 2 summarizes the risk of HBV reactivation faced by the planned use of biological therapies based on the baseline HBV status. Anti-CD20 antibodies, such as rituximab, deplete B lymphocytes by binding to CD20 surface markers. This depletion or dysfunction of B cells may impair the activation and proliferation of CD4-positive T cells specific to HBV, ultimately leading to HBVr infection [6,47]. Evens et al. conducted a meta-analysis of rituximab-associated HBVr cases reported in the United States Food and Drug Administration (FDA) Adverse Event Reporting System (FAERS) MedWatch database from 1997 to 2009. Their findings indicated that patients receiving rituximab-containing therapies had an over 5-fold-increased incidence of HBVr compared with those receiving non-rituximab treatments [48]. In another meta-analysis, conducted by Paul et al., the risk of reactivation was higher in patients who received rituximab-containing chemotherapy. Specifically, among 800 patients receiving rituximab-containing therapies from 10 studies, the reactivation risk was 10% (95% confidence interval [CI] 5.8–16%), compared with a 4.0% risk (95% CI 2.2–6.3%) in 580 patients from eight studies involving non-rituximab chemotherapy [49]. Other anti-CD20 monoclonal antibodies, such as ofatumumab, exert their effects through similar mechanisms [47,50]. Therefore, all patients undergoing anti-CD20 immunochemotherapy should be screened for HBVr. Patients who test positive for HBsAg or HBcAb should begin prophylactic anti-HBV nucleoside therapy (NAT) to prevent HBVr [47,51]. Data from the Phase III GOYA and GALLIUM studies indicate that prophylactic NAT can effectively prevent HBVr in patients with a history of hepatitis B undergoing treatment with obinutuzumab and rituximab, reducing the risk by 91% [50].

Studies using in vitro and animal models have demonstrated that TNF can inhibit HBV replication and stimulate HBV-specific T-cell responses, suggesting a vital role for TNF in the clearance of HBV from infected hepatocytes [52,53]. Consequently, the use of anti-TNF monoclonal antibodies may enable HBV to evade host antiviral defense mechanisms. A systematic review and analysis of case reports indicated that HBVr occurred in approximately 40% (35/89) of HBsAg-positive patients, and approximately 5% (9/168) of HBsAg-negative and anti-HBc-positive patients receiving anti-TNF therapy [54]. Recent meta-analyses have shown that HBsAg-negative/anti-HBc-positive patients receiving anti-TNF therapy without NAT prophylaxis have an HBVr risk of 1% [10]. It is widely recognized that individuals who test positive for HBsAg should receive prophylactic measures. Conversely, patients who are negative for HBsAg but positive for anti-HBc may be evaluated on a case-by-case basis, contingent on the given circumstances [55].

Table 2. Risk of HBV reactivation associated with planned biologic therapy based on baseline HBV status.

Category	HBsAg+ Risk	HBsAg-/Anti-HBc+ Risk
Anti-CD20	High	Moderate
Anti-TNF	High	Low
Anti-IL-6	High	Moderate
Other cytokine inhibitors	High	Moderate
Anti-CCR4	Moderate	Moderate
Immune checkpoint inhibitors	High	Low
JAK inhibitors	High	High

Definitions of risk level: high risk: anticipated incidence of HBVr > 10%; moderate risk: anticipated incidence of HBVr 1–10%; low risk: anticipated incidence of HBVr < 1%.

For cytokine inhibitors targeting factors other than TNF, a meta-analysis indicated that the overall risk of HBVr for HBsAg-positive patients not receiving prophylaxis is high (35.5%), while the risk for HBsAg-negative, anti-HBc-positive patients is moderate (7%) [10,56].

Mogamulizumab (Moga) is a humanized monoclonal antibody targeting C-C chemokine receptor type 4 (CCR4). In the United States, the FDA approved mogamulizumab in 2018 for the treatment of relapsed or refractory mycosis fungoides and Sézary syndrome [57]. In the FAERS database, eight documented cases of HBVr are linked to mogamulizumab use compared to the 2290 cases linked to other drugs. The calculated reporting odds ratio (ROR) was 143.67, with $p < 0.001$ and 95% CI of 71.17–290.04 [57].

Current research on the impact of other biological therapies that directly suppress T cells/B cells on HBVr, particularly in HBsAg-positive patients, remains sparse. A retrospective cohort study observed that four out of eight (50%) HBsAg-positive RA patients experienced HBVr following JAK inhibitor baricitinib therapy. In the same study, out of 207 HBsAg-negative/HBcAb-positive patients, 30 (14%) had HBVr [58]. Another retrospective cohort study revealed that among 116 patients with RA treated with the JAK inhibitor tofacitinib, only two HBsAg-positive patients who did not receive prophylactic treatment experienced HBVr, representing 50% of this subgroup [59].

A recent systematic review analyzed eight studies encompassing 1057 HBsAg-positive patients receiving immune checkpoint inhibitors and reported that HBVr occurred in 18 (1.7%) patients. Notably, the reactivation rate was approximately 11% among 56 patients who did not receive prophylactic HBV treatment [60].

A meta-analysis of four studies estimated the risk of HBVr associated with abatacept as approximately 40% in HBsAg-positive patients and 5% in HBsAg-negative anti-HBc-positive patients who did not receive NAT prophylaxis [60].

In conclusion, HBVr is a significant risk factor for biological therapies, particularly anti-CD20 and anti-TNF treatments. Prophylactic anti-HBV NAT is crucial for HBsAg-positive patients. Emerging biologics, including JAK and immune checkpoint inhibitors, also pose a risk for HBVr. Therefore, comprehensive HBV screening and prophylaxis are essential to ensure patient safety during biological therapy.

4. Hepatitis C Virus

4.1. Pathogenesis and Latency

Approximately 71 million people worldwide suffer from chronic Hepatitis C Virus (HCV) infection, with reactivation predominantly occurring in immunosuppressed individuals [61]. Typically, in patients with chronic HCV infection, the levels of HCV-RNA fluctuate within a stable range of approximately 0.5 $\log_{10}$ IU/mL, and HCVr is defined as an increase in HCV-RNA of $\geq$1 $\log_{10}$ IU/mL compared to baseline levels [61,62]. Approximately 23% of cancer treatment recipients experience HCV reactivation [62]. In comparison to HBVr, the clinical course associated with HCVr is generally milder [62].

4.2. Biologics and Reactivation

Patients with chronic HCV infection are at risk of viral reactivation or acute exacerbation following the initiation of anti-CD20 therapy [63]. Currently, there is insufficient evidence to suggest that anti-TNF treatment increases the risk of viral reactivation in patients with chronic hepatitis C. Nevertheless, caution should be exercised when administering these medications [63–65]. Patients should be closely monitored throughout the course of biological therapy.

5. Herpesviruses

5.1. Pathogenesis and Latency

Varicella zoster virus (VZV) belongs to the herpesvirus family and is prevalent worldwide. VZV is the causative agent of chickenpox (varicella) during primary infection and can later reactivate to cause shingles (herpes zoster [HZ]) [7,66]. Approximately 90% of the global population is estimated to become infected with VZV after reaching adulthood. Following primary infection, VZV establishes a latency period in the sensory ganglia, particularly the dorsal root and cranial nerve ganglia [67]. During latency, the virus remains dormant within neuronal cells and can persist for decades [67]. The reactivation of VZV can occur, and is often triggered by factors such as aging, immunosuppression, and stress, leading to HZ. The reactivation process involves the virus traveling along sensory nerves to the skin, causing the characteristic painful rash of shingles. Immunosuppressive therapies, including corticosteroids and biologics, can increase the risk of VZV reactivation by impairing the host immune response and allowing the dormant virus to become active again [67].

5.2. Biologics and Reactivation

Anti-TNF therapy can lead to the reactivation of VZV, but the associated risk is contentious. In a retrospective cohort study, several European studies indicated that anti-TNF treatment was significantly associated with an increased risk of HZ compared to non-biological disease-modifying antirheumatic drugs (DMARDs) [68–70]. In contrast, studies from the United States have not found an increased risk of HZ in patients receiving anti-TNF therapy compared to those receiving non-biologic DMARDs [71,72]. These conflicting results may be due to the differences in corticosteroid therapy practices between physicians in Europe and the United States [68,73].

Anti-CD20 monoclonal antibodies increase the risk of VZV reactivation. A large phase III randomized clinical trial involving previously untreated patients with follicular lymphoma investigated the anti-CD20 antibodies rituximab and obinutuzumab [74]. The study reported severe VZV infections in 8 of 597 patients (1.3%) in the rituximab group and in 6 of 595 patients (1%) in the obinutuzumab group. In another global phase III study, the Japanese subgroup reported VZV infections in 2 of 58 patients (3.4%) in the rituximab group and in 9 of 65 patients (13.8%) in the obinutuzumab group [75].

JAK inhibitors are associated with a significantly high risk of infectious complications, particularly HZ. In a phase III randomized controlled trial comparing ruxolitinib with the standard treatment for polycythemia vera, HZ occurred in 6% of patients treated with ruxolitinib, whereas no cases were reported in the control group [76]. Furthermore, a phase

IV post-marketing study involving 1144 patients revealed that HZ was the most common infectious complication, with an incidence rate of 8% [77]. Additionally, a meta-analysis conducted in 2017 investigating ruxolitinib-related infectious complications highlighted a substantially increased risk of HZ, with an odds ratio of 7.39 compared to controls [78].

Other therapies, such as abatacept and the CD38 agent (daratumumab) are also associated with HVZ reactivation. However, neither the crude incidence rate nor the adjusted hazard ratio differed significantly between the biological agents. [66,79].

In conclusion, biological therapies, particularly anti-CD20 monoclonal antibodies and JAK inhibitors, are associated with a risk of VZV reactivation. Anti-TNF therapies exhibit mixed results with respect to an increased risk of HZ. Other biologics such as abatacept and anti-CD38 drugs also pose a risk for reactivation. Comprehensive screening and preventive measures are essential to effectively manage these risks.

6. Cytomegalovirus (CMV)

6.1. Pathogenesis and Latency

CMV, a member of the herpesvirus family, is highly prevalent worldwide. CMV infection can be asymptomatic in healthy individuals, but may cause severe disease in immunocompromised patients and neonates. Reports estimate that 66–90% of adults have been infected with CMV worldwide, with the prevalence varying by region and socioeconomic status [80,81]. Following primary infection, CMV establishes latency in various cells including monocytes, macrophages, and endothelial cells. During latency, the virus remains dormant within these cells and persists throughout the lifetime of the host. The reactivation of CMV can occur especially in the context of immunosuppression, such as in organ transplant recipients, individuals with human immunodeficiency virus (HIV)/acquired immunodeficiency syndrome (AIDS), and patients undergoing chemotherapy. Reactivation involves the virus replicating and shedding, which can lead to symptomatic disease, including CMV retinitis, pneumonitis, colitis, and hepatitis [82,83].

6.2. Biologics and Reactivation

Alemtuzumab, a CD52-receptor-targeting antibody that depletes T and B cells, has been approved for the treatment of active relapsing–remitting multiple sclerosis and steroid-refractory severe acute graft-versus-host disease (aGVHD) [84,85]. Severe immunosuppression, which can lead to CMV reactivation, is the most potent side effect of alemtuzumab. A single-center study evaluated 54 patients with various hematological malignancies who underwent allogeneic hematopoietic cell transplantation and were treated with alemtuzumab for steroid-refractory grade III or IV aGVHD. Among the patients, 35 (65%) exhibited positive CMV status in the donor and/or recipient. This study included three cases of CMV-mediated colitis and one case of CMV pneumonia [85].

Patients treated with the anti-CCR4 antibody, mogamulizumab, exhibit increased risk of CMV reactivation. In the FAERS database, 17 cases of CMV-related infections have been reported in patients treated with mogamulizumab compared to the 12,849 cases reported in patients treated with other medications. The calculated ROR was 55.89 ($p < 0.001$; 95% CI: 34.31–91.06) [57].

Anti-CD20 monoclonal antibodies such as rituximab also increase the risk of CMV reactivation; however, the relative risk is lower than that of mogamulizumab and alemtuzumab [57]. Prospective studies on the effect of anti-TNF therapy on viral reactivation in patients with RA or Crohn's have not found any evidence of systemic CMV reactivation [86,87]. Clinical reports have also indicated only a few cases of severe CMV reactivation in patients undergoing anti-TNF therapy, including instances of CMV retinitis, CMV hepatitis, and disseminated infection [88].

7. Other Herpesviruses

Epstein–Barr virus (EBV) primarily targets the B lymphocytes, integrating into their genome and immortalizing the cells. During the latent phase of EBV infection, the immune system prevents viral reactivation through effective cellular immune responses [89]. Healthy individuals typically do not exhibit symptoms of EBV reactivation; however, immunocompromised individuals exhibit symptoms similar to those of initial EBV infection [89]. Although most EBV-infected individuals do not exhibit any symptoms, this virus can contribute to the onset of lymphoid and epithelial malignancies. Posttransplant lymphoproliferative disorder (PTLD) is a lymphoid or plasmacytic proliferative disease associated with immunosuppressive therapy following organ transplantation. EBV reactivation under immunosuppression is a significant risk factor for PTLD [90]. Belatacept (Nulojix) is a selective T-cell co-stimulation blocker that inhibits CD28-mediated T-cell co-stimulation by binding to CD80/86 ligands. It has been approved by the FDA for use in combination therapy to prevent kidney transplant rejection in patients seropositive for EBV. However, in a clinical trial, 0–4% of patients treated with belatacept developed PTLD [91].

Herpes simplex virus (HSV) and VZV belong to the same subgroup of herpesviruses (*Alphaherpesvirinae*) [67]. However, the risk of HSV reactivation is lower with biologics compared to VZV. Severe HSV diseases, including HSV encephalitis, HSV esophagitis, and disseminated cutaneous infections, have been reported in patients receiving TNFα-targeted therapies [92–95]. In a phase III randomized controlled trial investigating the effects of the anti-CD20 antibody, ocrelizumab, on primary progressive multiple sclerosis, oral HSV reactivation was more common in patients treated with ocrelizumab than in those receiving the placebo [96].

8. Polyomaviridae Family

8.1. Pathogenesis and Latency

Polyomaviridae is a family of small non-enveloped DNA viruses that infect various vertebrate hosts, including humans. Polyomaviruses are typically associated with asymptomatic infections; however, they can cause serious diseases under specific conditions. Human polyomaviruses, such as the BK virus (BKV) and JC virus (JCV), are widespread and usually acquired during childhood. Once acquired, these viruses can cause persistent infections in the host [97–99].

Polyomaviruses can remain latent in the kidneys, lymphoid tissues, and other organs throughout the host life. Reactivation of these viruses can occur in immunocompromised individuals, leading to various clinical manifestations. For instance, BKV reactivation is commonly associated with nephropathy in renal transplant patients, whereas JCV reactivation leads to progressive multifocal leukoencephalopathy (PML), a potentially fatal demyelinating disease of the central nervous system [98,99]. Immunosuppressive therapies, including those used for organ transplantation and autoimmune diseases, significantly increase the risk of polyomavirus reactivation.

8.2. Biologics and Reactivation

Treatment with biological agents is associated with the reactivation of JCV. Efalizumab, also known as raptiva, is a monoclonal antibody targeting CD11a that is used to treat psoriasis. However, as three of four patients receiving raptiva developed PML and died in previous studies, the drug has been withdrawn from the market [100,101]. Natalizumab (Tysabri) is a humanized monoclonal antibody targeting the alpha-4 integrin chain that is used to treat multiple sclerosis and Crohn's disease [101,102]. Approximately 400 documented cases of PML associated with natalizumab use were reported from 2005 to August 2013 [100,101]. Alemtuzumab targets the CD52 antigen, resulting in the destruction of B and T cells. It has been approved by the FDA for the treatment of B-cell chronic lymphocytic leukemia. Recently, 14 cases of PML have been reported in patients treated with alemtuzumab [103]. Some patients develop PML after receiving rituximab for the treatment of RA, non-Hodgkin lymphoma, and chronic lymphocytic leukemia [104,105]. Notably, alemtuzumab does not increase the risk of BKV infection, but it reduces immunosuppression; hence, switching to calcineurin inhibitor-free regimens is the most viable therapeutic option [8,106].

9. Retroviruses

Retroviruses are a unique class of RNA viruses characterized by their distinct replication mechanisms [107]. Their genomes consist of single-stranded RNA; however, after infecting the host, these viruses use reverse transcriptase to reverse-transcribe their RNA into DNA, which is subsequently integrated into the host genome. Among the retroviruses, HIV and human T-cell leukemia virus type 1 (HTLV-1) are mostly associated with human pathogenicity. Both primarily infect the CD4+ T lymphocytes [107–110].

Several weeks after the initial HIV infection, this virus enters the acute phase, followed by a latent period. Despite the absence of overt symptoms during this stage, HIV continues to replicate within the body and progressively weakens the immune system. When the number of CD4+ T cells decreases to a critical threshold, the patient's immune system becomes severely compromised, making them more susceptible to opportunistic infections and even cancer. This stage is referred to as the AIDS stage [111].

The overall safety profile of biological agents is generally favorable for HIV-infected individuals. A systematic review of 112 studies up to the year 2022 examined the use of biological therapies in 179 patients with HIV. The findings indicated that almost all categories of biological agents exhibit favorable safety profiles with minimal adverse events of mostly mild severity [112]. Therefore, HIV screening is necessary before initiating biological therapy, especially for patients engaging in high-risk behaviors. For HIV-positive patients, biological therapy should be administered only to those patients in a stable condition with CD4 cell counts > 200/mL [88].

Globally, HTLV-1 infects approximately 5–10 million people, with the vast majority being asymptomatic carriers. The most common diseases associated with HTLV-1 infection are adult T-cell leukemia/lymphoma (ATL), HTLV-1-associated myelopathy, and HTLV-1 uveitis [113–118]. TNF-α inhibitors do not affect the reactivation of HTLV-1 [119–122]. Some patients develop ATL after receiving the rituximab, cyclophosphamide, doxorubicin hydrochloride (hydroxydaunorubicin), vincristine sulfate (oncovin), and prednisone combination chemotherapy for B-cell lymphoma-associated hemophagocytic syndrome [123]. One patient with RA carrying the HTLV-1 virus was previously diagnosed with ATL after long-term treatment with TCZ [114,124]. Another case report also documented a patient with RA who experienced an exacerbation of HTLV-1-associated myelopathy and HTLV-1 uveitis following treatment with TCZ [125]. However, no direct evidence linking the development of HTLV-1-associated diseases to the use of biological agents has been reported to date.

10. Conclusions

Biological therapies are increasingly used for autoimmune diseases, cancer treatment, and organ transplantation. These therapies target specific molecular and cellular pathways, significantly improving patient prognosis. However, different biological therapies exhibit varying risks of reactivating latent pathogens owing to their distinct targets. Therefore, associations between different biological therapies and viral infections must be explored to better understand and manage the risk of pathogen reactivation. In clinical practice, potential risks and benefits should be considered to select the appropriate biological therapy. Furthermore, preventive measures must be developed to decrease the risk of latent pathogen reactivation.

Author Contributions: Conceptualization, K.K.; Writing—Original Draft Preparation, Y.Z. (Yuan Zong); Writing—Review and Editing, K.K., M.M., J.Z., M.Y., Y.Z. (Yaru Zou), and K.O.-M.; Funding Acquisition, K.K. All authors have read and agreed to the published version of the manuscript.

Funding: This work was supported by JSPS KAKENHI [grant number JP 20K09824], a grant for Rare and Intractable Diseases from the Ministry of Health, Labor, and Welfare of Japan [grant number 22FC0201], a Research Program on Emerging and Re-emerging Infectious Diseases grant from the Japan Agency for Medical Research and Development, AMED [grant numbers 23fk0108671h0001 and 23fk0108672h0001], and a High-Risk Emerging Infectious Diseases Research grant from Takeda Science Foundation [grant number FY2023].

Institutional Review Board Statement: Not applicable.

Informed Consent Statement: Not applicable.

Data Availability Statement: All data related to this study are presented and published here.

Conflicts of Interest: The authors declare no conflicts of interest.

References

1. Cai, Z.; Wang, S.; Li, J. Treatment of Inflammatory Bowel Disease: A Comprehensive Review. *Front. Med.* **2021**, *8*, 2681. [CrossRef] [PubMed]
2. Walsh, G. Biopharmaceuticals, an overview. In *Biopharmaceuticals, an Industrial Perspective*; Walsh, G., Murphy, B., Eds.; Springer: Dordrecht, The Netherlands, 1999; pp. 1–34.
3. Hansel, T.T.; Kropshofer, H.; Singer, T.; Mitchell, J.A.; George, A.J.T. The safety and side effects of monoclonal antibodies. *Nat. Rev. Drug Discov.* **2010**, *9*, 325–338. [CrossRef] [PubMed]
4. Taylor, P.C.; Adams, A.C.; Hufford, M.M.; de la Torre, I.; Winthrop, K.; Gottlieb, R.L. Neutralizing monoclonal antibodies for treatment of COVID-19. *Nat. Rev. Immunol.* **2021**, *21*, 382–393. [CrossRef]
5. Winthrop, K.L. The emerging safety profile of JAK inhibitors in rheumatic disease. *Nat. Rev. Rheumatol.* **2017**, *13*, 234–243. [CrossRef]
6. Anvari, S.; Tsoi, K. Hepatitis B Virus Reactivation with Immunosuppression: A Hidden Threat? *J. Clin. Med.* **2024**, *13*, 393. [CrossRef]
7. Gentile, G.; Foà, R. Viral infections associated with the clinical use of monoclonal antibodies. *Clin. Microbiol. Infect.* **2011**, *17*, 1769–1775. [CrossRef]
8. Noreña, I.; Fernández-Ruiz, M.; Aguado, J.M. Viral infections in the biologic therapy era. *Expert Rev. Anti-Infect. Ther.* **2018**, *16*, 781–791. [CrossRef] [PubMed]
9. Cantini, F.; Nannini, C.; Niccoli, L.; Petrone, L.; Ippolito, G.; Goletti, D. Risk of Tuberculosis Reactivation in Patients with Rheumatoid Arthritis, Ankylosing Spondylitis, and Psoriatic Arthritis Receiving Non-Anti-TNF-Targeted Biologics. *Mediat. Inflamm.* **2017**, *2017*, 8909834. [CrossRef]
10. Mastorino, L.; Dapavo, P.; Trunfio, M.; Avallone, G.; Rubatto, M.; Calcagno, A.; Ribero, S.; Quaglino, P. Risk of Reactivation of Latent Tuberculosis in Psoriasis Patients on Biologic Therapies: A Retrospective Cohort from a Tertiary Care Centre in Northern Italy. *Acta Derm. Venereol.* **2022**, *102*, adv00821. [CrossRef]
11. Torres, T.; Chiricozzi, A.; Puig, L.; Lé, A.M.; Marzano, A.V.; Dapavo, P.; Dauden, E.; Carrascosa, J.-M.; Lazaridou, E.; Duarte, G.; et al. Treatment of Psoriasis Patients with Latent Tuberculosis Using IL-17 and IL-23 Inhibitors: A Retrospective, Multinational, Multicentre Study. *Am. J. Clin. Dermatol.* **2024**, *25*, 333–342. [CrossRef]
12. Gupta, A.; Kaul, A.; Tsolaki, A.G.; Kishore, U.; Bhakta, S. Mycobacterium tuberculosis: Immune evasion, latency and reactivation. *Immunobiology* **2012**, *217*, 363–374. [CrossRef] [PubMed]
13. Langan, E.A.; Graetz, V.; Allerheiligen, J.; Zillikens, D.; Rupp, J.; Terheyden, P. Immune checkpoint inhibitors and tuberculosis: An old disease in a new context. *Lancet Oncol.* **2020**, *21*, e55–e65. [CrossRef] [PubMed]
14. Robert, M.; Miossec, P. Reactivation of latent tuberculosis with TNF inhibitors: Critical role of the beta 2 chain of the IL-12 receptor. *Cell. Mol. Immunol.* **2021**, *18*, 1644–1651. [CrossRef] [PubMed]
15. Sodenkamp, J.; Waetzig, G.H.; Scheller, J.; Seegert, D.; Grötzinger, J.; Rose-John, S.; Ehlers, S.; Hölscher, C. Therapeutic targeting of interleukin-6 trans-signaling does not affect the outcome of experimental tuberculosis. *Immunobiology* **2012**, *217*, 996–1004. [CrossRef]
16. Gómez-Reino, J.J.; Carmona, L.; Valverde, V.R.; Mola, E.M.; Montero, M.D. Treatment of rheumatoid arthritis with tumor necrosis factor inhibitors may predispose to significant increase in tuberculosis risk: A multicenter active-surveillance report. *Arthritis Rheum.* **2003**, *48*, 2122–2127. [CrossRef]
17. Long, R.; Gardam, M. Tumour necrosis factor-α inhibitors and the reactivation of latent tuberculosis infection. *Can. Med. Assoc. J.* **2003**, *168*, 1153.
18. Reichmann, M.T.; Marshall, B.G.; Cummings, F.; Elkington, P.T. Tuberculosis and TNF-inhibitors: History of exposure should outweigh investigations. *BMJ Case Rep.* **2014**, *2014*, bcr2013202127. [CrossRef]
19. Sánchez-Moya, A.I.; García-Doval, I.; Carretero, G.; Sánchez-Carazo, J.; Ferrandiz, C.; Herrera Ceballos, E.; Alsina, M.; Ferrán, M.; López-Estebaranz, J.L.; Gómez-García, F.; et al. Latent tuberculosis infection and active tuberculosis in patients with psoriasis: A study on the incidence of tuberculosis and the prevalence of latent tuberculosis disease in patients with moderate-severe psoriasis in Spain. BIOBADADERM registry. *J. Eur. Acad. Dermatol. Venereol.* **2013**, *27*, 1366–1374. [CrossRef]
20. Cataño, J.; Morales, M. Isoniazid toxicity and TB development during biological therapy of patients with psoriasis in Colombia. *J. Dermatol. Treat.* **2016**, *27*, 414–417. [CrossRef]
21. Gaffen, S.L.; Jain, R.; Garg, A.V.; Cua, D.J. The IL-23-IL-17 immune axis: From mechanisms to therapeutic testing. *Nat. Rev. Immunol.* **2014**, *14*, 585–600. [CrossRef]

22. Ruggiero, A.; Picone, V.; Martora, F.; Fabbrocini, G.; Megna, M. Guselkumab, Risankizumab, and Tildrakizumab in the Management of Psoriasis: A Review of the Real-World Evidence. *Clin. Cosmet. Investig. Dermatol.* **2022**, *15*, 1649–1658. [CrossRef]
23. Blauvelt, A.; Reich, K.; Papp, K.A.; Kimball, A.B.; Gooderham, M.; Tyring, S.K.; Sinclair, R.; Thaçi, D.; Li, Q.; Cichanowitz, N.; et al. Safety of tildrakizumab for moderate-to-severe plaque psoriasis: Pooled analysis of three randomized controlled trials. *Br. J. Dermatol.* **2018**, *179*, 615–622. [CrossRef]
24. Megna, M.; Patruno, C.; Bongiorno, M.R.; Gambardella, A.; Guarneri, C.; Foti, C.; Lembo, S.; Loconsole, F.; Fabbrocini, G. Lack of reactivation of tuberculosis in patients with psoriasis treated with secukinumab in a real-world setting of latent tuberculosis infection. *J. Dermatol. Treat.* **2022**, *33*, 2629–2633. [CrossRef]
25. Huang, Y.-W.; Tsai, T.-F. A drug safety evaluation of risankizumab for psoriasis. *Expert Opin. Drug Saf.* **2020**, *19*, 395–402. [CrossRef] [PubMed]
26. Tsai, T.F.; Chiu, H.Y.; Song, M.; Chan, D. A case of latent tuberculosis reactivation in a patient treated with ustekinumab without concomitant isoniazid chemoprophylaxis in the PEARL trial. *Br. J. Dermatol.* **2013**, *168*, 444–446. [CrossRef] [PubMed]
27. Lynch, M.; Roche, L.; Horgan, M.; Ahmad, K.; Hackett, C.; Ramsay, B. Peritoneal tuberculosis in the setting of ustekinumab treatment for psoriasis. *JAAD Case Rep.* **2017**, *3*, 230–232. [CrossRef] [PubMed]
28. Enzo, E.; Angelo, P. Latent tuberculosis reactivation in a patient with erythrodermic psoriasis under treatment with ustekinumab and a low dose steroid, despite isoniazid chemoprophylaxis. *Eur. J. Dermatol.* **2014**, *24*, 508–509.
29. Tsai, T.F.; Ho, V.; Song, M.; Szapary, P.; Kato, T.; Wasfi, Y.; Li, S.; Shen, Y.K.; Leonardi, C.; on behalf of the PHOENIX 1, PHOENIX 2, ACCEPT, PEARL; et al. The safety of ustekinumab treatment in patients with moderate-to-severe psoriasis and latent tuberculosis infection. *Br. J. Dermatol.* **2012**, *167*, 1145–1152. [CrossRef] [PubMed]
30. Manzanares, N.; Vilarrasa, E.; López, A.; Alonso, M.L.; Velasco, M.; Riera, J.; del Alcázar, E.; Carrascosa, J.M.; Azón, A.; Rivera, R.; et al. No tuberculosis reactivations in psoriasis patients initiating new generation biologics despite untreated latent tuberculosis infection: Multicenter case series of 35 patients. *J. Eur. Acad. Dermatol. Venereol.* **2024**, *38*, e26–e28. [CrossRef]
31. Choy, E.H.; De Benedetti, F.; Takeuchi, T.; Hashizume, M.; John, M.R.; Kishimoto, T. Translating IL-6 biology into effective treatments. *Nat. Rev. Rheumatol.* **2020**, *16*, 335–345. [CrossRef]
32. Tokuda, H.; Harigai, M.; Kameda, H.; Tomono, K.; Takayanagi, N.; Watanabe, A.; Tasaka, S.; Suda, T.; Tateda, K.; Kadota, J. Consensus statements for medical practice: Biological agents and lung disease [Abridged English translation by the Japanese Respiratory Society]. *Respir. Investig.* **2017**, *55*, 229–251. [CrossRef] [PubMed]
33. Koike, T.; Harigai, M.; Inokuma, S.; Ishiguro, N.; Ryu, J.; Takeuchi, T.; Takei, S.; Tanaka, Y.; Ito, K.; Yamanaka, H. Postmarketing surveillance of tocilizumab for rheumatoid arthritis in Japan: Interim analysis of 3881 patients. *Ann. Rheum. Dis* **2011**, *70*, 2148–2151. [CrossRef] [PubMed]
34. Schiff, M.H.; Kremer, J.M.; Jahreis, A.; Vernon, E.; Isaacs, J.D.; van Vollenhoven, R.F. Integrated safety in tocilizumab clinical trials. *Arthritis Res. Ther.* **2011**, *13*, R141. [CrossRef]
35. Itagaki, M.; Iketani, O.; Enoki, Y.; Chuang, V.T.G.; Taguchi, K.; Uno, S.; Uchida, S.; Namkoong, H.; Uwamino, Y.; Takano, Y.; et al. Analysis of Risk Factors for Developing Tuberculosis in Patients Who Received Prophylactic Latent Tuberculosis Infection Treatment with Experience of Biologic Medications. *Biol. Pharm. Bull.* **2023**, *46*, 1832–1837. [CrossRef]
36. Noh, S.; Dronavalli, G. Active TB after the Use of Tocilizumab for COVID-19 Infection. *Chest* **2021**, *160*, A289. [CrossRef]
37. Winthrop, K.L.; Park, S.H.; Gul, A.; Cardiel, M.H.; Gomez-Reino, J.J.; Tanaka, Y.; Kwok, K.; Lukic, T.; Mortensen, E.; Ponce de Leon, D.; et al. Tuberculosis and other opportunistic infections in tofacitinib-treated patients with rheumatoid arthritis. *Ann. Rheum. Dis.* **2016**, *75*, 1133–1138. [CrossRef]
38. Maiga, M.; Lun, S.; Guo, H.; Winglee, K.; Ammerman, N.C.; Bishai, W.R. Risk of Tuberculosis Reactivation with Tofacitinib (CP-690550). *J. Infect. Dis.* **2012**, *205*, 1705–1708. [CrossRef]
39. Bonelli, M.; Scheinecker, C. How does abatacept really work in rheumatoid arthritis? *Curr. Opin. Rheumatol.* **2018**, *30*, 295–300. [CrossRef]
40. Bernal-Fernandez, G.; Espinosa-Cueto, P.; Leyva-Meza, R.; Mancilla, N.; Mancilla, R. Decreased expression of T-cell costimulatory molecule CD28 on CD4 and CD8 T cells of mexican patients with pulmonary tuberculosis. *Tuberc. Res. Treat.* **2010**, *2010*, 517547. [CrossRef]
41. Martín Mola, E.; Balsa, A.; Martínez Taboada, V.; Sanmartí, R.; Marenco, J.L.; Navarro Sarabia, F.; Gómez-Reino, J.; Álvaro-Gracia, J.M.; Román Ivorra, J.A.; Lojo, L.; et al. Abatacept Use in Rheumatoid Arthritis: Evidence Review and Recommendations. *Reumatol. Clínica* **2013**, *9*, 5–17. [CrossRef]
42. Shi, Y.; Zheng, M. Hepatitis B virus persistence and reactivation. *BMJ* **2020**, *370*, m2200. [CrossRef] [PubMed]
43. Hsu, Y.C.; Huang, D.Q.; Nguyen, M.H. Global burden of hepatitis B virus: Current status, missed opportunities and a call for action. *Nat. Rev. Gastroenterol. Hepatol.* **2023**, *20*, 524–537. [CrossRef]
44. Clerico, M.; Dogliotti, I.; Ghione, P.; Zilioli, V.R.; Merli, F.; Botto, B.; Al Essa, W.; Battaglini, M.; Grimaldi, D.; Cervi, L.; et al. HBV Reactivation in Patients with Past Infection Affected by Non-Hodgkin Lymphoma and Treated with Anti-CD20 Antibody Based Immuno-Chemotherapy: A Multicenter Experience. *J. Pers. Med.* **2022**, *12*, 285. [CrossRef]
45. Chiu, H.Y.; Chiu, Y.M.; Chang Liao, N.F.; Chi, C.C.; Tsai, T.F.; Hsieh, C.Y.; Hsieh, T.Y.; Lai, K.L.; Chiu, T.M.; Wu, N.L.; et al. Predictors of hepatitis B and C virus reactivation in patients with psoriasis treated with biologic agents: A 9-year multicenter cohort study. *J. Am. Acad. Dermatol.* **2021**, *85*, 337–344. [CrossRef]

46. Lau, G.; Yu, M.-L.; Wong, G.; Thompson, A.; Ghazinian, H.; Hou, J.-L.; Piratvisuth, T.; Jia, J.-D.; Mizokami, M.; Cheng, G.; et al. APASL clinical practice guideline on hepatitis B reactivation related to the use of immunosuppressive therapy. *Hepatol. Int.* **2021**, *15*, 1031–1048. [CrossRef] [PubMed]
47. Cao, X.; Wang, Y.; Li, P.; Huang, W.; Lu, X.; Lu, H. HBV Reactivation during the Treatment of Non-Hodgkin Lymphoma and Management Strategies. *Front. Oncol.* **2021**, *11*, 68570. [CrossRef] [PubMed]
48. Evens, A.M.; Jovanovic, B.D.; Su, Y.C.; Raisch, D.W.; Ganger, D.; Belknap, S.M.; Dai, M.S.; Chiu, B.C.; Fintel, B.; Cheng, Y.; et al. Rituximab-associated hepatitis B virus (HBV) reactivation in lymphoproliferative diseases: Meta-analysis and examination of FDA safety reports. *Ann. Oncol.* **2011**, *22*, 1170–1180. [CrossRef]
49. Paul, S.; Dickstein, A.; Saxena, A.; Terrin, N.; Viveiros, K.; Balk, E.M.; Wong, J.B. Role of surface antibody in hepatitis B reactivation in patients with resolved infection and hematologic malignancy: A meta-analysis. *Hepatology* **2017**, *66*, 379–388. [CrossRef]
50. Kusumoto, S.; Arcaini, L.; Hong, X.; Jin, J.; Kim, W.S.; Kwong, Y.L.; Peters, M.G.; Tanaka, Y.; Zelenetz, A.D.; Kuriki, H.; et al. Risk of HBV reactivation in patients with B-cell lymphomas receiving obinutuzumab or rituximab immunochemotherapy. *Blood* **2019**, *133*, 137–146. [CrossRef]
51. Di Bisceglie, A.M.; Lok, A.S.; Martin, P.; Terrault, N.; Perrillo, R.P.; Hoofnagle, J.H. Recent US Food and Drug Administration warnings on hepatitis B reactivation with immune-suppressing and anticancer drugs: Just the tip of the iceberg? *Hepatology* **2015**, *61*, 703–711. [CrossRef]
52. Herbein, G.; O'Brien, W.A. Tumor necrosis factor (TNF)-alpha and TNF receptors in viral pathogenesis. *Proc. Soc. Exp. Biol. Med.* **2000**, *223*, 241–257. [CrossRef] [PubMed]
53. Kasahara, S.; Ando, K.; Saito, K.; Sekikawa, K.; Ito, H.; Ishikawa, T.; Ohnishi, H.; Seishima, M.; Kakumu, S.; Moriwaki, H. Lack of tumor necrosis factor alpha induces impaired proliferation of hepatitis B virus-specific cytotoxic T lymphocytes. *J. Virol.* **2003**, *77*, 2469–2476. [CrossRef]
54. Pérez-Alvarez, R.; Díaz-Lagares, C.; García-Hernández, F.; Lopez-Roses, L.; Brito-Zerón, P.; Pérez-de-Lis, M.; Retamozo, S.; Bové, A.; Bosch, X.; Sanchez-Tapias, J.M.; et al. Hepatitis B virus (HBV) reactivation in patients receiving tumor necrosis factor (TNF)-targeted therapy: Analysis of 257 cases. *Medicine* **2011**, *90*, 359–371. [CrossRef]
55. Terrault, N.A.; Lok, A.S.F.; McMahon, B.J.; Chang, K.M.; Hwang, J.P.; Jonas, M.M.; Brown, R.S., Jr.; Bzowej, N.H.; Wong, J.B. Update on prevention, diagnosis, and treatment of chronic hepatitis B: AASLD 2018 hepatitis B guidance. *Hepatology* **2018**, *67*, 1560–1599. [CrossRef] [PubMed]
56. Mezzacappa, C.; Lim, J.K. Management of HBV reactivation: Challenges and opportunities. *Clin. Liver Dis.* **2024**, *23*, e0143. [CrossRef]
57. Wang, S.; Jayarangaiah, A.; Malone, M.; Elrafei, T.; Steinberg, L.; Kumar, A. Risk of hepatitis B reactivation and cytomegalovirus related infections with Mogamulizumab: A retrospective study of international pharmacovigilance database. *eClinicalMedicine* **2020**, *28*, 100601. [CrossRef]
58. Harigai, M.; Winthrop, K.; Takeuchi, T.; Hsieh, T.-Y.; Chen, Y.-M.; Smolen, J.S.; Burmester, G.; Walls, C.; Wu, W.-S.; Dickson, C.; et al. Evaluation of hepatitis B virus in clinical trials of baricitinib in rheumatoid arthritis. *RMD Open* **2020**, *6*, e001095. [CrossRef]
59. Chen, Y.-M.; Huang, W.-N.; Wu, Y.-D.; Lin, C.-T.; Chen, Y.-H.; Chen, D.-Y.; Hsieh, T.-Y. Reactivation of hepatitis B virus infection in patients with rheumatoid arthritis receiving tofacitinib: A real-world study. *Ann. Rheum. Dis.* **2018**, *77*, 780–782. [CrossRef] [PubMed]
60. Papatheodoridis, G.V.; Lekakis, V.; Voulgaris, T.; Lampertico, P.; Berg, T.; Chan, H.L.Y.; Kao, J.-H.; Terrault, N.; Lok, A.S.; Reddy, K.R. Hepatitis B virus reactivation associated with new classes of immunosuppressants and immunomodulators: A systematic review, meta-analysis, and expert opinion. *J. Hepatol.* **2022**, *77*, 1670–1689. [CrossRef]
61. Mustafayev, K.; Torres, H. Hepatitis B virus and hepatitis C virus reactivation in cancer patients receiving novel anticancer therapies. *Clin. Microbiol. Infect.* **2022**, *28*, 1321–1327. [CrossRef]
62. Torres, H.A.; Hosry, J.; Mahale, P.; Economides, M.P.; Jiang, Y.; Lok, A.S. Hepatitis C virus reactivation in patients receiving cancer treatment: A prospective observational study. *Hepatology* **2018**, *67*, 36–47. [CrossRef] [PubMed]
63. Mahale, P.; Kontoyiannis, D.P.; Chemaly, R.F.; Jiang, Y.; Hwang, J.P.; Davila, M.; Torres, H.A. Acute exacerbation and reactivation of chronic hepatitis C virus infection in cancer patients. *J. Hepatol.* **2012**, *57*, 1177–1185. [CrossRef] [PubMed]
64. Asthana, A.K.; Lubel, J.S. Reactivation of latent viruses after treatment with biological therapies. *Virus Adapt. Treat.* **2014**, *6*, 1–10. [CrossRef]
65. Raychaudhuri, S.P.; Nguyen, C.T.; Raychaudhuri, S.K.; Gershwin, M.E. Incidence and nature of infectious disease in patients treated with anti-TNF agents. *Autoimmun. Rev.* **2009**, *9*, 67–81. [CrossRef] [PubMed]
66. Ho, D.Y.; Enriquez, K.; Multani, A. Herpesvirus Infections Potentiated by Biologics. *Infect. Dis. Clin. N. Am.* **2020**, *34*, 311–339. [CrossRef]
67. Tayyar, R.; Ho, D. Herpes Simplex Virus and Varicella Zoster Virus Infections in Cancer Patients. *Viruses* **2023**, *15*, 439. [CrossRef]
68. Strangfeld, A.; Listing, J.; Herzer, P.; Liebhaber, A.; Rockwitz, K.; Richter, C.; Zink, A. Risk of herpes zoster in patients with rheumatoid arthritis treated with anti-TNF-alpha agents. *JAMA* **2009**, *301*, 737–744. [CrossRef]
69. Galloway, J.B.; Mercer, L.K.; Moseley, A.; Dixon, W.G.; Ustianowski, A.P.; Helbert, M.; Watson, K.D.; Lunt, M.; Hyrich, K.L.; Symmons, D.P.M. Risk of skin and soft tissue infections (including shingles) in patients exposed to anti-tumour necrosis factor therapy: Results from the British Society for Rheumatology Biologics Register. *Ann. Rheum. Dis.* **2013**, *72*, 229. [CrossRef]

70. García-Doval, I.; Pérez-Zafrilla, B.; Descalzo, M.Á.; Roselló, R.; Hernández, M.V.; Gómez-Reino, J.J.; Carmona, L. Incidence and risk of hospitalisation due to shingles and chickenpox in patients with rheumatic diseases treated with TNF antagonists. *Ann. Rheum. Dis.* **2010**, *69*, 1751. [CrossRef]
71. McDonald, J.R.; Zeringue, A.L.; Caplan, L.; Ranganathan, P.; Xian, H.; Burroughs, T.E.; Fraser, V.J.; Cunningham, F.; Eisen, S.A. Herpes zoster risk factors in a national cohort of veterans with rheumatoid arthritis. *Clin. Infect. Dis.* **2009**, *48*, 1364–1371. [CrossRef]
72. Winthrop, K.L.; Baddley, J.W.; Chen, L.; Liu, L.; Grijalva, C.G.; Delzell, E.; Beukelman, T.; Patkar, N.M.; Xie, F.; Saag, K.G.; et al. Association Between the Initiation of Anti–Tumor Necrosis Factor Therapy and the Risk of Herpes Zoster. *JAMA* **2013**, *309*, 887–895. [CrossRef]
73. Baddley, J.W.; Cantini, F.; Goletti, D.; Gómez-Reino, J.J.; Mylonakis, E.; San-Juan, R.; Fernández-Ruiz, M.; Torre-Cisneros, J. ESCMID Study Group for Infections in Compromised Hosts (ESGICH) Consensus Document on the safety of targeted and biological therapies: An infectious diseases perspective (Soluble immune effector molecules [I]: Anti-tumor necrosis factor-α agents). *Clin. Microbiol. Infect.* **2018**, *24*, S10–S20. [CrossRef] [PubMed]
74. Marcus, R.; Davies, A.; Ando, K.; Klapper, W.; Opat, S.; Owen, C.; Phillips, E.; Sangha, R.; Schlag, R.; Seymour, J.F.; et al. Obinutuzumab for the First-Line Treatment of Follicular Lymphoma. *N. Engl. J. Med.* **2017**, *377*, 1331–1344. [CrossRef] [PubMed]
75. Ohmachi, K.; Tobinai, K.; Kinoshita, T.; Ishikawa, T.; Hatake, K.; Ichikawa, S.; Ohmine, K.; Kamitsuji, Y.; Choi, I.; Chou, T.; et al. Efficacy and safety of obinutuzumab in patients with previously untreated follicular lymphoma: A subgroup analysis of patients enrolled in Japan in the randomized phase III GALLIUM trial. *Int. J. Hematol.* **2018**, *108*, 499–509. [CrossRef]
76. Vannucchi, A.M.; Kiladjian, J.J.; Griesshammer, M.; Masszi, T.; Durrant, S.; Passamonti, F.; Harrison, C.N.; Pane, F.; Zachee, P.; Mesa, R.; et al. Ruxolitinib versus Standard Therapy for the Treatment of Polycythemia Vera. *N. Engl. J. Med.* **2015**, *372*, 426–435. [CrossRef]
77. Al-Ali, H.K.; Griesshammer, M.; le Coutre, P.; Waller, C.F.; Liberati, A.M.; Schafhausen, P.; Tavares, R.; Giraldo, P.; Foltz, L.; Raanani, P.; et al. Safety and efficacy of ruxolitinib in an open-label, multicenter, single-arm phase 3b expanded-access study in patients with myelofibrosis: A snapshot of 1144 patients in the JUMP trial. *Haematologica* **2016**, *101*, 1065–1073. [CrossRef] [PubMed]
78. Lussana, F.; Cattaneo, M.; Rambaldi, A.; Squizzato, A. Ruxolitinib-associated infections: A systematic review and meta-analysis. *Am. J. Hematol.* **2018**, *93*, 339–347. [CrossRef]
79. Yun, H.; Xie, F.; Delzell, E.; Chen, L.; Levitan, E.B.; Lewis, J.D.; Saag, K.G.; Beukelman, T.; Winthrop, K.; Baddley, J.W.; et al. Risks of herpes zoster in patients with rheumatoid arthritis according to biologic disease-modifying therapy. *Arthritis Care Res.* **2015**, *67*, 731–736. [CrossRef] [PubMed]
80. Kampouri, E.; Ibrahimi, S.S.; Xie, H.; Wong, E.R.; Hecht, J.B.; Sekhon, M.K.; Vo, A.; Stevens-Ayers, T.L.; Green, D.J.; Gauthier, J.; et al. Cytomegalovirus (CMV) Reactivation and CMV-Specific Cell-Mediated Immunity after Chimeric Antigen Receptor T-Cell Therapy. *Clin. Infect. Dis.* **2023**, *78*, 1022–1032. [CrossRef]
81. Zuhair, M.; Smit, G.S.A.; Wallis, G.; Jabbar, F.; Smith, C.; Devleesschauwer, B.; Griffiths, P. Estimation of the worldwide seroprevalence of cytomegalovirus: A systematic review and meta-analysis. *Rev. Med. Virol.* **2019**, *29*, e2034. [CrossRef]
82. Zhang, J.; Kamoi, K.; Zong, Y.; Yang, M.; Ohno-Matsui, K. Cytomegalovirus Anterior Uveitis: Clinical Manifestations, Diagnosis, Treatment, and Immunological Mechanisms. *Viruses* **2023**, *15*, 185. [CrossRef] [PubMed]
83. Dadwal, S.S.; Papanicolaou, G.A.; Boeckh, M. How I prevent viral reactivation in high-risk patients. *Blood* **2023**, *141*, 2062–2074. [CrossRef]
84. Eichau, S.; López Ruiz, R.; Castón Osorio, J.J.; Ramírez, E.; Domínguez-Mayoral, A.; Izquierdo, G. Primoinfección por citomegalovirus en un paciente con esclerosis múltiple recurrente-remitente tratado con alemtuzumab. *Neurología* **2020**, *35*, 440–443. [CrossRef]
85. Philippen, L.; Schub, N.; Günther, A.; Cario, G.; Schrappe, M.; Repp, R.; Gramatzki, M. Alemtuzumab treatment for steroid-refractory acute graft-versus-host disease leads to severe immunosuppression but not to relapse of malignant disease. *Bone Marrow Transplant.* **2024**, *59*, 153–155. [CrossRef] [PubMed]
86. Lavagna, A.; Bergallo, M.; Daperno, M.; Sostegni, R.; Costa, C.; Leto, R.; Crocellà, L.; Molinaro, G.; Rocca, R.; Cavallo, R.; et al. Infliximab and the risk of latent viruses reactivation in active Crohn's disease. *Inflamm. Bowel Dis.* **2007**, *13*, 896–902. [CrossRef] [PubMed]
87. Torre-Cisneros, J.; Del Castillo, M.; Castón, J.J.; Castro, M.C.; Pérez, V.; Collantes, E. Infliximab does not activate replication of lymphotropic herpesviruses in patients with refractory rheumatoid arthritis. *Rheumatology* **2005**, *44*, 1132–1135. [CrossRef] [PubMed]
88. Shale, M.J. The implications of anti-tumour necrosis factor therapy for viral infection in patients with inflammatory bowel disease. *Br. Med. Bull.* **2009**, *92*, 61–77. [CrossRef]
89. Bernal, K.D.E.; Whitehurst, C.B. Incidence of Epstein-Barr virus reactivation is elevated in COVID-19 patients. *Virus Res.* **2023**, *334*, 199157. [CrossRef]
90. Atallah-Yunes, S.A.; Salman, O.; Robertson, M.J. Post-transplant lymphoproliferative disorder: Update on treatment and novel therapies. *Br. J. Haematol.* **2023**, *201*, 383–395. [CrossRef]
91. Martin, S.T.; Powell, J.T.; Patel, M.; Tsapepas, D. Risk of posttransplant lymphoproliferative disorder associated with use of belatacept. *Am. J. Health-Syst. Pharm.* **2013**, *70*, 1977–1983. [CrossRef]

92. Bradford, R.D.; Pettit, A.C.; Wright, P.W.; Mulligan, M.J.; Moreland, L.W.; McLain, D.A.; Gnann, J.W.; Bloch, K.C. Herpes Simplex Encephalitis during Treatment with Tumor Necrosis Factor-α Inhibitors. *Clin. Infect. Dis.* **2009**, *49*, 924–927. [CrossRef] [PubMed]
93. Justice, E.A.; Khan, S.Y.; Logan, S.; Jobanputra, P. Disseminated cutaneous Herpes Simplex Virus-1 in a woman with rheumatoid arthritis receiving Infliximab: A case report. *J. Med. Case Rep.* **2008**, *2*, 282. [CrossRef] [PubMed]
94. Crusio, R.H.J.; Singson, S.V.; Haroun, F.; Mehta, H.H.; Parenti, D.M. Herpes simplex virus encephalitis during treatment with etanercept. *Scand. J. Infect. Dis.* **2014**, *46*, 152–154. [CrossRef]
95. Scheinpflug, K.; Schalk, E.; Reschke, K.; Franke, A.; Mohren, M. Diabetes Insipidus Due to Herpes Encephalitis in a Patient with Diffuse Large Cell Lymphoma. A Case Report. *Exp. Clin. Endocrinol. Diabetes* **2006**, *114*, 31–34. [CrossRef]
96. Montalban, X.; Hauser, S.L.; Kappos, L.; Arnold, D.L.; Bar-Or, A.; Comi, G.; Seze, J.D.; Giovannoni, G.; Hartung, H.-P.; Hemmer, B.; et al. Ocrelizumab versus Placebo in Primary Progressive Multiple Sclerosis. *N. Engl. J. Med.* **2017**, *376*, 209–220. [CrossRef] [PubMed]
97. Yang, J.F.; You, J. Regulation of Polyomavirus Transcription by Viral and Cellular Factors. *Viruses* **2020**, *12*, 1072. [CrossRef]
98. Hildreth, J.E.K.; Alcendor, D.J. JC Polyomavirus and Transplantation: Implications for Virus Reactivation after Immunosuppression in Transplant Patients and the Occurrence of PML Disease. *Transplantology* **2021**, *2*, 37–48. [CrossRef]
99. Zhou, X.; Zhu, C.; Li, H. BK polyomavirus: Latency, reactivation, diseases and tumorigenesis. *Front. Cell. Infect. Microbiol.* **2023**, *13*, 1263983. [CrossRef]
100. Chalkley, J.J.; Berger, J.R. Progressive multifocal leukoencephalopathy in multiple sclerosis. *Curr. Neurol. Neurosci. Rep.* **2013**, *13*, 408. [CrossRef]
101. Monaco, M.C.; Major, E.O. The link between VLA-4 and JC virus reactivation. *Expert. Rev. Clin. Immunol.* **2012**, *8*, 63–72. [CrossRef]
102. Chen, Y.; Bord, E.; Tompkins, T.; Miller, J.; Tan, C.S.; Kinkel, R.P.; Stein, M.C.; Viscidi, R.P.; Ngo, L.H.; Koralnik, I.J. Asymptomatic Reactivation of JC Virus in Patients Treated with Natalizumab. *N. Engl. J. Med.* **2009**, *361*, 1067–1074. [CrossRef] [PubMed]
103. Keene, D.L.; Legare, C.; Taylor, E.; Gallivan, J.; Cawthorn, G.M.; Vu, D. Monoclonal antibodies and progressive multifocal leukoencephalopathy. *Can. J. Neurol. Sci.* **2011**, *38*, 565–571. [CrossRef]
104. Fleischmann, R.M. Progressive multifocal leukoencephalopathy following rituximab treatment in a patient with rheumatoid arthritis. *Arthritis Rheum.* **2009**, *60*, 3225–3228. [CrossRef] [PubMed]
105. Bennett, C.L.; Focosi, D.; Socal, M.P.; Bian, J.C.; Nabhan, C.; Hrushesky, W.J.; Bennett, A.C.; Schoen, M.W.; Berger, J.R.; Armitage, J.O. Progressive multifocal leukoencephalopathy in patients treated with rituximab: A 20-year review from the Southern Network on Adverse Reactions. *Lancet Haematol.* **2021**, *8*, e593–e604. [CrossRef] [PubMed]
106. Cohen-Bucay, A.; Ramirez-Andrade, S.E.; Gordon, C.E.; Francis, J.M.; Chitalia, V.C. Advances in BK Virus Complications in Organ Transplantation and Beyond. *Kidney Med.* **2020**, *2*, 771–786. [CrossRef]
107. Yang, M.; Kamoi, K.; Zong, Y.; Zhang, J.; Ohno-Matsui, K. Human Immunodeficiency Virus and Uveitis. *Viruses* **2023**, *15*, 444. [CrossRef]
108. Zong, Y.; Kamoi, K.; Kurozumi-Karube, H.; Zhang, J.; Yang, M.; Ohno-Matsui, K. Safety of intraocular anti-VEGF antibody treatment under in vitro HTLV-1 infection. *Front. Immunol.* **2023**, *13*, 1089286. [CrossRef]
109. Yang, M.; Kamoi, K.; Zong, Y.; Zhang, J.; Zou, Y.; Ohno-Matsui, K. Ocular Manifestations of Human Immunodeficiency Virus Infection in the Combination Antiretroviral Therapy Era. *Pathogens* **2023**, *12*, 1417. [CrossRef]
110. Zong, Y.; Kamoi, K.; Ando, N.; Kurozumi-Karube, H.; Ohno-Matsui, K. Mechanism of Secondary Glaucoma Development in HTLV-1 Uveitis. *Front. Microbiol.* **2022**, *13*, 738742. [CrossRef]
111. Bekker, L.-G.; Beyrer, C.; Mgodi, N.; Lewin, S.R.; Delany-Moretlwe, S.; Taiwo, B.; Masters, M.C.; Lazarus, J.V. HIV infection. *Nat. Rev. Dis. Primers* **2023**, *9*, 42. [CrossRef]
112. Bang, A.S.; Hakimi, M.; Tahir, P.; Bhutani, T.; Leslie, K.S. Biologic Therapies in HIV/AIDS Patients with Inflammatory Diseases: A Systematic Review of the Literature. *AIDS Patient Care STDs* **2023**, *37*, 215–242. [CrossRef] [PubMed]
113. Kamoi, K.; Watanabe, T.; Uchimaru, K.; Okayama, A.; Kato, S.; Kawamata, T.; Kurozumi-Karube, H.; Horiguchi, N.; Zong, Y.; Yamano, Y.; et al. Updates on HTLV-1 Uveitis. *Viruses* **2022**, *14*, 794. [CrossRef]
114. Zhang, J.; Kamoi, K.; Zong, Y.; Yang, M.; Zou, Y.; Ohno-Matsui, K. Evaluating tocilizumab safety and immunomodulatory effects under ocular HTLV-1 infection in vitro. *Int. Immunopharmacol.* **2024**, *137*, 112460. [CrossRef]
115. Kamoi, K.; Horiguchi, N.; Kurozumi-Karube, H.; Hamaguchi, I.; Yamano, Y.; Uchimaru, K.; Tojo, A.; Watanabe, T.; Ohno-Matsui, K. Horizontal transmission of HTLV-1 causing uveitis. *Lancet Infect. Dis.* **2021**, *21*, 578. [CrossRef]
116. Kamoi, K.; Uchimaru, K.; Tojo, A.; Watanabe, T.; Ohno-Matsui, K. HTLV-1 uveitis and Graves' disease presenting with sudden onset of blurred vision. *Lancet* **2022**, *399*, 60. [CrossRef]
117. Kamoi, K. HTLV-1 in Ophthalmology. *Front. Microbiol.* **2020**, *11*, 388. [CrossRef] [PubMed]
118. Kamoi, K.; Mochizuki, M. HTLV infection and the eye. *Curr. Opin. Ophthalmol.* **2012**, *23*, 557–561. [CrossRef]
119. Fukui, S.; Nakamura, H.; Takahashi, Y.; Iwamoto, N.; Hasegawa, H.; Yanagihara, K.; Nakamura, T.; Okayama, A.; Kawakami, A. Tumor necrosis factor alpha inhibitors have no effect on a human T-lymphotropic virus type-I (HTLV-I)-infected cell line from patients with HTLV-I-associated myelopathy. *BMC Immunol.* **2017**, *18*, 7. [CrossRef] [PubMed]
120. Umekita, K.; Umeki, K.; Miyauchi, S.; Ueno, S.; Kubo, K.; Kusumoto, N.; Takajo, I.; Nagatomo, Y.; Okayama, A. Use of anti-tumor necrosis factor biologics in the treatment of rheumatoid arthritis does not change human T-lymphotropic virus type 1 markers: A case series. *Mod. Rheumatol.* **2015**, *25*, 794–797. [CrossRef]

121. Kurozumi-Karube, H.; Kamoi, K.; Ando, N.; Uchida, M.; Hamaguchi, I.; Ohno-Matsui, K. In vitro Evaluation of the Safety of Adalimumab for the Eye Under HTLV-1 Infection Status: A Preliminary Study. *Front. Microbiol.* **2020**, *11*, 522579. [CrossRef]
122. Uchida, M.; Kamoi, K.; Ando, N.; Wei, C.; Karube, H.; Ohno-Matsui, K. Safety of Infliximab for the Eye Under Human T-Cell Leukemia Virus Type 1 Infectious Conditions in vitro. *Front. Microbiol.* **2019**, *10*, 2148. [CrossRef]
123. Nagao, T.; Takahashi, N.; Saitoh, H.; Noguchi, S.; Guo, Y.-M.; Ito, M.; Watanabe, A.; Fujishima, N.; Kameoka, Y.; Tagawa, H.; et al. [Adult T-cell leukemia-lymphoma developed from an HTLV-1 carrier during treatment of B-cell lymphoma-associated hemophagocytic syndrome]. *Rinsho Ketsueki* **2012**, *53*, 2008–2012. [PubMed]
124. Nakamura, H.; Ueki, Y.; Saito, S.; Horai, Y.; Suzuki, T.; Naoe, T.; Eguchi, K.; Kawakami, A. Development of adult T-cell leukemia in a patient with rheumatoid arthritis treated with tocilizumab. *Intern. Med.* **2013**, *52*, 1983–1986. [CrossRef] [PubMed]
125. Terada, Y.; Kamoi, K.; Ohno-Matsui, K.; Miyata, K.; Yamano, C.; Coler-Reilly, A.; Yamano, Y. Treatment of rheumatoid arthritis with biologics may exacerbate HTLV-1-associated conditions: A case report. *Medicine* **2017**, *96*, e6021. [CrossRef] [PubMed]

International Journal of
Molecular Sciences

MDPI

Article

Shedding Light on Viral Shedding: Novel Insights into Nuclear Assembly, Cytoplasmic Transformation and Extracellular Vesicle Release of the BK Virus

Daniela Gerges [1,†], Karim Abd El-Ghany [1,2,†], Zsofia Hevesi [2], Monika Aiad [1], Haris Omic [1], Clemens Baumgartner [1,3], Wolfgang Winnicki [1], Michael Eder [1], Alice Schmidt [1], Farsad Eskandary [1,*,‡] and Ludwig Wagner [1,‡]

[1] Division of Nephrology and Dialysis, Department of Internal Medicine III, Medical University of Vienna, 1090 Vienna, Austria; daniela.gerges@meduniwien.ac.at (D.G.); karim_abdel-ghany@outlook.com (K.A.E.-G.); n01618704@students.meduniwien.ac.at (M.A.); haris.omic@meduniwien.ac.at (H.O.); clemens.baumgartner@meduniwien.ac.at (C.B.); wolfgang.winnicki@meduniwien.ac.at (W.W.); michael.eder@meduniwien.ac.at (M.E.); alice.schmidt@meduniwien.ac.at (A.S.); ludwig.wagner@meduniwien.ac.at (L.W.)

[2] Center for Brain Research, Medical University of Vienna, 1090 Vienna, Austria; zsofia.hevesi@meduniwien.ac.at

[3] Division of Endocrinology, Department of Internal Medicine III, Medical University of Vienna, 1090 Vienna, Austria

* Correspondence: farsad.eskandary@meduniwien.ac.at

† These authors contributed equally to this work as first authors.

‡ These authors contributed equally to this work as supervising last authors.

Abstract: Despite the high prevalence of BK polyomavirus (BKPyV) and the associated risk for BKPyV-associated nephropathy (BKPyVAN) in kidney transplant (KTX) recipients, many details on viral processes such as replication, maturation, assembly and virion release from host cells have not been fully elucidated. VP1 is a polyomavirus-specific protein that is expressed in the late phase of its replicative cycle with important functions in virion assembly and infectious particle release. This study investigated the localization and time-dependent changes in the distribution of VP1-positive viral particles and their association within the spectrum of differing cell morphologies that are observed in the urine of KTX patients upon active BKPyV infection. We found highly differing recognition patterns of two anti-VP1 antibodies with respect to intracellular and extracellular VP1 localization, pointing towards independent binding sites that were seemingly associated with differing stages of virion maturation. Cells originating from single clones were stably cultured out of the urine sediment of KTX recipients with suspected BKPyVAN. The cell morphology, polyploidy, virus replication and protein production were investigated by confocal microscopy using both a monoclonal (mAb 4942) and a polyclonal rabbit anti-VP1-specific antibody (RantiVP1 Ab). Immunoblotting was performed to investigate changes in the VP1 protein. Both antibodies visualized VP1 and the mAb 4942 recognized VP1 in cytoplasmic vesicles exhibiting idiomorphic sizes when released from the cells. In contrast, the polyclonal antibody detected VP1 within the nucleus and in cytoplasm in colocalization with the endoplasmic reticulum marker CNX. At the nuclear rim, VP1 was recognized by both antibodies. Immunoblotting revealed two smaller versions of VP1 in urinary decoy cell extracts, potentially from different translation start sites as evaluated by in silico analysis. Oxford Nanopore sequencing showed integration of BKPyV DNA in chromosomes 3, 4 and 7 in one of the five tested primary cell lines which produced high viral copies throughout four passages before transcending into senescence. The different staining with two VP1-specific antibodies emphasizes the modification of VP1 during the process of virus maturation and cellular exit. The integration of BKPyV into the human genome leads to high virus production; however, this alone does not transform the cell line into a permanently cycling and indefinitely replicating one.

Keywords: BK polyomavirus (BKPyV); tubular epithelial cell; BKPyV infection; capsid protein VP1; kidney transplant (KTX)

1. Introduction

BK polyomavirus (BKPyV) is a highly prevalent virus with oligo- or asymptomatic primary infection that occurs most often during childhood or adolescence, leaving individuals with lifelong protective immunity [1,2]. However, due to its permanent residence in the urogenital tract after primary infection, and certain virus-specific oncogenic potential, uncontrolled reactivation due to immunosuppression may cause BKPyV-associated nephropathy (BKPyVAN) and rarely even urogenital malignancy in recipients of kidney transplants (KTX) [3,4]. Although common, the exact mechanisms of infection, replication and potential integration into the host genome remain incompletely understood. Interindividual variation with respect to the extent of viral shedding during BKPyV infection and/or reactivation may be related to such genetic rearrangement within the viral genome but also upon site-specific integration into the host genome, temporarily enhancing viral fitness until viral clearance by immune reconstitution is re-established [5,6]. To shed light on possible clinical implications that arise from such modifications and to better define potential therapeutic strategies, a more deepened understanding of viral integration and shedding is key [7].

BKPyV primarily replicates in the epithelial cells of the urogenital tract. After infecting the cell, the BKPyV genome enters the host-cell nucleus, where early gene expression and replication of the mini-chromosome occur, followed by late gene expression. The virus utilizes the host's replisome to replicate its double-stranded DNA, and virion assembly subsequently takes place in the nucleus [8,9]. While integration in the host's genome has been reported, especially in patients with BKPyV-positive cancers, integration is not common, and it remains unclear which factors and alteration in gene expression are contributing [10–13].

The mode of viral shedding that has been most frequently described in the literature is cell lysis, where exuberant viral replication leads to cell rupture. It commonly occurs in the context of BKPyVAN in immunocompromised KTX patients. Infected tubular epithelial cells with high viral load undergo a lytic process and exfoliate from the tubular basement membrane and appear in the urine as so-called decoy cells [14–16]. However, recent reports indicate that non-lytic viral shedding might be possible [17]. BKPyV-infected tubular epithelial cells appear to release infectious particles via extracellular vesicles (EVs), which can traffic tens of viral particles [18]. Other members of the *Polyomaviridiae* family include the JC polyomavirus (JCPyV), simian virus 40 (SV40) and the Merkel cell polyomavirus. Such EVs have also been described for the genetically related JC polyomavirus (JCPyV), which utilizes EVs to target host cells and spread infection [19]. Regarding SV40, it is clinical practice that a SV40-specific antibody against the large T antigen is utilized to stain kidney biopsy samples for the immunohistological diagnosis of BKPyVAN. In SV40, various viral proteins such as VP4 appear to regulate exocytosis [20,21]. VP4 was initially discovered as a VP2-gene-derived product translated from an alternate translation start site downstream from that of the genuine VP2 [20].

Release enveloped in host membranes is thought to enhance virion stability and immune evasion [22], ensuring an uncontained spread of infection, unnoticed by the host's immune response [23,24].

In this work, we investigated the replication and means of viral spreading of BKPyV within tubular cells, specifically highlighting possible modifications of the virus and its proteins as it traverses the cell. Our primary aim was to establish patient-specific cell culture models derived from urinary cell sediment that show active BKPyV replication, which might provide insight into replicative patterns based on individual genetic alterations of the virus that might also include potential chromosomal insertion sites. To detect differences in viral packaging, nuclear egress patterns and intracellular accumulation of virus-like particles, we included the use of two specific antibodies targeting the main viral protein VP1 that were characterized in their distinct staining behavior throughout intra- and extracellular virus production and release.

2. Results

2.1. Patient Demographics

Twenty-one patients of European descent were screened for this study, of which stable cell cultures from urinary sediment were established in nine patients (Tables 1 and 2). Two of them showed active ongoing BKPyV replication until senescence (Table 3). Median age was 58 years (27–78 years), and two (22.2%) were female. Prior KTX had occurred in two (22.2%) patients, and all others were recipients of a first KTX. Eight patients were on tacrolimus bid (median tacrolimus target-trough level at the time of sampling: 5.2 ng/mL) and one patient (#7) received 4-weekly belatacept infusions (5 mg/kg at the time of sampling). Seven patients received a reduced dose of mycophenolate mofetil (MMF, 500 mg, bid) due to ongoing BKPyV replication. In one case, MMF was paused due to BKPyVAN, while another patient (#8) received azathioprine (150 mg daily). At the time of sampling, plasma BKPyV load and urine BKPyV load ranged between 1.4×10^2 and 3.6×10^4 copies and 7.0×10^4 and 3.0×10^{10}, respectively. KTX biopsy results were available in all patients, and BKPyVAN diagnosis was established according to the Banff Working Group Classification of Definitive BKPyVAN: five patients were classified as PVN class 1, one patient as PVN class 2 and three patients as PVN class 3 [25]. Immunohistochemistry for SV40 was performed on six of nine tissue samples. Only one biopsy sample stained positive for SV40 (Tables 1 and 2).

Table 1. Patient demographics.

Characteristics	All Patients (n = 9)
Age, y	58 (36–58)
Male	7 (77.8)
First KTX	7 (77.8)
Second KTX	2 (22.2)
Duration after last KTX, mo	10 (9–15)
Plasma BKPyV load, copies per mL	11,566 (11,459–11,674)
Urine BKPyV load, copies per mL	7.02×10^9 (5.65×10^9–8.38×10^9)
Immunosuppression	
TAC	8 (88.9)
MMF	7 (77.8)
Azathioprine	1 (11.1)
Belatacept	1 (11.1)
Prednisolone	9 (100)
BKPyVAN [25]	
PVN class 1	5 (55.6)
PVN class 2	1 (11.1)
PVN class 3	3 (33.3)
SV40	
Positive	1 (11.1)
Negative	5 (55.6)
Not performed	3 (33.3)

Values are median with 1st and 3rd quartile in brackets (Q1–Q3). Immunosuppression is given at time of sampling. BKPyV = BK polyomavirus; BKPyVAN = BK polyomavirus associated nephropathy; KTX = kidney transplantation; mo = months; mL = milliliters; SV40 = simian vacuolating virus 40; y = years.

Table 2. Patient demographics. Age in years, numbers (#) of KTX; duration after last KTX in months (duration); immunosuppressive regimen with tacrolimus and mycophenolate mofetil; percentage of decoy cells in urine (HE), urinary BKPyV load in copies per mL, plasma BKPyV load in copies per mL, SV40 positivity in KTX biopsy and PVN class according to Nickeleit V et al. specified by KTX biopsy [25].

ID	Age (y)	Sex	# of KTX	Duration (mo)	TAC	MMF	HE (%)	uBKPyV (Copies/mL)	pBKPyV (Copies/mL)	SV40	PVN Class
1	52	m	1	13	Y	Y	93	8.7×10^5	1.4×10^2	-	1
2	55	f	2	75	Y	Y	0	5.8×10^6	2.0×10^3	-	3
3	27	m	1	15	Y	Y	30	1.7×10^7	1.5×10^4	NA	1
4	72	m	1	10	Y	Y	10	4.3×10^9	3.6×10^4	-	3
5	60	f	2	24	Y	Y	0	5.2×10^8	8.3×10^3	NA	2
6	78	m	1	9	Y	Y	50	3.0×10^{10}	2.0×10^4	+	3
7	56	m	1	8	N	N	0	7.0×10^4	2.2×10^2	NA	1
8	63	m	1	10	Y	N	30	1.8×10^9	4.5×10^3	-	1
9	58	m	1	8	Y	Y	30	1.9×10^9	1.6×10^4	-	1

f = female; HE = hematoxylin eosin; ID = identification; m = male; mL = milliliters; MMF = mycophenolate mofetil; mo = months; NA = not available; pBKPyV = plasma BKPyV load; PVN = polyomavirus nephropathy; SV40 = simian vacuolating virus 40; TAC = tacrolimus; uBKPyV = urinary BKPyV load; y = year, # = number, + = positive, - = negative.

Table 3. Viral infection model showing integration into host DNA, VP1-positivity with both polyclonal Ranti-VP1 and the mAb 4942, cells type identification to either urinary bladder epithelia (KRT13) or the kidney (TMEM27), the infective potential of cell lines (infectivity) and number of passages after which cells exhibited growth arrest (senescence).

ID	Integration	RantiVP1	mAb 4942	KRT13	TMEM27	Infectivity	Senescence
1	NP	+	-	-	+	NP	3
2	NI	+	-	-	+/-	NP	3
3	NI	+	-	-	+	NP	3
4	NI	+	-	-	+/-	+	5
5	NI	-	-	-	+	++	15
6	Chr 3,4,7	+	++	-	+	Infected	5
7	NA	+	++	-	+/-	Infected	4
8	NP	-	-	-	+	+	14
9	NP	+/-	-	-	+/-	+	5

Chr = chromosome; integration = viral integration into host DNA; RantiVP1 = polyclonal rabbit antibody; mAB 4942 = commercial VP1-specific antibody; NA = not applicable due to insufficient sequencing coverage; NI = no integration detected; NP = not performed; KRT13 = cytokeratin 13; TMEM27 = collectrin; + = infective, ++ = highly infective; infected = infected itself with BKPyV.

2.2. BKPyV Replicates in Non-Immortalized Tubular Cell Lines

In order to procure cells capable of BKPyV replication and to facilitate further investigation into its biology, urinary sediment cells obtained from 21 patients diagnosed with BKPyVAN were cultured. Among these, nine cell lines were successfully established, exhibiting variations in cell morphology and growth kinetics (Supplementary Figures S1 and S2), yet all reaching confluence. Although the replicative potential differed among cell lines, the percentage of tetraploid cells (M-Phase) in each population ranged between 17 and 25% (Figure 1A–C). As detailed in Table 3, identification of the clonally expanding cells of renal tubular epithelial origin was analyzed by positive immunofluorescence for TMEM27 as well as negative labelling for the urinary bladder-specific cytokeratin 13 (KRT13). The cells continued to expand and grow over a period of 15 days to 2 months even after repeated cryopreservation. They were split between 3 and 15 times, and stable BKPyV replication was established in a total in six out of nine cell lines (Table 3). Two of them showed high BKPyV secretion (#6 and #7) from clonal expansion until they reached senescence, and four of them were re-infected.

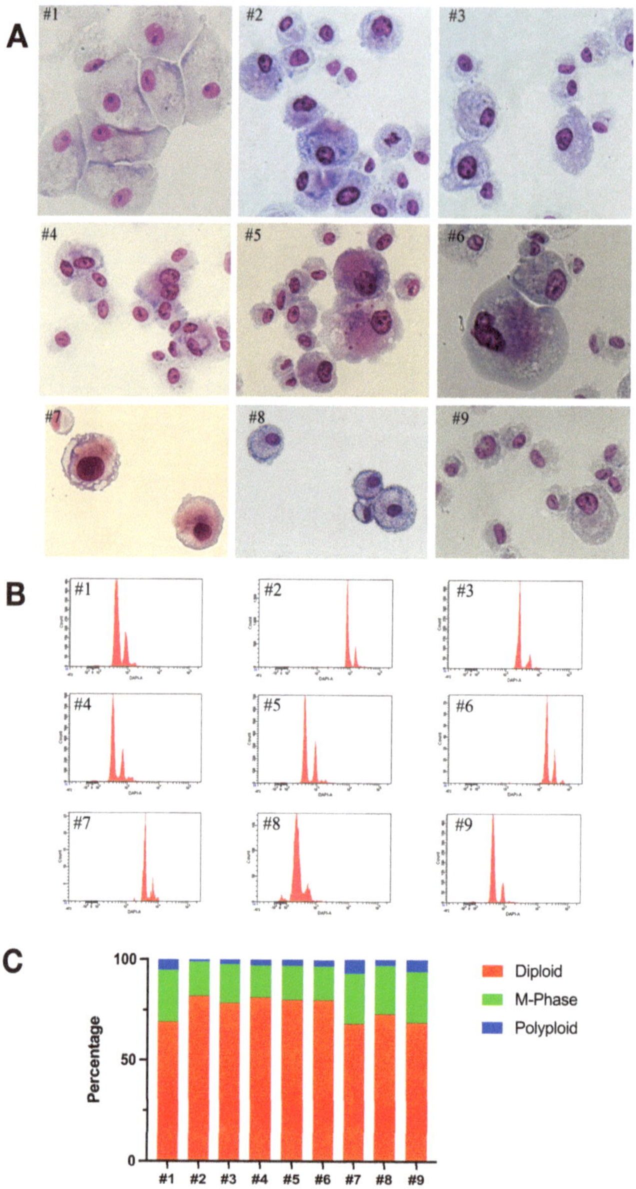

Figure 1. Hematoxylin/Eosin staining, FACS analysis and distribution of DNA content of primary cell lines. (**A**) HE staining exhibited diploid cells, cells with enlarged nucleus (DNA synthesis phase) and polyploidy as depicted in patients #4 and #6, (orig. magnification 400×). (**B**) Flow cytometry (FACS) analysis of cellular DNA content was performed on all nine established primary tubular cell lines. The cells were permeabilized and stained with DAPI prior to analysis. (**C**) Distribution of DNA content in diploid, tetraploid (M-phase), and polyploid cells, expressed as a percentage of the total cell population, was determined by DAPI staining of permeabilized cell line cells. The data are presented for each of the nine patients.

Following the establishment of nine stably cultured urinary cell isolates from nine different individuals that exhibited active in vitro replication of BKPyV, it was of further interest to characterize BKPyV's intracellular maturation, assembly and shedding utilizing an anti-VP1-specific antibody for further visualization.

2.3. Vesicles Enable BKPyV Shedding from Infected Tubular Cells

Our next goal was to describe the mode of infection and viral shedding of BKPyV within the tubular cell cultures in order to differentiate between low and high replicators. Therefore, the supernatant of the four BKPyV producing cultured patient-derived decoy cells was utilized to infect those cell lines, which were lacking the formation and secretion of mature BKPyV (#4, #5, #8, #9). As mentioned above, two different antibodies recognizing the VP1 antigen were applied to visualize the viral distribution and shedding in these cell culture specimens.

In order to characterize the localization of VP1 within our BKPyV-infected tubular cell culture model, we applied two different anti-VP1 antibodies with variations in their epitope specificities. We had observed that our polyclonal antibody might recognize also earlier translational products of VP1 involved in different stages of viral particle maturation, whereas the monoclonal antibody (mAb 4942) would be more specific for a mature conformational VP1 epitope. Staining with the polyclonal RantiVP1 Ab (green) revealed localization of nuclear virus assembly factories and the perinuclear region, suggesting a possible concentration in the endoplasmic reticulum (ER) (Figure 2).

Figure 2. Confocal immunofluorescence for VP1 protein. Immunofluorescence for VP1 in patient #1 derived cell line using RantiVP1 (green) and DAPI (purple). Nuclear punctate staining was observed, representing "nuclear virus assembly factories", and shadowed staining in the cytoplasm, indicating VP1 translation and production after passage #1.

In contrast, the staining pattern of mAb 4942 predominantly exhibited a granular pattern within the cytoplasm and immediate nuclear periphery (Figure 3). It also showed high specificity in the detection of decoy cells [26]. Additionally, it revealed extracellular virus-loaded vesicles following cell rupture, which were observed in typical idiomorphic sizes and BKPyV-loaded vesicles immobilized on cytopreparations (Figure 4). The staining pattern also varied significantly with the cell cycle stage of the individual cells (Figure 5).

Staining of both antibodies appeared to overlap at the nuclear rim, as a site of viral accumulation before egress through the nuclear membrane as shown in Figures 3 and 4 in yellow. The staining pattern of mAb 4942 may suggest that this antibody recognizes the VP1 protein in the later stages of viral production and egress.

To further clarify the different recognition patterns, RantiVP1 was tested for colocalization with a marker protein of the ER, CNX. As indicated by the shaded staining pattern, VP1 was recognized by RantiVP1 at the ER production stage and clearly colocalized with the ER marker protein CNX, as shown in Figure 6.

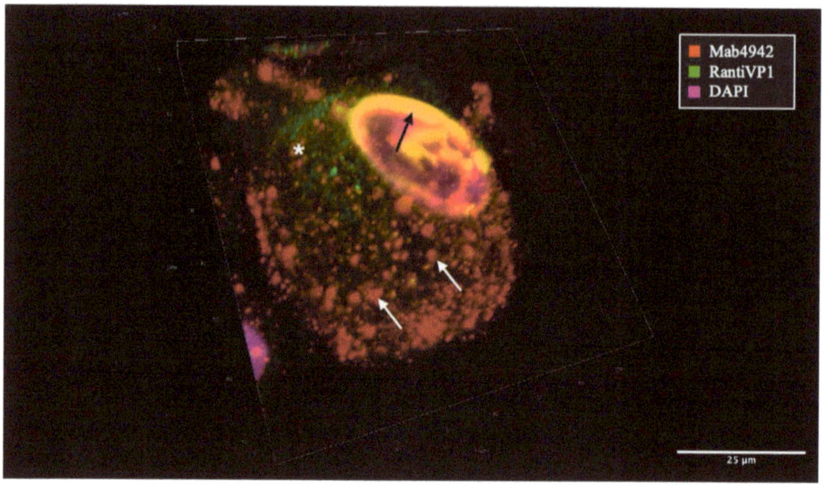

Figure 3. Merged image showing colocalization of VP1 as detected by mAb 4942 (red) and the RantiVP1 (green) resulting in yellow coloration. Confocal fluorescence microscopy revealed the process of virus particle formation using two distinct VP1-specific antibodies (mAb 4942 = red, RantiVP1 = green). BKPyV-VP1 encapsulated within small vesicles traversing the cytoplasm, ultimately accumulating at the cellular periphery was mainly stained by mAb 4942 (red) (annotated by white arrows). Conversely, the VP1-specific rabbit antibody (RantiVP1) highlighted the VP1 protein localization in the cytosol and in proximity to the nucleus (white asterisk) at a different stage of VP1 production and maturation. Virus formation within nuclear regions is depicted as accumulation at the nuclear rim by both antibodies resulting in yellow staining (annotated by black arrow).

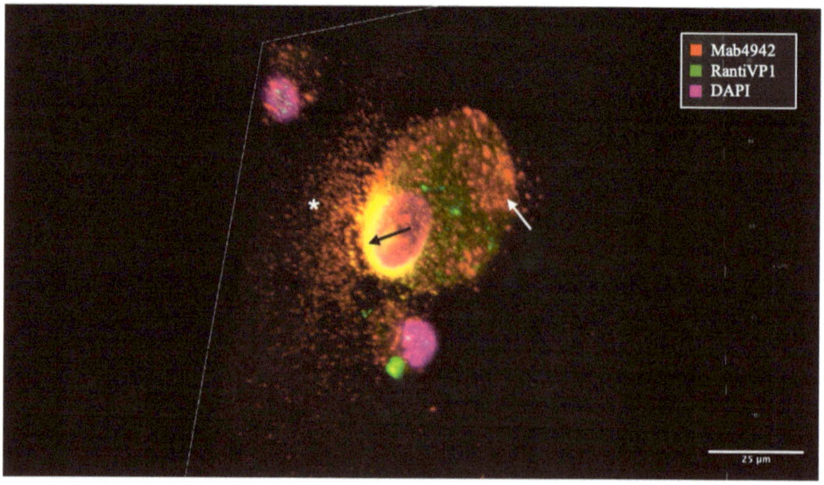

Figure 4. VP1 staining encapsulated within small vesicles released from a virus-producing cell of patient #7's primary cell line. VP1-containing vesicles as stained by mAb 4942 were distributed throughout the cytoplasm (white arrow). At the nuclear rim, the BKPyV-VP1 was prominently recognized by both antibodies mAb 4942 and RantiVP1 resulting in a high degree of overlap and yellow coloration (indicated by small, long-tailed black arrow). Extracellular VP1-loaded vesicles stained with mAb 4942 after cell rupture, in idiomorphic sizes, immobilized on the cytopreparation, marked by a white asterisk.

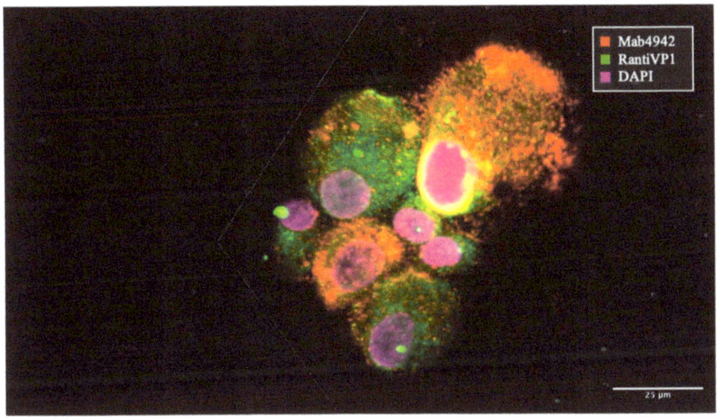

Figure 5. BKPyV-infected cells in different stages of the cell cycle. Two cells exhibited the presence of mature enveloped virus particles dispersed throughout the cytoplasm, predominantly of uniform size (mAb 4942 = red). The cell positioned at the top of the image displayed virus accumulation at the nuclear rim, where it was co-recognized by both antibodies (yellow). In contrast, in four other cells, VP1 staining, predominantly recognized by RantiVP1 antibodies (green) was observed, with a minor presence of virus particles stained by mAb 4942.

To visualize the cytoplasmic membrane of VP1-positive cell lines, dual-color staining for AQP1 and VP1 using mAb 4942 was performed. In Figure 7, cells with VP1-positive vesicles appeared to have an intact cytoplasmic membrane. The cell with an enlarged nucleus (white arrow) showed reduced VP1 staining.

As the distribution of the BKPyV genotype in our patient cohort may reflect that of Europeans, the anti-BKPyV VP1 reactivity of the rabbit polyclonal antibody against both genotype I and genotype IV had to be confirmed and established; it is shown in Figure 8.

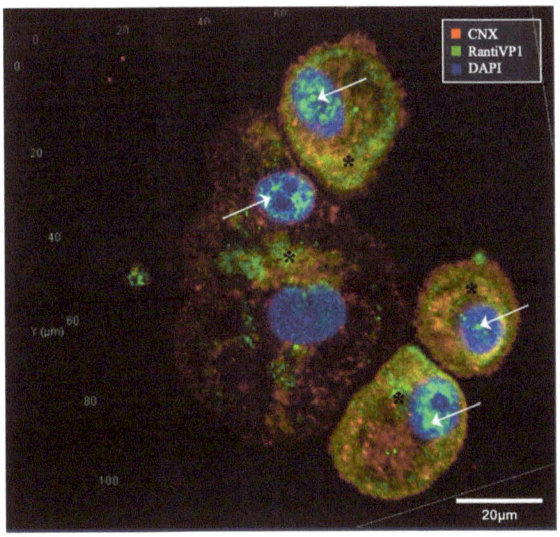

Figure 6. Confocal BKPyV-VP1 staining and colocalization with ER marker CNX. The VP1 detected by the RantiVP1 (green) colocalizes (yellow, indicated by *) with the ER marker CNX (red) in younger cells of the cell cycle. VP1 staining at the nuclear virus assembly factories is present in the smaller cells and in one of the nuclei of the large tetraploid cells (indicated by arrow).

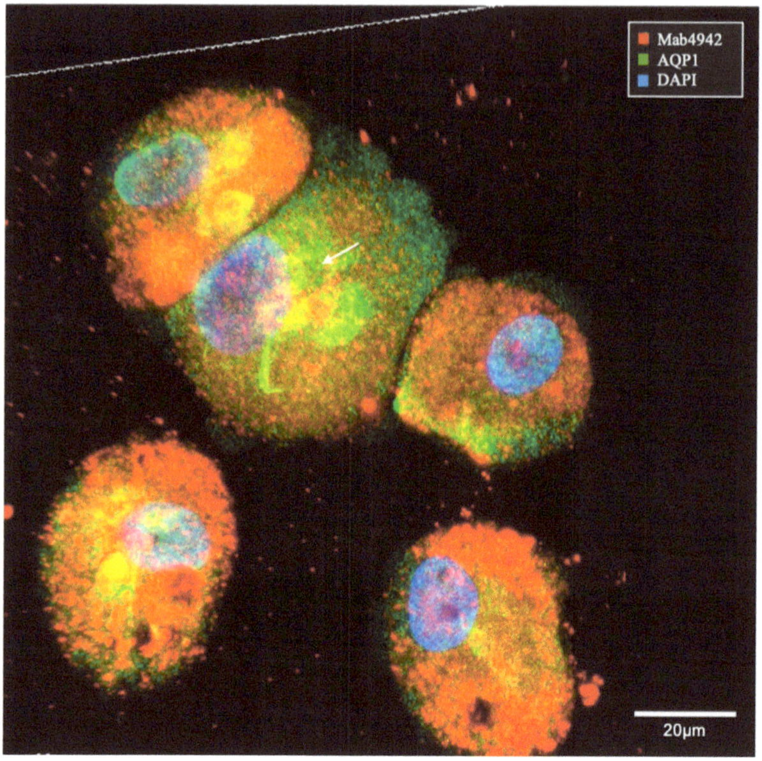

Figure 7. Confocal VP1 staining and visualization of the plasma membrane. VP1 positivity was detected by mAb 4942 (red), and the cytoplasmic membrane is visualized by AQP1 (green). The cell with an enlarged nucleus, marked by a white arrow, showed reduced VP1 staining.

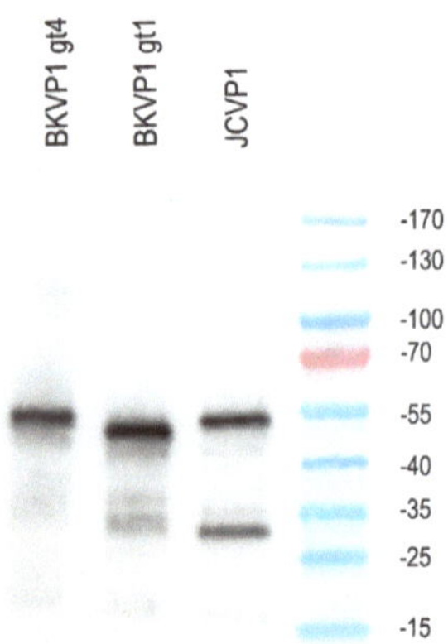

Figure 8. Comparative analysis between BKPyV VP1 genotype I and genotype IV using recombinant protein expressed in *E. coli* using RantiVP1. JCPyV VP1 protein recognition is depicted in the lane next to the molecular weight marker protein. VP1 proteins produced in *E. coli* from the JCPyV and BKPyV, followed by purification utilizing the His tag and Ni NTA resin. The slightly different sizes of the recombinant variants are attributed to the additional amino acids representing the Tag depending on the expression plasmid used (see Supplementary Data Sets S1 and S2).

2.4. Decoy Cells Exhibit Truncated Forms of the VP1-Gene Product

Recent findings by Henriksen et al. indicated that SV40 and BKPyV produce a truncated VP2/3, denoted VP4 [27], and this led us to perform tests to investigate whether a similar truncated form of VP1 might exist. To investigate possible changes in the VP1 protein, immunoblotting was conducted on both decoy cells and mature virions isolated from patients' urine.

Immunoblotting of decoy cells with the polyclonal RantiVP1 antibody confirmed VP1 at its established 41 kDa weight and also identified two additional reactive bands at 36–37 and 17 kDa. The highest expression levels were observed in extracts from decoy cells. Thus, it appears that two additional *VP1*-gene derived products (VP1a and VP1b at 36–7 kDa and 17 kDa, respectively) were expressed in decoy cells (Figure 8).

In contrast, immunoblotting of mature BKPyV virions isolated from patients' urine, with high viral load (>10^7 copies/mL), showed only a single VP1 band at 41 kDa. This indicates that the VP1 variants, VP1a and VP1b, identified in patient decoy cells are likely not present in mature secreted virions (Figure 9).

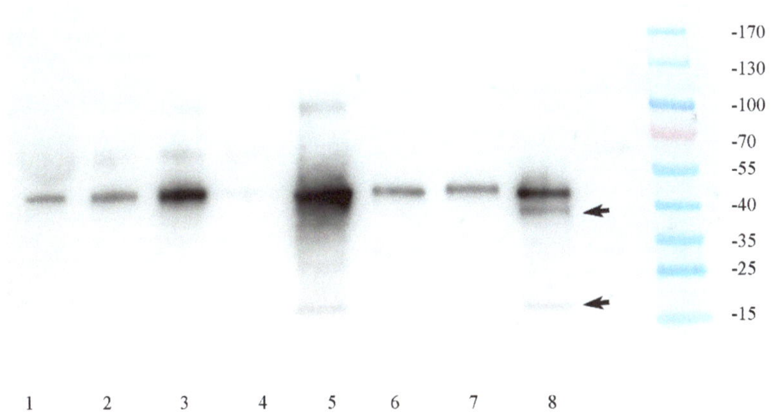

Figure 9. SDS-Page and immunoblot analysis of BKPyV virions as pelleted from urine by ultracentrifugation and protein extracts from urine sediment with decoy cells showed genuine VP1 and VP1 variants. Lanes 1–3 display urinary virion extracts pelleted by ultracentrifugation from patients with a urine viral load of 10^7–10^9 copies/mL. No additional VP1-specific smaller bands were detectable in any of the urine virion isolates. From Lanes 4 to 8, extracts of 2×10^4 urinary cells were loaded into each lane. Lane 4 shows cell extract from a patient recovered from disease but still the presence of faint VP1 bands. Lane 5 represents urine sediment obtained from a patient at the time of diagnosis, exhibiting over 50% decoy cells, with two variant bands detected: one at approximately 100 kDa representing aggregates and another at 16–17 kDa. Conversely, Lanes 6 and 7 depict patients in the recovery phase, showcasing less than 10% decoy cells among urine sediment cells. Lane 8 illustrates the VP1 banding pattern of a patient with 30% decoy cells in the sediment, wherein the 16–17 kDa band resembles that of Lane 5, albeit with an additional band approximately 2–3 kDa lower in size than the genuine VP1 protein.

2.5. Identification of Possible Translation Products from VP1 mRNAs by In Silico Analysis

The VP1 cDNA sequence was analyzed using an initiation codon identification server. A total of six potential translation products were identified. As indicated in Figure 10, the product with the highest probability represented the genuine 362-amino-acid-long VP1 protein. However, among the other potential products, one consisted of 307 amino acids with a calculated molecular weight of 34.23 kDa, and another consisted of 173 amino acids with a calculated molecular weight of 18 kDa (Figure 10). These two variants could correspond to the experimentally identified products VP1a and VP1b, respectively, as observed in the immunoblots. The size variants observed in the immunoblot, when compared to the in silico analysis, can only be explained by additional post-translational modifications. These modifications may include proteolytic cleavage or glycosylation and phosphorylation. Other variants identified by in silico analysis when including BKPyV genotype I and IV were not present in immunoblotting experiments.

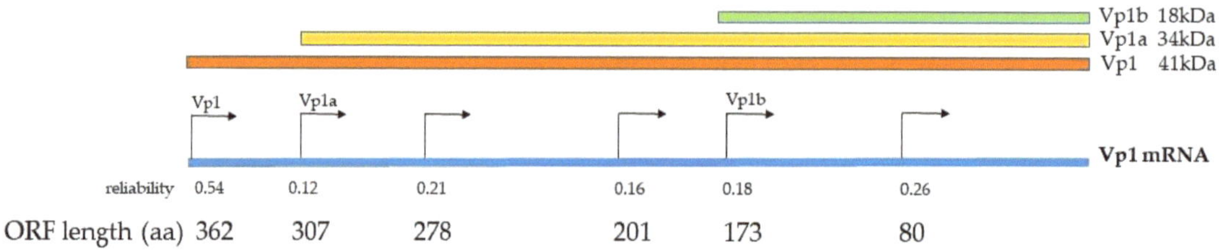

Figure 10. In silico analysis for potential translation products originating from the BKPyV-VP1 mRNA revealed several possible products. The highest reliability score of 0.54 was assigned to the genuine VP1 protein, which is 362 amino acids in length. Additionally, two other potential products, VP1a and VP1b, with lengths of 307 and 173 amino acids, respectively, were also identified. These reached reliability scores of 0.12 and 0.18.

2.6. BKPyV DNA Integrates into the Host Cell

As a final step, we thought it would be of interest to investigate whether integration into the host genome of BKPyV had occurred in our cell culture models. BKPyV DNA has been reported to occasionally integrate into the host genome, affecting all chromosomes except the Y chromosome, primarily in patients with BKPyV-related tumors [12,13]. We investigated whether such integration was also present in our study's cell lines. Of particular interest was whether differences in gene integration existed between cells that produce low numbers of viral copies and those that secreted high levels of mature virus particles. Therefore, we chose to compare the two cell lines with permanent BKPyV-replication and production (cell lines #6 and #7) to three of the primarily non-infected cell lines (cell lines #2, #4 and #5 from an early non-infected passage) through ONT long read sequencing.

No integration was detected in patient cell lines #2, #4 and #5. Sample #7 had insufficient sequencing coverage to confirm integration. However, sample #6 exhibited significant viral genome accumulation on chromosomes 3, 4 and 7 (Table 4, Figure 11). At these sites the BKPyV genome could be identified and presented in relation to the genetic surroundings using the IGV server (Figure 11, Table 4). On chromosome 3, the BKPyV genome was found to be integrated into the TAFA1 gene, which plays a role in chemotaxis. On chromosome 4, the BKPyV genome was integrated into a noncoding region between the Pyroglutamylated RFamide peptide receptor (QRFPR) and Annexin A5 (ANXA5) genes. On chromosome 7, the BKPyV genome was found integrated into an intron of the inner mitochondrial membrane peptidase subunit 2 (2IMMP2L) gene.

Table 4. Exact BKPyV integration into chromosomes 3, 4 and 7. Location of integration (location), genetic relation, disruption of host genes and supporting reads identified by ONT in patient #6.

Chromosome	Location	Genetic Relation	Host Genes	Supporting Reads
3	68666797	Intragenic	TAFA1	2123
4	122666278	Intergenic	QRFPR-ANXA5	171
7	110766605	Intragenic	IMMP2L	58

ANXA5 = Annexin A5; IMMP2L = inner mitochondrial membrane peptidase subunit 2; QRFPR = Pyroglutamylated RFamide peptide receptor; TAFA1 = Chemokine like family member 1

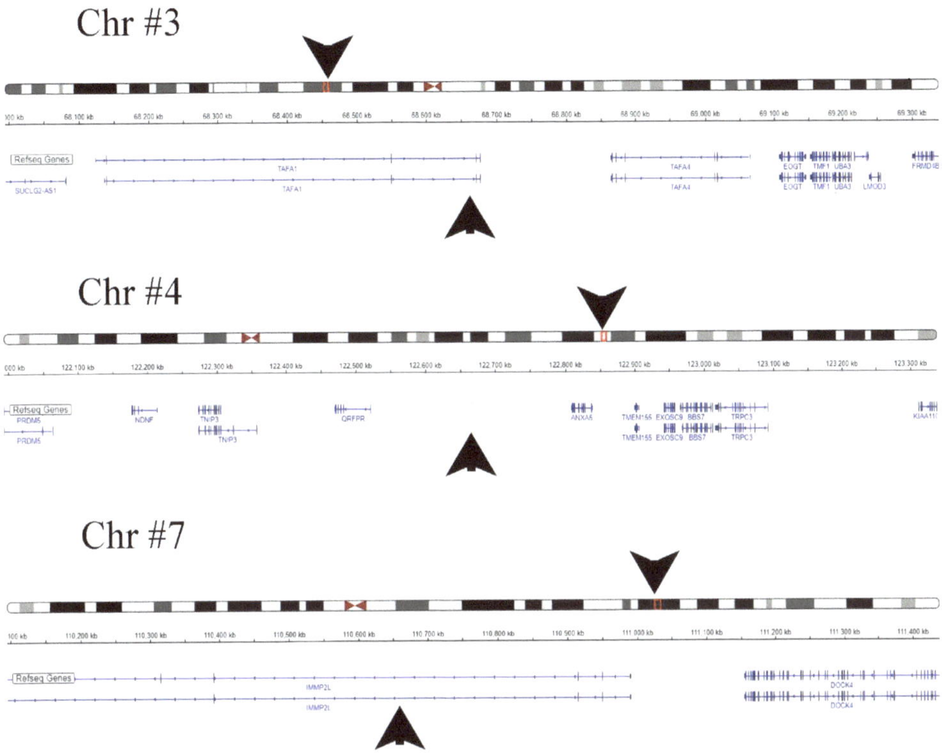

Figure 11. Schematic representation of BKPyV integration sites on chromosomes 3, 4 and 7, identified by ONT in patient #6. By passage three, the patient's cell line secreted virions at a logarithmic count of 10^{10}/mL. Post-passage four, a decline in cell growth rate occurred. Chromosomal integration sites are marked by red box, with black arrows indicating their positions relative to adjacent genes, which are listed in Table 4. The illustration was created using the publicly accessible IGV hg18 database.

3. Discussion

Infections with BKPyV are highly prevalent in the general population, affecting over 90% of individuals, typically without clinical manifestation. However, in the immunocompromised, especially KTX patients, BKPyV may lead to organ failure and graft loss of the transplanted organ as well as cause tumors. To date, the precise mechanisms of BKPyV infection and propagation, including its cell-to-cell transmission and potential disparities in viral shedding between immunocompetent and immunosuppressed hosts, remain inadequately characterized.

In this study, we successfully established a BKPyV infection model in newly cultured tubular progenitor cell lines. These cells were directly isolated from the source tissue via urine samples and importantly were neither immortalized nor transformed in any way. This natural propagation provides a more accurate representation of BKPyV's behavior in native cellular environments, as shown in detail earlier [28]. Furthermore, using two VP1-specific antibodies with different recognition patterns, we could discern viral changes during its maturation and to our best knowledge visualize BKPyV shedding via EVs, confirming data of earlier authors [18]. Additionally, in one case we proved BKPyV integration into three chromosomes of the host genome.

In the generation of the cell lines, we demonstrated that BKPyV from culture supernatant of urinary decoy cells can be used to establish infection, while this could not be achieved using isolated virions from urine. In doing so, we established a BKPyV infection model in non-immortalized tubular progenitor primary cells lines from KTX patients, which has some novelty. Experiments on infectivity and virus tropism were conducted by other authors using primary human epithelial cells, epithelial cells, endothelial cells and fibroblasts derived from the respiratory and urinary tracts. These experiments were performed to assess the susceptibility and specific cellular targets of the virus across different tissue types [10]. Further experiments were performed using immortalized tubular cell lines expressing human telomerase reverse transcriptase [29]. Our tissue culture model might originate from the changes in tubular cell dynamics and presence of tubular progenitor cells in urine due to BKPyV infection. As kidney injury diminishes and viral load decreases, the renal tubules heal and undergo replication, causing replicating tubular epithelial cells to appear in urine [30–32]. Our research shows these progenitor cells expanded effectively from a clonal phase and continued BKPyV replication and secretion.

Following virus replication and translation, BKPyV undergoes post-translational modifications in the cytoplasm, which are critical for virion assembly and future infectivity [22]. While cell lysis is one known mechanism of BKPyV spread, envelopment of the virus into the lipid bilayer of the host's cell membrane represents a different mode of virus egress, which bears advantages for the virus [18,24,33]: (1) Protection: the lipid envelope provides an additional layer of protection for the viral genome, keeping it from degradation by host nucleases and the hostile environment [34]. (2) Cell entry: Enveloped virions can enter host cells by different mechanisms than non-enveloped virions. The lipid envelope can facilitate fusion with cell membranes or promote endocytosis, potentially increasing the efficiency of viral entry [35]. (3) Evasion of the immune system: The envelope consists of the host's cell membrane; therefore, viruses traveling within EVs evade recognition and neutralization by the host's immune system [24]. Vice versa, the elimination of viruses in vesicles out of a cell might serve as a defense strategy for individual cells; however, it comes at the cost of a considerable risk to the entire organism due to the reasons mentioned above. Up to now, the precise mechanisms governing EV formation at the virus–cell interaction level remain incompletely elucidated [36]. In our study, mAb 4942 successfully detected mature virions packaged in vesicles.

Pertaining to our potential identification of two additional modified *VP1*-gene products, similar observations have been made by other authors with respect to the *VP2*-gene product. For VP2, a truncated protein named VP4 has been described, which arises due to alternate translational start sites of VP2, and it appears that VP4 might be involved in progeny release [27]. In this context, several different polyomavirus proteins have been described as viroporins such as agnoprotein, VP2, VP3 and VP4 [37–39]. These proteins form small channels in host cell membranes and can alter cellular processes, including calcium signaling and cell cycle regulation, which are crucial for viral replication and release. VP4 has been shown to be of importance for viral entry into the host cell and to trigger the timely lytic release of BKPyV's and SV40's progeny [20,27]. In our study, the origin of the two additionally identified *VP1*-gene products—whether through alternative translation variants or protease cleavage—and their potential beneficial function to BKPyV must be left unanswered. Further studies will have to investigate if VP1a/VP1b exhibit functions similar to VP4.

Regarding virus genome integration, earlier studies confirm fragmentary integration of BKPyV at multiple sites into the host genome allowing the virus to persist in host cells unrecognized by the immune system and undergo transcription and translation of viral proteins [12,13]. This phenomenon has been observed in the monkey kidney-derived COS cell line, which is capable of transcribing and translating the large T antigen protein of SV40 but does not produce capsid proteins or mature virions [26]. Our data confirmed the integration of BKPyV DNA into three different chromosomes in one cell line, but no driver insertion was found which has been linked to carcinoma formation in the past [12]. Certain genes could be disrupted in this cell line, but this is not necessary for the maintenance of a cell line culture.

A significant observation that is important for clinical considerations is the verified integration of the BKPyV genome into three different human chromosomes in one of the cell lines. Despite this integration, there was no activation of pertinent cell cycling pathways, and the cell line underwent growth arrest after passage 4. This suggests that BKPyV integration alone does not directly induce malignant transformation. It is likely that a specific genomic environment is required to facilitate a driver-type integration that leads to oncogenesis. These findings highlight the complexity of BKPyV-associated cellular transformation and underscore the need for further research to elucidate the precise conditions and mechanisms that enable viral integration to contribute to cancer development.

In conclusion, the maturation process of the BKPyV under these culture conditions is orchestrated towards envelopment into vesicles for egress, a mechanism that enhances infectivity and allows evasion of the immune system. Two additional *VP1*-gene-derived products exist, which are only found within the host cell during virus replication of so-called decoy cells. Utilizing freshly established tubular progenitor cells derived from patients with manifest BKPyVAN provides a valuable platform for studying BKPyV biology. Furthermore, these cells offer the opportunity to identify genomic integration of the virus, facilitating deeper investigations into the mechanisms of BKPyV infection and the search for inhibitory compounds.

4. Materials and Methods

4.1. Cell Culture and Cell Lines

Kidney transplant patients with a positive serum BKPyV copy number under observation at the outpatient clinic of the Division of Nephrology and Dialysis at the Medical University of Vienna starting May 2023 till end of January 2024 were sequentially included. Individuals providing urine volumes with less than 10^5 cells in total were excluded from this study. Out of 982 screened KTX patients for BKPyV-uria, 75 were eligible for this study. However, 54 were excluded as they showed less than 10^5 cells in the provided urine sample.

Twenty-one patients were enrolled after providing informed consent. The immunosuppression was reduced due to BKPyVAN. From nine of them—classified as PVN Class 1 (n = 5), PVN Class 2 (n = 1) and PVN Class 3 (n = 3)—clonal cell outgrowth could be achieved [25]. KTX biopsy results were available for all patients. Clinical routine charts provided laboratory parameters, urine and serum BKPyV copy numbers and details about the immunosuppressive regimen.

The morning urine (80 mL) of these patients was centrifuged at $2000\times g$ rounds per minute (rpm) for 10 min. The resulting cell pellet was resuspended in RPMI 1640 culture medium containing 10% fetal bovine serum, penicillin and streptomycin. The cells were seeded in Cellstar 60 × 15 mm tissue-culture-treated dishes (Greiner bio-one) and incubated at 37 °C, 5% CO_2 and 95% humidity.

The medium was changed after 3 days to proximal tubular cell culture media (REBM/REGMTM) obtained from Lonza (Lonza Group AG, Basel, Switzerland). Normocin (InvivoGen, San Diego, CA, USA; Cat. # ant-nr-05) was added as an antibiotic. After 6–7 days, small clones comprising 3–9 cells in adherent growth mode were observed. These clones increased in size due to cell proliferation. As indicated in Supplementary Figure S1, when expansive colonies with confluence and rapid growth were reached, a first cell passage was performed using trypsin-EDTA 1x in calcium/magnesium-free PBS (Biowest, Nuaillé, France, L0940-100) without phenol red. Similar methods for culturing urinary renal epithelial cells have been described by other authors [30].

This study was approved by the Ethics Committee of the Medical University of Vienna under the number 1065/2021.

4.2. BKPyV Copy Number Evaluation

BKPyV copy number was evaluated as detailed by Pajenda et al. [40]. In brief, to evaluate BKPyV copy number, cDNA or DNA was mixed with Universal Master Mix (Applied Biosystems, Waltham, MA, USA) and the BKPyV probe set (Pa03453401_s1, Applied Biosystems, Waltham, MA, USA). The PCR reaction was performed on a QuantStudio 6 Flex qPCR machine (Applied Biosystems, Waltham, MA, USA). A sample was considered positive if it reached the amplification threshold by cycle 38.

A tenfold serial dilution of a plasmid containing the BKPyV-specific qPCR target sequence (Pa03453401_s1) was used as a standard series. Copy numbers of positive samples were calculated according to this standard series. As an alternative and for clinical samples, the test kit from GeneProof®, Brno, Czech Republic (GeneProof BK Virus (BKV) PCR Kit—IVDR) was utilized.

4.3. Obtaining BK Polyoma Virus by Ultracentrifugation

Urine samples of patients with BKPyVAN were pelleted using an Optima Max-XP tabletop ultracentrifuge (Beckman Coulter, Brea, CA, USA). Therefore, one milliliter of each sample was applied to an open-top thick wall polycarbonate centrifuge tube, and symmetrically aligned tubes were centrifuged at 100,000 rpm for 2:30 h in a TLA-120.2 fixed-angle rotor. Times for acceleration and deceleration between zero and 5000 rpm were set to 2:30 min and 6:00 min, respectively.

4.4. Cell Line Infecton Using BKPyV-Containing Medium from Decoy Cell Culture

To prepare the virus-containing medium, decoy cells obtained from urine of patients were incubated at a concentration of 5×10^4 cells/mL in culture medium (REBM/REGMTM) for 3 days. The culture fluid was then centrifuged for 10 min at $3000\times g$ to pellet cellular fragments. The clarified supernatant was tested for BKPyV copy number (10^8 copies/mL) and stored in aliquots at 4 °C for immediate use. For longer storage periods, the aliquots were kept at −80 °C. This tissue culture medium containing high copy numbers of BKPyV was used for infecting renal epithelial cell lines.

Semiconfluent (50–60%) renal epithelial cell lines from patients #4, #5, #8 and #9 were exposed to virus-containing medium (2×10^7 copies/mL) for 24 h. Following this incubation period, the cell cultures were passaged, and the culture medium was replaced with fresh culture medium.

4.5. Immunoreagent

A polyclonal antibody against BKPyV-VP1 was generated by immunizing a female rabbit (*Oryctolagus cuniculus*) with recombinant His-tagged VP1 protein expressed in *Escherichia coli* (*E. coli*) and purified using Ni-NTA agarose (Qiagen, Hilden, Germany), as described in Pajenda et al. [40]. The amino acid sequence of the protein and its coding gene sequence are provided in Data Set S1 in the attached file. Serum was collected after five injections of the purified VP1 protein. Anti-VP1 IgG was subsequently purified from the serum using either a protein G column or ammonium sulfate precipitation. The polyclonal antibody was used in 0.8–1 µg/mL. In addition, a second mouse monoclonal VP1-specific antibody (mAb, 4942) was purchased from ThermoFisher (Invitrogen, Carlsbad, CA, USA, Catalog # MA5-33242) and used in 1–2 µg/mL. This mAb was generated by the use of BKPyV-VP1 as immunogen and has been validated and tested for staining BKPyV-infected decoy cells by immunofluorescence in an earlier work [26]. The collectrin (TMEM27)-specific antibody PA5-88721 (Invitrogen) was used in a dilution ratio of 1:100. The Cytokeratin 13-specific mAb CL488-66684 (Proteintech, Rosemont, IL, USA) was diluted at a ratio of 1:400 in PBS. The mouse monoclonal antibody specific to calnexin (CNX) was purchased (Catalog # MA3-027) from ThermoFisher, Waltham, MA, USA. The anti-aquaporin 1 (AQP1) rabbit polyclonal Ab (AB2219, diluted) was purchased from Millipore, Burlington, MA, USA.

4.6. Immunofluorescence by Confocal Microscopy

Confluent cells were liberated from the tissue culture dish using trypsin as described above. Following inactivation of trypsin with culture medium, a cell suspension of 10^5 cells/mL was generated. Out of this, 60 µL was applied to the funnel of the cytocentrifuge, which was run at 1200 rpm for 3 min. The resultant cytopreparation was air-dried for 1 h followed by fixation in acetone for 4 min. A water repellent cycle was drawn around the cell-containing area, and the primary antibody was applied. The mAb 4942 anti-human BK-VP1 was diluted at 1:30, and the rabbit anti-human BK-VP1 was diluted at 1/1000 and incubated for 2 h at room temperature or at 4 °C overnights. Following a washing step in PBS for 10 min under constant stirring, the second antibody was applied to either goat anti-mouse Alexa Fluor 594 (diluted at 1:700 in PBS) or goat anti-rabbit Alexa Fluor 488 (diluted at 1:700 in PBS). DAPI for nuclear staining was contained in the secondary antibodies. After a second washing step in PBS, slides were mounted with Vectashield mounting medium for immunofluorescence (Vecotor Laboratories, Burlingham, CA, USA) and covered with a microscope glass coverslip for confocal microscopy. A 3D confocal digital assembly was achieved by performing Z-stack recording comprising 40 individual scans, with a scan density of 0.2 µm, capturing imaging data from the top to the bottom of the nuclear staining. The recorded images were subsequently processed into the 3D format using ZEN blue (Version 3.4 for Windows 10, Carl Zeiss Microscopy, LLC, White Plains, NY, USA). Samples from BKPyV-negative patients served as negative controls.

4.7. Immunoblotting

The cell pellet (2000 rpm) was resuspended in PBS and pelleted again already in the sample grinding tubes containing the DNA-binding/grinding resin. Following a brief centrifugation and removal of the supernatant, the extraction solution (1x sample buffer containing DTT and protease inhibitors, cOmplete Tablets, Mini, REF 04 693 124 001) was added, and cells were broken up using the pestle (Cytiva, 80-6483-37). The resultant cell lysate was spun at 12,000× *g* for 7 min, and the precleared supernatant was loaded onto a BioRad PAGE mini gel (Hercules, CA, USA) for protein separation. Following protein separation, the gel was transferred onto nitrocellulose by semidry blotting. The blotted membrane was blocked in blocking buffer (Pierce™ Protein-Free T20 Blocking Buffer, Appleton, WI, USA, 37571) for 20 min, and rabbit anti-human BK-VP1 was added (diluted at 1:1500) and incubated over night at 4 °C. After washing the blot with TPBS twice for 10 min, the goat anti-rabbit HRP conjugate (Dako P0448, Agilent, Sana Clara, CA, USA, diluted at 1:10,000 in PBS substituted with 10% blocking buffer) was incubated for 60 min at room temperature under constant shaking. Finally, the blot was developed for specific antibody binding after washing again with TPBS twice for 10 min using chemiluminescence reagent (Merck KGaA, Darmstadt, Germany) and recorded by the lumi-imaging device Fusion FX Vilber Lourmat (Vilber, Eberhardzell, Germany).

4.8. Virus DNA Isolation

For virus DNA isolation from culture supernatant, the InnuPREP Sewage Water DNA/RNA Kit (Innuscreen Gmbh, Berlin, Germany) was used. In brief, 1 mL of supernatant was transferred to a 2.5 mL tube, followed by the addition of 100 µL of VCR-1 and 100 µL of VCR-2 reagent. The mixture was then incubated at room temperature for 10 min. Subsequently, centrifugation at 13,000 rpm was performed for 10 min. The pellet was then washed with 1 mL of RNase-free water and then dissolved in 150 µL of Lysis Solution RL, which was followed by the addition of 10 µL of Proteinase K. The tube was placed in a thermomixer and shaken continuously for 15 min at 60 °C. Subsequently, 135 µL of binding solution SBS and 10 µL of MAG suspension M were added. After thorough mixing and incubation for 5 min at room temperature with constant rotation, the suspension underwent two washes with 300 µL of wash solution HS and two washes with 300 µL of wash solution LS. Following the final washing step, the solution was removed as completely as possible, while ensuring the retention of the beads. The beads were then dried at 50 °C for 15 min. Finally, the DNA was eluted with 15 µL of RNase-free water.

4.9. DNA Isolation from Human Cell Lines and Sequencing by Oxford Nanopore Technology

Cell line cells were suspended in PBS and DNA isolation was performed using the QIAamp® DNA Blood Mini Kit 50 protocol (Qiagen, Hilden, Germany), following the guidelines outlined in the test manual. The column-purified DNA was subsequently submitted for Oxford Nanopore Technology (ONT) sequencing at the Vienna Biocenter core facility. After undergoing quality checks, the obtained reads were utilized for bioinformatics analysis, specifically focusing on the integration of the virus into the human host genome.

4.10. Bioinformatics Identification of BK Polyoma Virus Integration

Basecalling of raw data (pod5-format) from the promethION sequencing device was performed with ONT's basecaller Dorado [https://github.com/nanoporetech/dorado]. Samples were aligned against the viral reference sequence available at https://www.ncbi.nlm.nih.gov/nuccore/OQ230873.1, using Minimap2 v2.24 [https://github.com/lh3/minimap2], (accessed on 20 February 2024), (minimap2 -t 16 -ax mapont $viralReference $reads). Subsequently, only reads that aligned to the viral reference sequence with a mapping quality of at least 60 were retained after filtering ("filteredViralReads"). Filtering and format conversion were performed using samtools v1.18 [https://samtools.github.io/], (accessed on 20 February 2024). In order to identify putative insertion sites, the viral reads were aligned against the canonical chromosomes of the human genome available at https://www.ncbi.nlm.nih.gov/datasets/genome/GCF_000001405.40/, using the splice-aware alignment mode of Minimap2 (accessed on 20 February 2024), with the following command: minimap2 -t 16 -ax splice $humanReference $filteredViralReads. The IGV hg18 database for localizing the genetic environment of integration was accessed on 20 February 2024 (https://igv.org/app/).

4.11. In Silico Analysis of Potential Translation Initiation Sites on the VP1 mRNA

The freely available calculator from the Helix Research Institute (https://atgpr.dbcls.jp) was used to calculate the translation initiation probability. The 1089 nucleotide sequence encoding the VP1 genotype I and IV cDNA was entered into the server to identify potential open reading frames with possible translation start sites.

Supplementary Materials: The supporting information can be downloaded at: https://www.mdpi.com/article/10.3390/ijms25169130/s1.

Author Contributions: Conceptualization, W.W., F.E. and L.W.; Methodology, F.E. and L.W.; Validation, D.G. and F.E.; Formal analysis, M.A.; Investigation, D.G., K.A.E.-G. and L.W.; Resources, W.W., A.S., F.E. and L.W.; Data curation, K.A.E.-G., Z.H., M.A., C.B. and L.W.; Writing—original draft, D.G., K.A.E.-G., F.E. and L.W.; Writing—review & editing, D.G., K.A.E.-G., H.O., C.B., W.W., M.E., A.S., F.E. and L.W.; Visualization, K.A.E.-G.; Supervision, L.W.; Funding acquisition, D.G., W.W. and F.E. All authors have read and agreed to the published version of the manuscript.

Funding: This research was funded by the Medical Scientific Fund of the Mayor of the City of Vienna, grant number 22100, 21122 and 22099.

Institutional Review Board Statement: This study was approved by the Ethics Committee of the Medical University of Vienna under the number 1065/2021.

Informed Consent Statement: Informed consent was obtained from all study subjects involved in the study.

Data Availability Statement: The original contributions presented in the study are included in the article/Supplementary Materials, further inquiries can be directed to the corresponding author/s.

Conflicts of Interest: The authors declare no conflict of interest.

References

1. Egli, A.; Infanti, L.; Dumoulin, A.; Buser, A.; Samaridis, J.; Stebler, C.; Gosert, R.; Hirsch, H.H. Prevalence of polyomavirus BK and JC infection and replication in 400 healthy blood donors. *J. Infect. Dis.* **2009**, *199*, 837–846. [CrossRef]
2. Kean, J.M.; Rao, S.; Wang, M.; Garcea, R.L. Seroepidemiology of human polyomaviruses. *PLoS Pathog.* **2009**, *5*, e1000363. [CrossRef]
3. Kotton, C.N.; Kamar, N.; Wojciechowski, D.; Eder, M.; Hopfer, H.; Randhawa, P.; Sester, M.; Comoli, P.; Tedesco Silva, H.; Knoll, G.; et al. The Second International Consensus Guidelines on the Management of BK Polyomavirus in Kidney Transplantation. *Transplantation* **2024**. *online ahead of print*. [CrossRef] [PubMed]
4. Nickeleit, V.; Singh, H.K.; Dadhania, D.; Cornea, V.; El-Husseini, A.; Castellanos, A.; Davis, V.G.; Waid, T.; Seshan, S.V. The 2018 Banff Working Group classification of definitive polyomavirus nephropathy: A multicenter validation study in the modern era. *Am. J. Transplant.* **2021**, *21*, 669–680. [CrossRef] [PubMed]
5. Buck, C.B. Exposing the Molecular Machinery of BK Polyomavirus. *Structure* **2016**, *24*, 495. [CrossRef] [PubMed]
6. Jin, Y.; Zhou, Y.; Deng, W.; Wang, Y.; Lee, R.J.; Liu, Y.; Elias, N.; Hu, Y.; Luo, M.H.; Liu, R.; et al. Genome-wide profiling of BK polyomavirus integration in bladder cancer of kidney transplant recipients reveals mechanisms of the integration at the nucleotide level. *Oncogene* **2021**, *40*, 46–54. [CrossRef]
7. Mbianda, C.; El-Meanawy, A.; Sorokin, A. Mechanisms of BK virus infection of renal cells and therapeutic implications. *J. Clin. Virol.* **2015**, *71*, 59–62. [CrossRef]
8. Dalianis, T.; Hirsch, H.H. Human polyomaviruses in disease and cancer. *Virology* **2013**, *437*, 63–72. [CrossRef]
9. Hirsch, H.H.; Steiger, J. Polyomavirus BK. *Lancet Infect. Dis.* **2003**, *3*, 611–623. [CrossRef]
10. An, P.; Sáenz Robles, M.T.; Cantalupo, P.G.; Naik, A.S.; Sealfon, R.; Imperiale, M.J.; Pipas, J.M. Cultured Renal Proximal Tubular Epithelial Cells Resemble a Stressed/Damaged Kidney While Supporting BK Virus Infection. *J. Virol.* **2023**, *97*, e0034323. [CrossRef]
11. An, P.; Cantalupo, P.G.; Zheng, W.; Saenz-Robles, M.T.; Duray, A.M.; Weitz, D.; Pipas, J.M. Single-Cell Transcriptomics Reveals a Heterogeneous Cellular Response to BK Virus Infection. *J. Virol.* **2021**, *95*, e02237-20. [CrossRef] [PubMed]
12. Wang, Y.; Liu, Y.; Deng, W.; Fu, F.; Yan, S.; Yang, H.; Liu, R.; Geng, J.; Xu, J.; Wu, Y.; et al. Viral integration in BK polyomavirus-associated urothelial carcinoma in renal transplant recipients: Multistage carcinogenesis revealed by next-generation virome capture sequencing. *Oncogene* **2020**, *39*, 5734–5742. [CrossRef] [PubMed]
13. Wang, Y.; Yan, S.; Liu, Y.; Yan, Z.; Deng, W.; Geng, J.; Li, Z.; Xia, R.; Zeng, W.; Zhao, T.; et al. Dynamic viral integration patterns actively participate in the progression of BK polyomavirus-associated diseases after renal transplantation. *Am. J. Transplant.* **2023**, *23*, 1694–1708. [CrossRef] [PubMed]
14. Nickeleit, V.; Singh, H.K.; Kenan, D.J.; Mieczkowski, P.A. The two-faced nature of BK polyomavirus: Lytic infection or non-lytic large-T-positive carcinoma. *J. Pathol.* **2018**, *246*, 7–11. [CrossRef]
15. Sekito, T.; Araki, M.; Yoshinaga, K.; Maruyama, Y.; Sadahira, T.; Nishimura, S.; Wada, K.; Watanabe, M.; Watanabe, T.; Tanabe, K.; et al. Presence of decoy cells for 6 months on urine cytology efficiently predicts BK virus nephropathy in renal transplant recipients. *Int. J. Urol.* **2021**, *28*, 1240–1246. [CrossRef]
16. Yan, L.; Guo, H.; Han, L.; Huang, H.; Shen, Y.; He, J.; Liu, J. Sternheimer-Malbin Staining to Detect Decoy Cells in Urine of 213 Kidney Transplant Patients. *Transplant. Proc.* **2020**, *52*, 823–828. [CrossRef]
17. Evans, G.L.; Caller, L.G.; Foster, V.; Crump, C.M. Anion homeostasis is important for non-lytic release of BK polyomavirus from infected cells. *Open Biol.* **2015**, *5*, 150041. [CrossRef]
18. Handala, L.; Blanchard, E.; Raynal, P.-I.; Roingeard, P.; Morel, V.; Descamps, V.; Castelain, S.; Francois, C.; Duverlie, G.; Brochot, E.; et al. BK Polyomavirus Hijacks Extracellular Vesicles for *En Bloc* Transmission. *J. Virol.* **2020**, *94*, e01834-19. [CrossRef]
19. Morris-Love, J.; O'Hara, B.A.; Gee, G.V.; Dugan, A.S.; O'Rourke, R.S.; Armstead, B.E.; Assetta, B.; Haley, S.A.; Atwood, W.J. Biogenesis of JC polyomavirus associated extracellular vesicles. *J. Extracell. Biol.* **2022**, *1*, e43. [CrossRef]
20. Daniels, R.; Sadowicz, D.; Hebert, D.N. A very late viral protein triggers the lytic release of SV40. *PLoS Pathog.* **2007**, *3*, e98. [CrossRef]
21. Santiana, M.; Altan-Bonnet, N. Insane in the Membrane: Glial Extracellular Vesicles Transmit Polyomaviruses. *mBio* **2019**, *10*, e01024-19. [CrossRef]
22. Giannecchini, S. Evidence of the Mechanism by Which Polyomaviruses Exploit the Extracellular Vesicle Delivery System during Infection. *Viruses* **2020**, *12*, 585. [CrossRef]
23. Kerviel, A.; Zhang, M.; Altan-Bonnet, N. A New Infectious Unit: Extracellular Vesicles Carrying Virus Populations. *Annu. Rev. Cell Dev. Biol.* **2021**, *37*, 171–197. [CrossRef] [PubMed]
24. Morris-Love, J.; Gee, G.V.; O'Hara, B.A.; Assetta, B.; Atkinson, A.L.; Dugan, A.S.; Haley, S.A.; Atwood, W.J. JC Polyomavirus Uses Extracellular Vesicles to Infect Target Cells. *mBio* **2019**, *10*, e00379-19. [CrossRef] [PubMed]
25. Nickeleit, V.; Singh, H.K.; Randhawa, P.; Drachenberg, C.B.; Bhatnagar, R.; Bracamonte, E.; Chang, A.; Chon, W.J.; Dadhania, D.; Davis, V.G.; et al. The Banff Working Group Classification of Definitive Polyomavirus Nephropathy: Morphologic Definitions and Clinical Correlations. *J. Am. Soc. Nephrol.* **2018**, *29*, 680–693. [CrossRef] [PubMed]
26. Pajenda, S.; Hevesi, Z.; Eder, M.; Gerges, D.; Aiad, M.; Koldyka, O.; Winnicki, W.; Wagner, L.; Eskandary, F.; Schmidt, A. Lessons from Polyomavirus Immunofluorescence Staining of Urinary Decoy Cells. *Life* **2023**, *13*, 1526. [CrossRef] [PubMed]

27. Henriksen, S.; Hansen, T.; Bruun, J.A.; Rinaldo, C.H. The Presumed Polyomavirus Viroporin VP4 of Simian Virus 40 or Human BK Polyomavirus Is Not Required for Viral Progeny Release. *J. Virol.* **2016**, *90*, 10398–10413. [CrossRef] [PubMed]
28. An, P.; Saenz Robles, M.T.; Duray, A.M.; Cantalupo, P.G.; Pipas, J.M. Human polyomavirus BKV infection of endothelial cells results in interferon pathway induction and persistence. *PLoS Pathog.* **2019**, *15*, e1007505. [CrossRef]
29. Zhao, L.; Imperiale, M.J. A Cell Culture Model of BK Polyomavirus Persistence, Genome Recombination, and Reactivation. *MBio* **2021**, *12*, e0235621. [CrossRef]
30. Bach, C.; Knaup, K.X.; Herrmann, M.; Krumbiegel, M.; Pfister, F.; Buttner-Herold, M.; Steffen, M.; Zecher, D.; Lopau, K.; Schneider, K.; et al. A noninvasive diagnostic approach to retrospective donor HLA typing in kidney transplant patients using urine. *Transpl. Int.* **2021**, *34*, 1226–1238. [CrossRef]
31. Knafl, D.; Winnicki, W.; Mazal, P.; Wagner, L. Urinary nephrospheres indicate recovery from acute kidney injury in renal allograft recipients—A pilot study. *BMC Nephrol.* **2019**, *20*, 251. [CrossRef] [PubMed]
32. Gerges, D.; Hevesi, Z.; Schmidt, S.H.; Kapps, S.; Pajenda, S.; Geist, B.; Schmidt, A.; Wagner, L.; Winnicki, W. Tubular epithelial progenitors are excreted in urine during recovery from severe acute kidney injury and are able to expand and differentiate in vitro. *PeerJ* **2022**, *10*, e14110. [CrossRef] [PubMed]
33. Helle, F.; Handala, L.; Bentz, M.; Duverlie, G.; Brochot, E. Intercellular Transmission of Naked Viruses through Extracellular Vesicles: Focus on Polyomaviruses. *Viruses* **2020**, *12*, 1086. [CrossRef] [PubMed]
34. Sin, J.; McIntyre, L.; Stotland, A.; Feuer, R.; Gottlieb, R.A. Coxsackievirus B Escapes the Infected Cell in Ejected Mitophagosomes. *J. Virol.* **2017**, *91*, e01347-17. [CrossRef] [PubMed]
35. Santiana, M.; Ghosh, S.; Ho, B.A.; Rajasekaran, V.; Du, W.L.; Mutsafi, Y.; De Jésus-Diaz, D.A.; Sosnovtsev, S.V.; Levenson, E.A.; Parra, G.I.; et al. Vesicle-Cloaked Virus Clusters Are Optimal Units for Inter-organismal Viral Transmission. *Cell Host Microbe* **2018**, *24*, 208–220.e8. [CrossRef] [PubMed]
36. Jia, X.; Yin, Y.; Chen, Y.; Mao, L. The Role of Viral Proteins in the Regulation of Exosomes Biogenesis. *Front. Cell Infect. Microbiol.* **2021**, 11. [CrossRef] [PubMed]
37. Suzuki, T.; Orba, Y.; Okada, Y.; Sunden, Y.; Kimura, T.; Tanaka, S.; Nagashima, K.; Hall, W.W.; Sawa, H. The human polyoma JC virus agnoprotein acts as a viroporin. *PLoS Pathog.* **2010**, *6*, e1000801. [CrossRef] [PubMed]
38. Raghava, S.; Giorda, K.M.; Romano, F.B.; Heuck, A.P.; Hebert, D.N. The SV40 late protein VP4 is a viroporin that forms pores to disrupt membranes for viral release. *PLoS Pathog.* **2011**, *7*, e1002116. [CrossRef] [PubMed]
39. Giorda, K.M.; Raghava, S.; Zhang, M.W.; Hebert, D.N. The viroporin activity of the minor structural proteins VP2 and VP3 is required for SV40 propagation. *J. Biol. Chem.* **2013**, *288*, 2510–2520. [CrossRef] [PubMed]
40. Pajenda, S.; Gerges, D.A.; Freire, R.; Wagner, L.; Hevesi, Z.; Aiad, M.; Eder, M.; Schmidt, A.; Winnicki, W.; Eskandary, F.A. Acute Kidney Injury and BK Polyomavirus in Urine Sediment Cells. *Int. J. Mol. Sci.* **2023**, *24*, 17511. [CrossRef] [PubMed]

International Journal of
Molecular Sciences

MDPI

Article

Reactive Oxygen Species Induction by Hepatitis B Virus: Implications for Viral Replication in p53-Positive Human Hepatoma Cells

Yuna Jeong [1,†], Jiwoo Han [1,†] and Kyung Lib Jang [1,2,3,*]

1 Department of Integrated Biological Science, The Graduate School, Pusan National University, Busan 46241, Republic of Korea; da3967@naver.com (Y.J.); hanjiwoo@pusan.ac.kr (J.H.)
2 Department of Microbiology, College of Natural Science, Pusan National University, Busan 46241, Republic of Korea
3 Microbiological Resource Research Institute, Pusan National University, Busan 46241, Republic of Korea
* Correspondence: kljang@pusan.ac.kr; Tel.: +82-51-510-2178
† These authors contributed equally to this work.

Abstract: Hepatitis B virus (HBV) infects approximately 300 million people worldwide, causing chronic infections. The HBV X protein (HBx) is crucial for viral replication and induces reactive oxygen species (ROS), leading to cellular damage. This study explores the relationship between HBx-induced ROS, p53 activation, and HBV replication. Using HepG2 and Hep3B cell lines that express the HBV receptor NTCP, we compared ROS generation and HBV replication relative to p53 status. Results indicated that HBV infection significantly increased ROS levels in p53-positive HepG2-NTCP cells compared to p53-deficient Hep3B-NTCP cells. Knockdown of p53 reduced ROS levels and enhanced HBV replication in HepG2-NTCP cells, whereas p53 overexpression increased ROS and inhibited HBV replication in Hep3B-NTCP cells. The ROS scavenger N-acetyl-L-cysteine (NAC) reversed these effects. The study also found that ROS-induced degradation of the HBx is mediated by the E3 ligase Siah-1, which is activated by p53. Mutations in p53 or inhibition of its transcriptional activity prevented ROS-mediated HBx degradation and HBV inhibition. These findings reveal a p53-dependent negative feedback loop where HBx-induced ROS increases p53 levels, leading to Siah-1-mediated HBx degradation and HBV replication inhibition. This study offers insights into the molecular mechanisms of HBV replication and identifies potential therapeutic targets involving ROS and p53 pathways.

Keywords: HBx; Hepatitis B virus; proteasome; reactive oxygen species; Siah-1; p53

1. Introduction

The infection of the hepatitis B virus (HBV)—a member of the *Hepadnaviridae* family—poses a significant public health challenge, affecting approximately 350 million individuals globally with chronic HBV infection [1]. This condition is associated with a high risk of developing severe liver diseases, including acute and chronic hepatitis, cirrhosis, and hepatocellular carcinoma [1]. The HBV replicates a partially double-stranded circular DNA genome of approximately 3200 base pairs through reverse transcription of its pregenomic RNA (pgRNA) [2]. This genome includes four major open reading frames (ORFs) that encode surface antigens (preS1, preS2, and S proteins), core antigens (preC and C proteins), polymerase (P protein), and X protein (HBx). The HBx is a 17-kDa multifunctional oncoprotein that influences signal transduction, transcription, and mitochondrial function and is found in the cytoplasm, nucleus, and mitochondria [3,4]. The HBx is recognized as a positive regulator of HBV replication, as demonstrated in various models including in vitro infection systems, human hepatocyte chimeric mice, and murine hydrodynamic injection studies [5–7]. The HBx activates viral promoters to synthesize HBV mRNA and pgRNA from a covalently closed circular DNA template [2]. Additionally, HBx indirectly promotes HBV replication by disrupting cellular signaling pathways such as the cytosolic calcium signaling pathway and the phosphatidylinositol 3-kinase/Akt pathway [8,9].

Reactive oxygen species (ROS) are highly reactive molecules derived from oxygen. Under normal circumstances, cells generate ROS continuously through processes like protein folding in the endoplasmic reticulum, oxidative phosphorylation in mitochondria, and the catabolism of lipids and amino acids in the cytosol [10]. It has been demonstrated that ROS levels are frequently elevated in the liver and blood of HBV-infected patients [11,12]. Moreover, in vitro HBV infection has been shown to induce ROS production in cells [13]. While viral proteins like the HBsAg and the HBcAg may contribute to this process [14], HBx is considered the primary driver of oxidative stress in the course of HBV infection. For instance, the HBx interacts with the human voltage-dependent anion channel in mitochondria, affecting its transmembrane potential [15,16] and decreasing mitochondrial enzymes involved in electron transport, resulting in elevated ROS levels in mitochondria [17]. Excess ROS from chronic HBV infection can damage proteins, lipids, and DNA, potentially leading to liver fibrosis and hepatocellular carcinoma [18,19]. Although ROS plays a well-established role in HBV pathogenesis, their impact on HBV replication is not yet fully understood.

Several studies have suggested that the HBx levels are primarily regulated by the E3 ligase seven in absentia homolog 1 (Siah-1), which facilitates ubiquitin (Ub)-dependent proteasomal degradation of the HBx. Additionally, p53 can stimulate the transcriptional activation of *Siah-1* via p53 response elements located in its promoter [20]. Given that p53 is a crucial mediator of DNA damage induced by excess ROS [21], it is valuable to investigate whether HBx-induced ROS triggers *Siah-1* expression through p53 activation. This could lead to the downregulation of the HBx via a negative feedback loop, thereby inhibiting HBV replication. For this study, we used an optimized in vitro HBV replication system [22,23] to accurately evaluate the effect of HBx-induced ROS on HBV replication. For this purpose, we first investigated whether the HBV induces varying levels of ROS generation depending on the p53 status in human hepatoma cells. Following this, we examined the mechanism by which the HBV amplifies ROS in a p53-dependent manner. Subsequently, we explored whether HBV-induced ROS negatively regulates HBV replication in hepatoma cells. Furthermore, we assessed the role of the HBx in the p53-dependent ROS amplification and the ROS-mediated inhibition of the HBV in these cells. Finally, we investigated whether ROS upregulates *Siah-1* expression by elevating p53 levels and subsequently reduces HBx levels through Siah-1-mediated proteasomal degradation.

2. Results

2.1. Exploring the Relationship between ROS Induction and HBV Replication in p53-Positive Human Hepatoma Cells

Initially, we explored whether HBV infection leads to differential ROS generation in human hepatoma cells based on the p53 status. HepG2 cells express wild-type (WT) p53, whereas Hep3B cells do not, providing a unique platform for comparative studies on the role of p53 in HBV-related molecular mechanisms [24]. For in vitro HBV replication studies, HepG2 and Hep3B cell lines were stably transfected with a plasmid encoding the HBV receptor, sodium taurocholate co-transporting polypeptide (NTCP) [25], to establish HepG2-NTCP and Hep3B-NTCP cell lines, respectively [26]. These cell lines support the entire HBV life cycle, from entry to the release of infectious virions. HBV particles were produced from a 1.2-mer HBV replicon [26,27]. Replication of the HBV in both HepG2-NTCP and Hep3B-NTCP cells was confirmed by western blot analysis of viral proteins such as the HBx and HBsAg in cell lysates (Figure 1a), detection of extracellular hepatitis B envelope antigen (HBeAg) by ELISA (Figure 1b), and quantification of virus particles in cell culture media using conventional PCR and quantitative real-time PCR (qPCR) (Figure 1c). Large amounts of large (L)- and middle (M)-HBsAg, along with trace amounts of small (S)-HBsAg, were detected in the infected cells (Figure 1a), consistent with previous findings [26]. In contrast to HepG2-NTCP cells, small quantities of HBsAg derived from the inserted viral genome [28] were present in the cell lysates of uninfected Hep3B-NTCP cells (Figure 1a, lane 4), while neither conventional PCR nor qPCR detected evidence of HBV replication in these cells (Figure 1c, lane 4). These findings suggest that the HBV replicates successfully in HepG2-NTCP and Hep3B-NTCP cells under our experimental conditions. In line with prior studies that identified p53 as a negative regulator of HBV replication [26,29], all three experimental methods revealed higher levels of HBV replication in Hep3B-NTCP cells (Figure 1a–c).

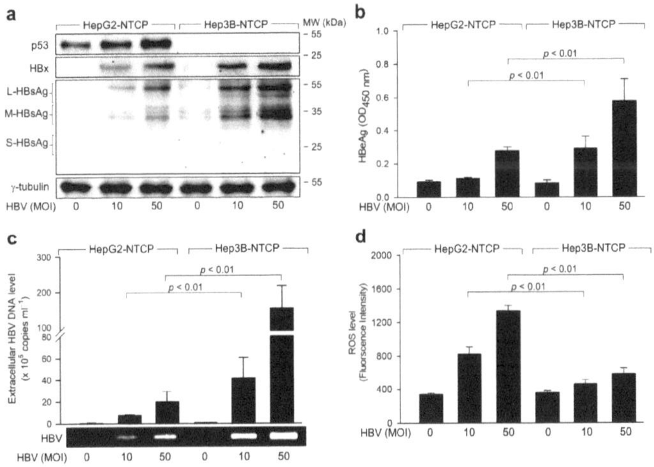

Figure 1. Exploring the relationship between ROS induction and HBV replication in p53-positive human hepatoma cells: (**a**) HepG2-NTCP and Hep3B-NTCP cells were infected with HBV at 10 or 50 MOI per cell for 24 h in DMEM with 2% DMSO and 4% PEG 8000. The cells were washed twice with DMEM without serum and incubated for an additional three days in DMEM with 3% FBS, 2% DMSO, and 4% PEG 8000. Cell lysates were analyzed by western blotting to measure the levels of indicated proteins. (**b**) Levels of HBeAg derived from the cells prepared in (**a**) were detected by ELISA. Results are presented as mean ± standard deviation from five independent experiments (n = 5). (**c**) Levels of extracellular HBV DNA from (**a**) were determined using both conventional PCR and quantitative real-time PCR (qPCR) (n = 4). (**d**) Intracellular ROS levels were assessed with chloromethyl dichlorodihydrofluorescein diacetate (CM-H2DCFDA; Invitrogen), a widely used H_2O_2-specific probe, in intact cells [30]. The conversion of CM-H2DCFDA to the green fluorescent product, DCF, was measured using a microplate reader with excitation and emission wavelengths of 485 nm and 535 nm, respectively (n = 4).

We next focused on determining whether the HBV induces differential ROS generation based on the p53 status during its replication in human hepatoma cells. In alignment with prior research [31], HBV infection elevated intracellular ROS levels in both cell lines (Figure 1d). Notably, the effect of HBV infection on ROS levels was markedly more pronounced in HepG2-NTCP cells (approximately 4-fold increase) compared to Hep3B-NTCP cells (approximately 1.5-fold increase). These observations align with a previous study [32,33] demonstrating the role of p53 in amplifying ROS levels during HBV infection. Moreover, the inverse relationship between HBV and ROS levels hints at a possible role for ROS in negatively regulating HBV replication2.2. HBV Activates p53 to Amplify ROS during Infection in Human Hepatoma Cells

We further investigated the mechanism by which HBV triggers ROS amplification in a p53-dependent manner. In line with earlier findings [20,34], HBV infection led to an upregulation of p53 levels in HepG2-NTCP cells (Figure 1a). To ascertain the role of p53 activation in ROS amplification, we employed p53 shRNA to knock down p53 in HepG2-

NTCP cells and used ectopic p53 expression to overexpress p53 in Hep3B-NTCP cells. In this study, scrambled short hairpin RNA (SC shRNA) served as a negative control, ensuring that the effects observed are due to specific gene silencing rather than nonspecific shRNA actions. Concurrently, p53 shRNA was utilized to specifically knock down the p53 gene, enabling the assessment of the role of p53 in cellular processes. The comparison between these two treatments allowed for clear differentiation of specific gene silencing effects from potential off-target effects or general RNA interference toxicity. As a result, reduction in p53 resulted in decreased ROS levels in HBV-infected HepG2-NTCP cells, while increased p53 expression elevated ROS levels in HBV-infected Hep3B-NTCP cells (Figure 2a). This underscores the involvement of p53 in ROS amplification during HBV infection in human hepatoma cells. Conversely, reducing p53 levels boosted both intracellular viral proteins and extracellular HBV virions in HBV-infected HepG2-NTCP cells, while increased p53 expression curtailed HBV replication in Hep3B-NTCP cells (Figure 2b,c). These findings confirm the role of p53 as a negative modulator of HBV replication in hepatocytes [26,29]. Based on these observations, we conclude that p53 activation plays a dual role: it is both implicated in ROS amplification and contributes to the inhibition of HBV replication in p53-positive human hepatoma cells during HBV infection.

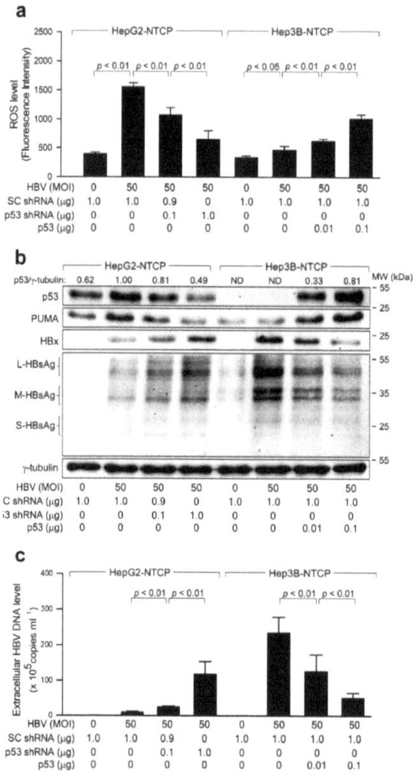

Figure 2. HBV activates p53 to amplify ROS during infection in human hepatoma cells. (**a**) HepG2-NTCP and Hep3B-NTCP cells were transfected with p53 shRNA and p53 expression plasmid for 24 h. Cells were then infected with HBV at 50 MOI for an additional 3 days. Intracellular ROS levels were analyzed as in Figure 1d ($n = 4$). (**b**) The levels of the specified proteins were measured by western blotting. Each protein band's normalized intensity was calculated relative to the housekeeping protein γ-tubulin to measure protein expression levels across samples. (**c**) Extracellular HBV DNA levels were quantified using qPCR ($n = 4$).

2.2. HBV Induces ROS Generation to Inhibit Virus Replication via a Negative Feedback Loop

To determine if HBV-induced ROS regulates HBV replication in both cell lines, we assessed the impact of the antioxidant N-acetyl-L-cysteine (NAC) on HBV replication. Known as a potent ROS scavenger, NAC effectively neutralized ROS during HBV infection in a dose-dependent manner, thereby nullifying the ability of the HBV to increase ROS and p53 levels in HepG2-NTCP cells (Figure 3a,b). Furthermore, NAC treatment led to dose-dependent activation of HBV replication in HepG2-NTCP cells, evidenced by the incremental rise in intracellular viral proteins and extracellular virus particles (Figure 3b,c). In Hep3B-NTCP cells, NAC treatment did not affect the levels of intracellular HBV proteins or extracellular HBV virions, although it effectively quelled HBV-induced ROS (Figure 3d–f). From these results, we deduce that HBV-induced ROS acts as an inhibitor of HBV replication in human hepatoma cells, but this inhibition is contingent upon the presence of p53.

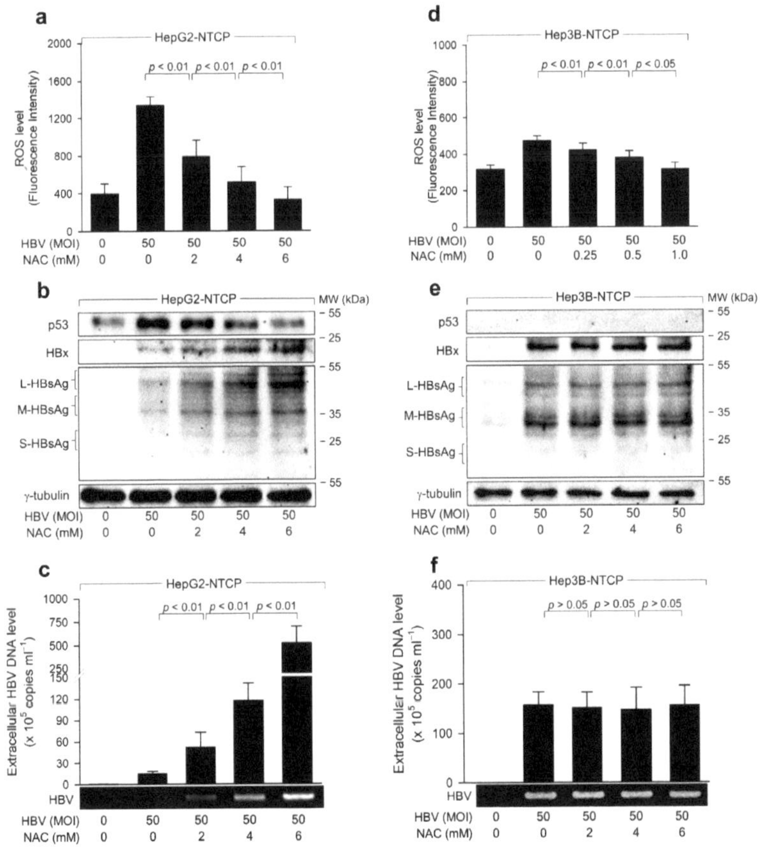

Figure 3. HBV induces ROS generation to inhibit virus replication via a negative feedback loop. (**a**,**d**) HepG2-NTCP and Hep3B-NTCP cells were infected with HBV for 48 h. Cells were then treated with NAC at the specified concentrations for 24 h before harvesting. ROS levels in cells were measured using CM-H2DCFDA (n = 5). (**b**,**e**) Levels of p53, HBx, HBsAg, and γ-tubulin in cell lysates were analyzed by western blotting. (**c**,**f**) Extracellular HBV DNA levels were quantified using conventional PCR and qPCR (n = 3).

2.3. The Transcriptional Activity of p53 Is Crucial for the ROS-Induced Suppression of HBV Replication

We next explored whether the potential of ROS to inhibit HBV replication depends on p53 transcriptional activity. Two p53 mutants, p53 R175H and p53 R248Q, cannot activate target gene transcription [35]. The protein levels of p53 upregulated modulator of apoptosis (PUMA) were upregulated during HBV infection in Hep3B-NTCP cells in the presence of WT p53, whereas HBV infection either without p53 or with the mutated p53 forms minimally affected PUMA levels (Figure 4a). Under these conditions, WT p53 amplified ROS and suppressed HBV replication in Hep3B-NTCP cells, while neither R175H nor p53 R248Q displayed these effects (Figure 4a–c), suggesting that p53 transcriptional activity is pivotal for ROS amplification and subsequent HBV inhibition in human hepatoma cells.

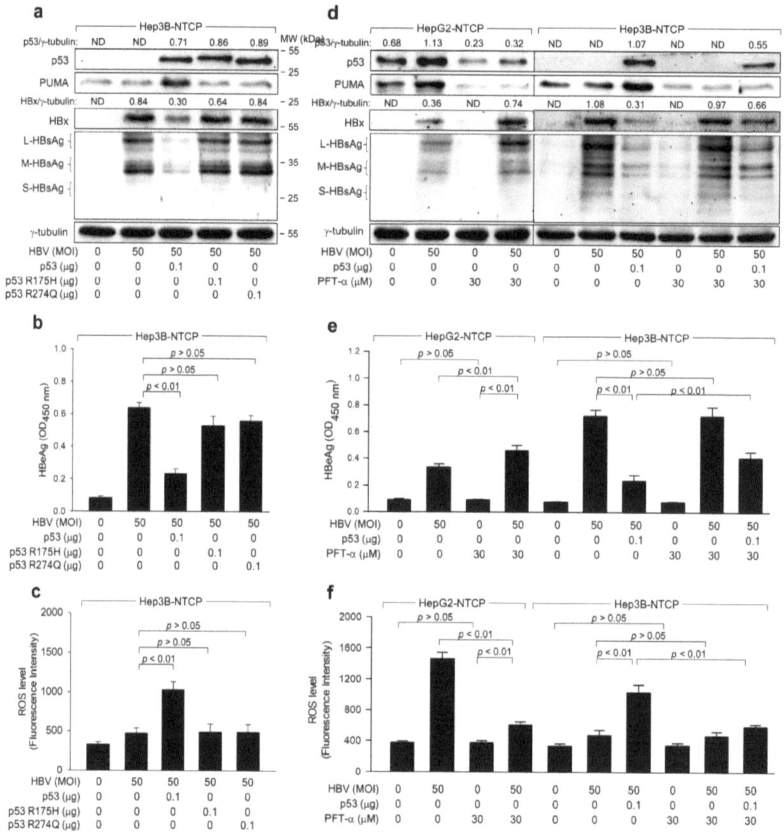

Figure 4. The transcriptional activity of p53 is crucial for the ROS-induced suppression of HBV replication. (**a–c**) Hep3B-NTCP cells were transfected with an expression plasmid encoding wild type (WT) p53, p53 R175H, and p53 R248Q for 24 h and then infected with HBV 50 MOI for an additional 3 days. (**d–f**) HepG2-NTCP cells and Hep3B-NTCP cells were infected with HBV, followed by treatment with PFT-α at the indicated concentrations. For lanes 7 and 10, Hep3B-NTCP cells were transfected with indicated plasmid before infection. (**a,d**) Levels of the intracellular proteins were determined by western blotting. (**b,e**) Levels of extracellular HBeAg were determined by ELISA ($n = 3$). (**c,f**) Levels of ROS were determined after treatment with CM-H2DCFDA ($n = 3$). (**a,d**) Each protein band's normalized intensity was calculated relative to the housekeeping protein γ-tubulin to measure protein expression levels across samples.

To corroborate the importance of p53 transcriptional activity for ROS-induced HBV inhibition, we used the specific p53 inhibitor, pifithrin-α (PFT-α), which curtails the transcriptional activation of p53 target genes [36]. PFT-α treatment diminished PUMA levels in HepG2-NTCP cells, regardless of HBV infection status, thereby nearly abolishing the effect of the HBV on PUMA levels (Figure 4d). This indicates successful inhibition of p53 transcriptional activity by PFT-α during HBV infection in HepG2-NTCP cells. PFT-α also reduced p53 levels in uninfected and HBV-infected HepG2-NTCP cells (Figure 4d), as it interrupted p53 amplification through a positive feedback loop [37]. However, PFT-α could not entirely negate the effect of the HBV on p53 levels (Figure 4d), likely because initial ROS induction and subsequent p53 upregulation during HBV infection are not contingent on p53 transcriptional activity. In HepG2-NTCP cells, PFT-α stimulated HBV replication, as evidenced by heightened intracellular HBx and HBsAg and extracellular HBeAg levels (Figure 4d,e). Moreover, PFT-α reduced ROS levels by approximately one-third during HBV infection in these cells, whereas the effect was minimal in uninfected cells, possibly due to ROS detection limit (Figure 4f). Conversely, PFT-α did not induce noticeable changes in ROS production and HBV replication in Hep3B-NTCP cells (Figure 4d–f). Ectopic p53 expression restored the potential of PFT-α to diminish ROS levels and stimulate HBV replication in Hep3B-NTCP cells (Figure 4d–f). Based on these observations, we conclude that p53 transcriptional activity is indispensable for HBV-induced ROS generation and the subsequent p53-mediated inhibition of HBV replication in human hepatoma cells.

2.4. ROS Reduces HBx Levels to Suppress HBV Replication in Human Hepatoma Cells

Next, we examined the potential role of HBx in p53-dependent ROS amplification and ROS-mediated HBV inhibition in hepatoma cells. HepG2-NTCP cells were infected with either WT HBV or its HBx-null variant. Notably, WT HBV displayed a significantly higher replication rate than HBx-null HBV, which is evident from elevated levels of intracellular HBsAg and extracellular virions (Figure 5a,b). The potential of WT HBV to induce ROS and elevate p53 levels during infection in HepG2-NTCP cells was also more pronounced compared to HBx-null HBV (Figure 5a,c). Furthermore, ectopic *HBx* expression largely complemented the defects observed in HBx-null HBV (Figure 5d–f), confirming the pivotal roles of the HBx in HBV replication [6,26,27], ROS generation [13,17], and p53 activation [26]. Intriguingly, HBx-null HBV still managed to induce ROS during replication in HepG2-NTCP cells (Figure 5c), implying potential contributions from other viral factors such as HBsAg and HBcAg in this mechanism. Treatment with NAC significantly attenuated the potential of WT HBV to elevate p53 and ROS levels during infection in HepG2-NTCP cells (Figure 5a,c). Likewise, NAC abolished the potential of HBx-null HBV to augment ROS and p53 levels in HepG2-NTCP cells with *HBx* expression (Figure 5d,f). Furthermore, NAC promoted the replication of HBx-null HBV in the presence of the HBx and WT HBV in HepG2-NTCP cells (Figure 5a,b,d,e). Conversely, in the absence of the HBx, NAC had a marginal effect on HBx-null HBV replication, even though it effectively neutralized the virus-induced alterations of p53 and ROS levels (Figure 5a–c). These findings suggest that HBx predominantly mediates the p53-dependent inhibition of HBV replication provoked by ROS.

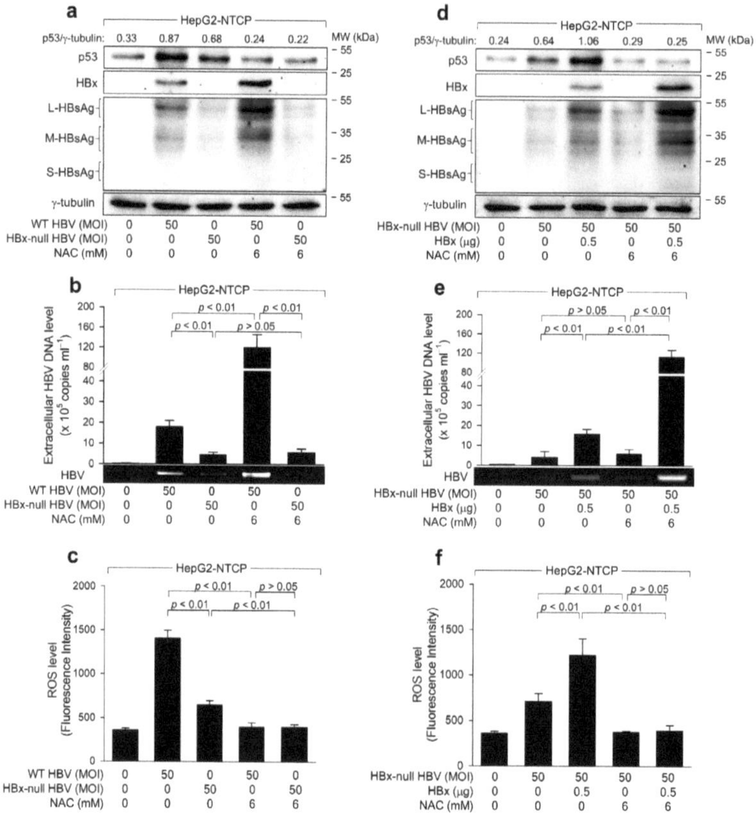

Figure 5. ROS reduces HBx levels to suppress HBV replication in human hepatoma cells. (**a**) HepG2-NTCP cells were infected with either WT HBV or HBx-null HBV for 3 days. If required, cells were treated with NAC for 24 h before harvesting. For lanes 3 and 5 in (**d–f**), pCMV-HA-HBx was transfected 24 h before HBV infection. (**a**,**d**) The levels of the indicated proteins were measured by western blotting. (**b**,**e**) Extracellular HBV DNA levels were quantified using conventional PCR and qPCR (n = 3). (**c**,**f**) ROS levels were measured after treatment with CM-H2DCFDA (n = 3). (**a**,**d**) The normalized intensity of each protein band was calculated relative to the housekeeping protein γ-tubulin to compare protein expression levels across samples.

2.5. ROS Drives Siah-1-Mediated Proteasomal Degradation of HBx in a p53-Reliant Manner

To elucidate the direct roles of the HBx in ROS generation and subsequent p53 activation, we transiently transfected HBx in HepG2 and Hep3B cells. Ectopic *HBx* expression alone elevated ROS and p53 levels in HepG2 cells (Figure 6a,b). In Hep3B cells, although the HBx slightly upregulated ROS levels, the effect was less pronounced than in HepG2 cells (Figure 6c,d). This suggests that HBx initially induces ROS generation via a p53-independent mechanism, which is further amplified mainly through p53 upregulation, consistent with a previous study [32]. Notably, p53 knockdown resulted in elevated HBx levels during HBV infection in HepG2-NTCP cells, whereas p53 overexpression led to reduced HBx levels in Hep3B-NTCP cells (Figure 2b), aligning with previous findings on the negative regulatory effect of p53 on the HBx during HBV infection [20,26,29,38]. Moreover, ROS depletion using NAC increased HBx levels during HBV infection and ectopic expression in HepG2-NTCP cells (Figures 5a and 6b). However, NAC had minimal effect on HBx levels during HBV replication and ectopic expression in Hep3B-NTCP cells where no p53 signal was detected (Figures 3e and 6d). These data strongly link HBx-induced ROS generation to p53-mediated HBx downregulation in hepatoma cells.

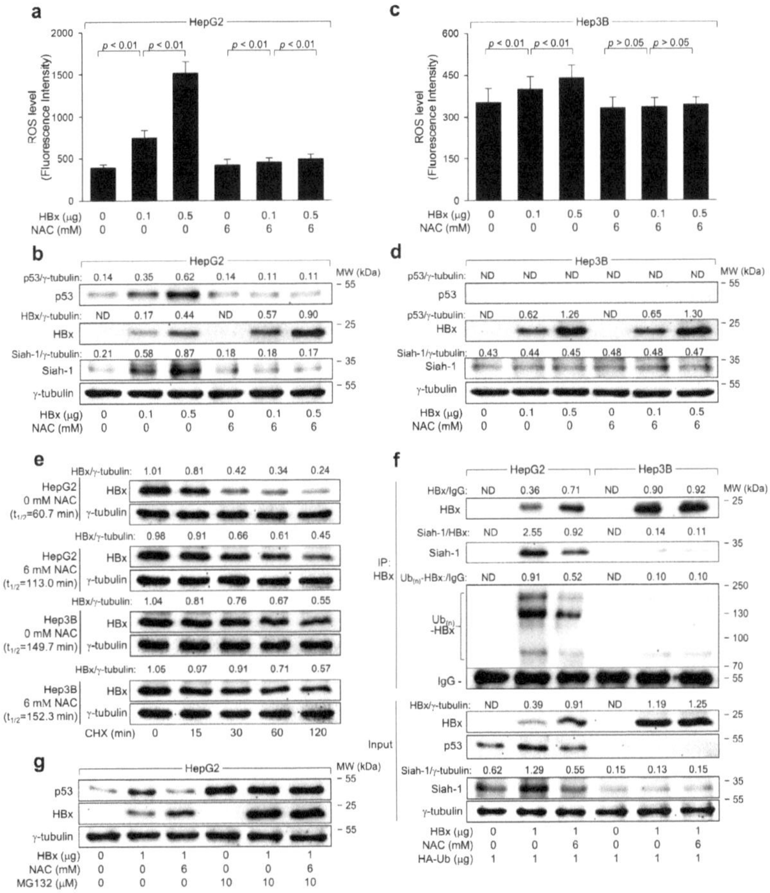

Figure 6. ROS drives Siah-1-mediated proteasomal degradation of HBx in a p53-reliant manner. HepG2 and Hep3B cells were transfected with an *HBx* expression plasmid for 48 h in the presence or absence of NAC. (**a**,**c**) ROS levels were determined ($n = 4$). (**b**,**d**) Levels of the indicated proteins were determined by western blotting. (**e**) Cells were treated with 50 μg/mL of cycloheximide (CHX) for the indicated time to inhibit further protein synthesis before harvesting, followed by western blotting. Each band was quantified using ImageJ image-analysis software (Version 2.1.0, NIH) to determine the half-life ($t_{1/2}$) of HBx protein (**f**) HA-Ub expression plasmid was added in the transfection cocktails. Ubiquitin-conjugated HBx products were immunoprecipitated with anti-HA antibodies and pulled down with magnetic beads. Western blotting was performed to measure levels of HBx, p53, Siah-1, and Ub-complexed HBx. The input lanes show the levels of the proteins. (**g**) Cells were mock-treated or treated with 10 μM MG132 for 4 h before protein sampling. (**b**,**d**,**e**,**f**) Each protein band's normalized intensity was calculated relative to the housekeeping protein γ-tubulin to measure protein expression levels across samples.

Previous reports indicate that p53 transcriptionally activates the *Siah-1* gene, which encodes an E3 ligase for the HBx [20,23,39]. Indeed, ectopic *HBx* expression, independent of other viral proteins, upregulated Siah-1 levels via p53 upregulation in HepG2 cells but not in Hep3B cells (Figure 6b,d). Additionally, NAC treatment reduced p53 and Siah-1 levels in HepG2 but not in Hep3B cells (Figure 6b,d). These findings strongly suggest that Siah-1 is involved in ROS-mediated HBx downregulation in a p53-dependent manner. To substantiate that ROS reduces HBx levels via p53-dependent Siah-1 upregulation, we evaluated if ROS affects HBx protein stability in a p53-dependent manner. We treated HBx-expressing HepG2 and Hep3B cells with or without NAC, followed by cycloheximide (CHX) treatment to halt protein synthesis. The half-life ($t_{1/2}$) of the HBx in HepG2 cells was 60.7 min and increased to 113.0 min with NAC, while in Hep3B cells, it was 149.7 min and minimally affected by NAC (Figure 6e), confirming the role of ROS in HBx destabilization via p53.

Additionally, we investigated whether ROS enhances Siah-1-mediated ubiquitination of the HBx through p53 dependence. We co-transfected *HBx* and HA-tagged ubiquitin into both cell lines, with or without NAC treatment, and immunoprecipitated the ubiquitin-complexed HBx. Co-immunoprecipitation revealed that Siah-1 interacted with the HBx, inducing its ubiquitination in HepG2 cells (Figure 6f, lane 2). In contrast, this interaction and subsequent HBx ubiquitination were compromised in Hep3B cells due to the absence of p53-mediated Siah-1 upregulation (Figure 6f, lane 5). NAC treatment reduced HBx-Siah-1 interaction in HepG2 cells, decreasing HBx ubiquitination and elevating its protein levels, whereas these effects were largely absent in Hep3B cells (Figure 6f, lanes 3 and 6). MG132, a proteasomal inhibitor, attenuated HBx and NAC-induced p53 and HBx upregulation in HepG2 cells (Figure 6g). In conclusion, these findings establish that Siah-1 drives ROS-induced ubiquitination and proteasomal degradation of the HBx in a p53-dependent manner.

3. Discussion

HBV infection typically induces oxidative stress, marked by increased levels of ROS, such as H_2O_2, in the liver and blood of patients, and both host and viral factors are known to be included in this process [11,12]. Virus-specific cytotoxic T lymphocytes, which are mainly responsible for liver injury by destroying infected hepatocytes and promoting the production of inflammatory cytokines, significantly contribute to ROS generation in the liver [40]. Additionally, the HBV contributes to the production of ROS in hepatocytes, and this effect is primarily mediated by the HBx [13–15,17]. In this study, we employed an optimized HBV infection system, which allows successful HBV infection [22]. Infection of HepG2-NTCP cells with the HBV at an indicated multiplicity of infection (MOI) for 4 days enabled quantitative analysis of HBV replication rates through measurement of HBV proteins in the cell lysates and extracellular HBV particles in the culture media (Figure 1). Our findings revealed a significant difference in ROS levels between HepG2 and Hep3B cells, suggesting a potential role for p53 in ROS modulation. Given the inherent limitations of using Hep3B cells, including the possible impact of genomic integration, we further sought to clarify the role of p53 in regulating ROS. To this end, we manipulated p53 expression by knocking down its expression in HepG2 cells and ectopically expressing it in Hep3B cells (Figure 2a). These manipulations consistently demonstrated that ROS levels are influenced by the status of p53, thereby reinforcing the link between p53 and ROS regulation. To conclusively determine the influence of p53 on ROS and HBV replication, we employed p53 inhibitors, specifically the p53 dominant negative mutant R175H and R248Q, and PFT-α, to disrupt normal p53 function. The application of these inhibitors resulted in the attenuation of the p53 effects on ROS and subsequent HBV replication (Figure 4), thereby confirming the interdependence of p53, ROS production, and HBV replication dynamics. These results collectively highlight the critical role of p53 in modulating ROS levels and its subsequent impact on HBV replication, providing valuable insights into the cellular mechanisms governing viral replication in liver cells.

The HBx appears to play a predominant role in ROS production during HBV infection. When localized within mitochondria, the HBx alters mitochondrial membrane potential, triggering ROS generation [15,16]. Additionally, the HBx modulates the permeability of the mitochondrial membrane and disrupts the mitochondrial respiratory chain [15,41]. HBx-induced ROS may induce DNA damage, such as double-strand DNA breaks, resulting in activation of the ATM-Chk2 pathway and subsequent stabilization of p53. Previous reports have demonstrated that the HBx induces the accumulation of γ-H2AX [42] and activates ATM and Chk2 via phosphorylation at the Ser-1981 and Thr-68 residues, respectively, leading to activation of p53 via phosphorylation at the Ser-15 and Ser-20 residues. Consistently, the present study showed that ectopic *HBx* expression alone is sufficient to elevate both ROS and p53 levels in hepatoma cells (Figure 6). While HBx-null HBV also induces ROS generation and activates p53 (Figure 5a), presumably due to the possible roles of HBsAg and HBcAg [43], the magnitude of these effects appears considerably weaker compared to that observed in the presence of the HBx (Figures 1 and 6).

The slight increase in ROS levels in Hep3B-NTCP cells following HBV infection (Figure 1) suggests that HBV can trigger ROS production through a p53-independent mechanism. The involvement of the HBx in this phenomenon was validated through ectopic expression experiments in Hep3B cells, where the HBx alone induced ROS elevation independent of other viral proteins (Figure 6c). This effect was likely due to the potential of the HBx to induce ROS generation in mitochondria through p53-independent mechanisms [15,16,41]. However, the ROS levels generated by the HBx in the absence of p53 were considerably lower than those obtained in the presence of p53 (Figure 6a,c), underscoring the crucial role of p53 in amplifying ROS in HBV-infected cells. Many p53-inducible genes encode redox-active proteins involved in ROS generation, such as quinone oxidoreductase, proline oxidase, and manganese superoxide dismutase [44,45]. Therefore, p53 may amplify its protein levels through a positive feedback loop involving ROS and the ATM-ChK2 pathway.

The R175 residue located within the zinc-binding site adjacent to the DNA-binding interface of p53 is crucial for maintaining structural stability, while R248 is involved in direct interaction with DNA [35]. Mutations in these residues result in the loss of transcriptional activator function and interfere with p53 tetramer formation as dominant negatives. Consistently, both *p53 R175H* and *p53 R248Q* expressed in Hep3B-NTCP cells did not exhibit transcriptional activity, as demonstrated by the protein levels of PUMA (Figure 4a). Neither mutation induced ROS amplification nor inhibited HBV replication (Figure 4a–c), indicating the importance of p53 transcriptional activity in regulating ROS production and HBV replication. Additionally, treatment with PFT-α, which decreases nuclear p53 stability [46] or its DNA binding activity [47], stimulated HBV replication but decreased ROS levels in both HepG2-NTCP and Hep3B-NTCP cells (Figure 4d–f). This suggests that transcriptional activity is required for p53 to regulate ROS generation and viral replication. Interestingly, the potential of p53 to inhibit HBV replication remained active even when treatment with PFT-α completely inhibited the ability of p53 to activate the expression of its target genes. This suggests that p53 can function as a negative regulator of HBV replication through another mechanism that does not involve its transcriptional activity.

The role of p53 as a negative regulator of HBV replication is relatively well-established [26,32,48]. The present study provides several lines of evidence supporting that HBx-induced ROS inhibits HBV replication by downregulating HBx levels in a p53-dependent manner. First, lower levels of intracellular HBx and HBsAg and extracellular HBeAg and HBV particles were produced during HBV infection in HepG2-NTCP cells where the HBx effectively upregulated p53 and ROS levels (Figure 1). Second, p53 knockdown downregulated ROS levels but stimulated HBV replication in HepG2-NTCP cells, whereas ectopic p53 expression increased ROS levels but inhibited HBV replication in Hep3B-NTCP cells (Figure 2). Third, HBV replication was enhanced when ROS levels were lowered by treatment with NAC (Figure 3a–c), as demonstrated in a previous study suggesting the negative role of ROS in HBV replication [49]. Notably, this effect was not observed during WT HBV infection in Hep3B-NTCP cells and HBx-null HBV infection in HepG2-NTCP and Hep3B NTCP cells, where p53 was largely unaffected or absent (Figures 3 and 5), indicating the significant role of p53 activation in this process.

It has been shown that p53 reduces HBx levels through ubiquitin-dependent proteasomal degradation, inhibiting HBV replication [20,50]. Siah-1, acting as an E3 ligase for the HBx, facilitates the ability of p53 to downregulate HBx levels. [20,39]. Given that *Siah-1* expression is recognized to be transcriptionally activated by p53 [20,51], it is conceivable that the p53 activated by HBx-induced ROS could stimulate the expression of *Siah-1*. Consistently, ectopic *HBx* expression upregulated both p53 and Siah-1 levels in HepG2 cells, whereas treatment with NAC nullified the ability of the HBx to induce this effect (Figure 6b). In addition, treatment with NAC decreased the binding of Siah-1 to the HBx and HBx ubiquitination, resulting in the upregulation of HBx levels in a p53-dependent manner (Figure 6). Given the potential of the HBx to enhance HBV replication [6,52,53], as also shown in the current study (Figure 5), the Siah-1-mediated ubiquitination and proteasomal degradation of the HBx likely play a crucial role in ROS-induced inhibition of HBV replication.

This study demonstrated that the HBx potentially contributes to ROS-induced inhibition of HBV replication by being downregulated through a negative feedback loop. The reasons why the HBx limits its expression during HBV replication via this feedback loop involving ROS and p53 are not yet clear. One possibility is that the HBV modulates HBx protein levels to remain within a specific threshold during persistent infection. Alternatively, the ROS-induced inhibition of HBV replication may serve as an innate host defense mechanism against HBV infection, highlighting p53's role as the guardian of the genome. By demonstrating the inhibitory effect of ROS on HBV replication in a p53-dependent manner and elucidating the involvement of Siah-1 in HBx degradation, this research underscores the therapeutic potential of targeting ROS signaling pathways for HBV treatment. Further studies on ROS-inducing agents could pave the way for innovative strategies to combat HBV and enhance patient outcomes. A limitation of the current study is that it does not address the long-term effects of chronic infection, as the current in vitro system is not suitable for studying prolonged infection periods necessary to confirm chronic infection. To address the role of p53 in ROS generation and HBV replication during chronic hepatitis B, in vivo experiments based on the p53 status of patients should be performed.

4. Materials and Methods

4.1. Plasmid

The plasmid pCMV-3 × HA1-HBx (HA-HBx) carries the *HBx* gene (genotype D) situated downstream from three influenza virus haemagglutinin (HA) tags [54]. The 1.2-mer WT HBV replicon, which includes 1.2 copies of the genotype D HBV genome, and a version without the *HBx* gene were provided by W. S. Ryu (Yonsei University, Seoul, Republic of Korea) [27]. The plasmid RC210241, which codes for the human *NTCP*, was sourced from OriGene (Rockville, MD, USA, Cat No. 003049). Scrambled (SC) and p53-targeting short hairpin RNA (shRNA) were acquired from Santa Cruz Biotechnology (Santa Cruz, CA, USA, Cat No. sc-37007 and sc-29435, respectively). The pCMV p53-WT plasmid was a gift from Chang-Woo Lee (Sungkyunkwan University, Suwon, Republic of Korea). The pCMV p53 R175H and pCMV p53 R274Q plasmids were provided by Bum-Joon Park (Pusan National University, Busan, Republic of Korea). The plasmid pHA-Ub that encodes HA-tagged ubiquitin was generously supplied by Y. Xiong (University of North Carolina at Chapel Hill, NC, USA).

4.2. Cell Culture and Transfection

The HepG2 (Cat No. 88065) and Hep3B (Cat No. 88064) cell lines were gained from the Korean Cell Line Bank (KCLB, Seoul, Republic of Korea). To create stable cell lines, HepG2-NTCP and Hep3B-NTCP cells were transfected with *NTCP* expression plasmid and then selected using 500 μg/mL G418 sulfate (Sigma-Aldrich, St. Louis, MO, USA, Cat No. A1720). These cells were cultured in Dulbecco's Modified Eagle Medium (DMEM) by WelGENE (Gyeongsan, Republic of Korea, Cat No. LM001-05), with 10% fetal bovine serum (FBS, Capricorn Scientific, Ebsdorfergrund, Germany, Cat No. FBS-22A), 100 μg/mL streptomycin (United States Biological, Salem, MA, USA, Cat No. 21865), and 100 units/mL penicillin G (Sigma-Aldrich, Cat No. P3032). The culture conditions were maintained at 37 °C in a 5% CO_2-humidified atmosphere. For transient expression studies, 2×10^5 cells per well in a 6-well plate were transfected with a transfection reagent (Thermo Fisher Scientific, Waltham, MA, USA, Cat No. R0532). The cells were subjected to various treatments, including CHX(Sigma-Aldrich, Cat No. C7698), N-acetylcysteine (NAC, Sigma-Aldrich, Cat No. A7250), (PFT-α, Sigma-Aldrich, Cat No. P4359), and MG132 (Millipore, Burlington, MA, USA, Cat No. 474790), under specific experimental conditions.

4.3. HBV Cell Culture System

For HBV stock preparations, Hep3B-NTCP cells were transfected with the 1.2-mer HBV replicon plasmid as described above. The culture medium was collected to prepare the HBV stocks. The concentration of HBV in these stocks was assessed by qPCR, details of which are provided in the following section. For the infection experiments, HBV was introduced into cells seeded in 6-well plates at a MOI of 50, based on a slightly modified protocol from an established HBV cell culture protocol [22,23]. Initially, 2×10^5 cells per well were exposed to 1×10^6 genome equivalents (GEQs) of HBV and incubated for 24 h in serum-free DMEM supplemented with 2% DMSO (Sigma-Aldrich, Cat No. D8418) and 4% PEG 8000 (Sigma-Aldrich, Cat No. D4463). After two PBS washes, the cells were further cultured in DMEM enriched with 3% FBS, 4% PEG 8000, and 2% DMSO for 72 h.

4.4. Quantitative Real-Time PCR of HBV DNA

The extracellular HBV levels were measured using a qPCR protocol [26]. Initially, extracellular HBV DNA was extracted from the collected culture supernatant employing the QIAamp DNA Mini Kit (Qiagen, Hilden, Germany, Cat No. 51306). For conventional PCR analysis, the genomic DNA underwent amplification using 2 × Taq PCR Master Mix 1 (BioFACT, Daejeon, Republic of Korea, Cat No. ST301-19h) with the primers HBV 1399F (5′-TGG TAC CTG CGC GGG ACG TCC TT-3′) and HBV 1632R (5′-AGC TAG CGT TCA CGG TGG TCT CC-3′). Subsequent qPCR analysis was performed using SYBR premix Ex Taq II (Takara Bio, Kusatsu, Japan, Cat No. RR82LR) and the primers HBV 379F (5′-GTG TCT GCG GCG TTT TAT CA-3′) and HBV 476R (5′-GAC AAA CGG GCA ACA TAC CTT-3′) on a Rotor-Gene qPCR machine (Qiagen).

4.5. Measurement of Intracellular ROS Levels

Intracellular ROS levels were determined using the H_2O_2-specific probe, CM-H2DCFDA (Invitrogen, Waltham, MA, USA, Cat No. C6827), which is commonly used for detecting ROS in living cells [30]. In the procedure, 1×10^5 cells per well were incubated in 12-well plates with 10 μM CM-H2DCFDA in serum-free medium for 30 min. Cells were washed with PBS and collected using Trypsin-EDTA (Gibco, Cleveland, TN, USA, Cat No. 25200-072) for analysis. The fluorescence intensity of DCF was analyzed using a fluorescence microplate reader (Mithras LB940, Berthold Technologies, Bad Wildbad, Germany) set to excitation and emission wavelengths of 485 nm and 535 nm, respectively.

4.6. Western Blot Analysis

Cells were ruptured using a lysis buffer composed of 0.1% SDS, 50 mM Tris-HCl (pH 8.0), 1% NP-40, and 150 mM NaCl, which also contained protease inhibitors from Roche (Basel, Switzerland, Cat No. 11836153001). The protein levels in the lysates were determined with a Bio-Rad protein assay kit (Hercules, CA, USA, Cat No. 5000006). Proteins were separated based on sizes by SDS-PAGE. Transferred proteins on a nitrocellulose membrane (Amersham, UK, Cat No. 10600003) were probed with primary antibodies such as Siah-1 (Abcam, Cambridge, UK, Cat No. ab2237, 1:2000), p53 (Santa Cruz Biotechnology, Cat No. sc-126, 1:1000), HBsAg (Santa Cruz Biotechnology, Cat No. sc-53300, 1:400), γ-tubulin (Santa Cruz Biotechnology, Cat No. sc-17787, 1:500), PUMA (Cell signaling, 4976S, 1:2000), HBx (Millipore, Cat No. MAB8419, 1:500), and HA (Santa Cruz Biotechnology, Cat No.

sc-7392, 1:500). These were then incubated with HRP-conjugated secondary antibodies, specifically anti-mouse (Bio-Rad, Cat No. BR170-6516, 1:3000), anti-rabbit IgG (H + L)-HRP (Bio-Rad, Cat No. BR170-6515, 1:3000), or anti-goat IgG (H + L)-HRP (Thermo Scientific, Cat No. 31400, 1:10,000). Protein detection was performed using an ECL kit (Advansta, San Jose, CA, USA, Cat No. K-12043-D20) and imaged with the ChemiDoc XRS system (Bio-Rad).

4.7. Immunoprecipitation

Immunoprecipitation (IP) assay was performed using a kit from Thermo Fisher Scientific (Classic Magnetic IP/Co-IP, Cat No. 88804), following the manufacturer's instructions. Initially, 4×10^5 cells were seeded per 60 mm-diameter dish and transfected with designated expression plasmids for 48 h as specified. Cell lysates were then treated with an anti-HBx antibody (Millipore, Cat No. 8419) at 4 °C overnight to facilitate immune complex formation. To isolate the immune complexes, the mixture was incubated with protein A/G beads, which bind to the Fc region of the antibody, allowing the antibody-protein complexes to be precipitated out of the lysate. A magnetic stand (Pierce, Waltham, MA, USA) was used to elute the protein complexes from the beads. The eluted proteins are analyzed by SDS-PAGE, followed by western blotting to detect the presence of the target protein and any interacting partners.

4.8. Quantification of Western Blot Images

Western blot images were captured using a standardized protocol to ensure consistent lighting and exposure across all samples [55]. Band intensity quantification was performed using ImageJ software (version 2.1.0, National Institutes of Health, Bethesda, MD, USA). Each band of interest was outlined with a uniformly sized rectangular selection tool to measure the integrated density, which includes the area and mean gray value of the selected region. Background intensity was determined by measuring a similarly sized area adjacent to the bands. This background value was subtracted from the corresponding band intensity to correct for ambient noise and nonspecific signals. The normalized intensity of each protein band was then calculated relative to the intensity of the housekeeping protein γ-tubulin, providing a relative value of protein expression level in each sample.

4.9. Statistical Analysis

The reported data represent the average values and standard deviations calculated from a minimum of three separate experiments. A two-tailed Student's *t*-test was performed using SigmaPlot (ver 12.5). Statistical significance was determined based on the P values; specifically, a p value of 0.05 or less indicates significant differences between experimental groups. Conversely, a p value greater than 0.05 was regarded as not statistically significant, suggesting no meaningful difference between the groups under comparison. This statistical method helps to ensure that the observed effects are not due to random variation, providing a reliable measure of the reproducibility and accuracy of the experimental results.

Author Contributions: Conceptualization, Y.J., J.H. and K.L.J.; methodology, Y.J. and J.H.; software, J.H.; validation, Y.J., J.H. and K.L.J.; formal analysis, Y.J. and J.H.; investigation, Y.J. and J.H.; data curation, Y.J., J.H. and K.L.J.; writing—original draft preparation, Y.J. and J.H.; writing—review and editing, K.L.J.; supervision, K.L.J.; project administration, K.L.J.; funding acquisition, K.L.J. All authors have read and agreed to the published version of the manuscript.

Funding: This work was supported by a two-year Research Grant from Pusan National University.

Institutional Review Board Statement: Not applicable.

Informed Consent Statement: Not applicable.

Data Availability Statement: The data presented in this study are available from the corresponding author upon reasonable request.

Acknowledgments: We thank W.-S. Ryu (Yonsei University, Seoul, Republic of Korea), C.-W. Lee (Sungkyunkwan University, Suwon, Republic of Korea), and Y. Xiong (University of North Carolina at Chapel Hill, NC, USA) for kindly providing the 1.2-mer-HBV replicons, pCMV p53, and pHA-Ub, respectively.

Conflicts of Interest: The authors declare no conflicts of interest.

References

1. Ott, J.J.; Stevens, G.A.; Groeger, J.; Wiersma, S.T. Global epidemiology of hepatitis B virus infection: New estimates of age-specific HBsAg seroprevalence and endemicity. *Vaccine* **2012**, *30*, 2212–2219. [CrossRef] [PubMed]
2. Locarnini, S.; Zoulim, F. Molecular genetics of HBV infection. *Antivir. Ther.* **2010**, *15* (Suppl. S3), 3–14. [CrossRef] [PubMed]
3. Liu, S.; Koh, S.S.; Lee, C.G. Hepatitis B Virus X Protein and Hepatocarcinogenesis. *Int. J. Mol. Sci.* **2016**, *17*, 940. [CrossRef] [PubMed]
4. Levrero, M.; Zucman-Rossi, J. Mechanisms of HBV-induced hepatocellular carcinoma. *J. Hepatol.* **2016**, *64* (Suppl. S1), S84–S101. [CrossRef] [PubMed]
5. Tsuge, M.; Hiraga, N.; Akiyama, R.; Tanaka, S.; Matsushita, M.; Mitsui, F.; Abe, H.; Kitamura, S.; Hatakeyama, T.; Kimura, T.; et al. HBx protein is indispensable for development of viraemia in human hepatocyte chimeric mice. *J. Gen. Virol.* **2010**, *91 Pt 7*, 1854–1864. [CrossRef] [PubMed]
6. Keasler, V.V.; Hodgson, A.J.; Madden, C.R.; Slagle, B.L. Enhancement of hepatitis B virus replication by the regulatory X protein in vitro and in vivo. *J. Virol.* **2007**, *81*, 2656–2662. [CrossRef] [PubMed]
7. Keasler, V.V.; Hodgson, A.J.; Madden, C.R.; Slagle, B.L. Hepatitis B virus HBx protein localized to the nucleus restores HBx-deficient virus replication in HepG2 cells and in vivo in hydrodynamically-injected mice. *Virology* **2009**, *390*, 122–129. [CrossRef] [PubMed]
8. Bouchard, M.J.; Wang, L.H.; Schneider, R.J. Calcium signaling by HBx protein in hepatitis B virus DNA replication. *Science* **2001**, *294*, 2376–2378. [CrossRef]
9. Rawat, S.; Bouchard, M.J. The hepatitis B virus (HBV) HBx protein activates AKT to simultaneously regulate HBV replication and hepatocyte survival. *J. Virol.* **2015**, *89*, 999–1012. [CrossRef]
10. Finkel, T. Oxidant signals and oxidative stress. *Curr. Opin. Cell Biol.* **2003**, *15*, 247–254. [CrossRef]
11. Fujita, N.; Sugimoto, R.; Ma, N.; Tanaka, H.; Iwasa, M.; Kobayashi, Y.; Kawanishi, S.; Watanabe, S.; Kaito, M.; Takei, Y. Comparison of hepatic oxidative DNA damage in patients with chronic hepatitis B and C. *J. Viral Hepat.* **2008**, *15*, 498–507. [CrossRef]
12. Bolukbas, C.; Bolukbas, F.F.; Horoz, M.; Aslan, M.; Celik, H.; Erel, O. Increased oxidative stress associated with the severity of the liver disease in various forms of hepatitis B virus infection. *BMC Infect. Dis.* **2005**, *5*, 95. [CrossRef]
13. Ren, J.-H.; Chen, X.; Zhou, L.; Tao, N.-N.; Zhou, H.-Z.; Liu, B.; Li, W.-Y.; Huang, A.-L.; Chen, J. Protective Role of Sirtuin3 (SIRT3) in Oxidative Stress Mediated by Hepatitis B Virus X Protein Expression. *PLoS ONE* **2016**, *11*, e0150961. [CrossRef]
14. Ivanov, A.V.; Valuev-Elliston, V.T.; Tyurina, D.A.; Ivanova, O.N.; Kochetkov, S.N.; Bartosch, B.; Isaguliants, M.G. Oxidative stress, a trigger of hepatitis C and B virus-induced liver carcinogenesis. *Oncotarget* **2017**, *8*, 3895–3932. [CrossRef] [PubMed]
15. Rahmani, Z.; Huh, K.W.; Lasher, R.; Siddiqui, A. Hepatitis B virus X protein colocalizes to mitochondria with a human voltage-dependent anion channel, HVDAC3, and alters its transmembrane potential. *J. Virol.* **2000**, *74*, 2840–2846. [CrossRef] [PubMed]
16. Clippinger, A.J.; Bouchard, M.J. Hepatitis B virus HBx protein localizes to mitochondria in primary rat hepatocytes and modulates mitochondrial membrane potential. *J. Virol.* **2008**, *82*, 6798–6811. [CrossRef] [PubMed]
17. Lee, Y.I.; Hwang, J.M.; Im, J.H.; Lee, Y.I.; Kim, N.S.; Kim, D.G.; Yu, D.Y.; Moon, H.B.; Park, S.K. Human hepatitis B virus-X protein alters mitochondrial function and physiology in human liver cells. *J. Biol. Chem.* **2004**, *279*, 15460–15471. [CrossRef]
18. Xie, H.; Xie, D.; Zhang, J.; Jin, W.; Li, Y.; Yao, J.; Pan, Z.; Xie, D. ROS/NF-kappaB Signaling Pathway-Mediated Transcriptional Activation of TRIM37 Promotes HBV-Associated Hepatic Fibrosis. *Mol. Ther. Nucleic Acids* **2020**, *22*, 114–123. [CrossRef]
19. Foo, J.; Bellot, G.; Pervaiz, S.; Alonso, S. Mitochondria-mediated oxidative stress during viral infection. *Trends Microbiol.* **2022**, *30*, 679–692. [CrossRef]
20. Yeom, S.; Kim, S.S.; Jeong, H.; Jang, K.L. Hepatitis B virus X protein activates E3 ubiquitin ligase Siah-1 to control virus propagation via a negative feedback loop. *J. Gen. Virol.* **2017**, *98*, 1774–1784. [CrossRef]
21. Achanta, G.; Huang, P. Role of p53 in sensing oxidative DNA damage in response to reactive oxygen species-generating agents. *Cancer Res.* **2004**, *64*, 6233–6239. [CrossRef] [PubMed]
22. Yoon, H.; Han, J.; Jang, K.L. Hepatitis B Virus X Protein Stimulates Hepatitis C Virus (HCV) Replication by Protecting HCV Core Protein from E6AP-Mediated Proteasomal Degradation. *Microbiol. Spectr.* **2022**, *10*, e0143222. [CrossRef] [PubMed]
23. Michailidis, E.; Pabon, J.; Xiang, K.; Park, P.; Ramanan, V.; Hoffmann, H.H.; Schneider, W.M.; Bhatia, S.N.; de Jong, Y.P.; Shlomai, A.; et al. A robust cell culture system supporting the complete life cycle of hepatitis B virus. *Sci. Rep.* **2017**, *7*, 16616. [CrossRef] [PubMed]
24. Qiu, G.H.; Xie, X.; Xu, F.; Shi, X.; Wang, Y.; Deng, L. Distinctive pharmacological differences between liver cancer cell lines HepG2 and Hep3B. *Cytotechnology* **2015**, *67*, 1–12. [CrossRef] [PubMed]
25. Li, W. The hepatitis B virus receptor. *Annu. Rev. Cell Dev. Biol.* **2015**, *31*, 125–147. [CrossRef] [PubMed]

26. Lim, H.Y.; Han, J.; Yoon, H.; Jang, K.L. Tumor Suppressor p53 Inhibits Hepatitis B Virus Replication by Downregulating HBx via E6AP-Mediated Proteasomal Degradation in Human Hepatocellular Carcinoma Cell Lines. *Viruses* **2022**, *14*, 2313. [CrossRef] [PubMed]
27. Cha, M.Y.; Ryu, D.K.; Jung, H.S.; Chang, H.E.; Ryu, W.S. Stimulation of hepatitis B virus genome replication by HBx is linked to both nuclear and cytoplasmic HBx expression. *J. Gen. Virol.* **2009**, *90 Pt 4*, 978–986. [CrossRef] [PubMed]
28. Knowles, B.B.; Howe, C.C.; Aden, D.P. Human hepatocellular carcinoma cell lines secrete the major plasma proteins and hepatitis B surface antigen. *Science* **1980**, *209*, 497–499. [CrossRef]
29. Joschko, N.M. The Role of the Tumour Suppressor p53 in the Elimination of Hepatitis B Virus-infected Hepatocytes-Implications for Viral Clearance. Ph.D. Thesis, Universität Heidelberg, Heidelberg, Germany, 2012.
30. Oparka, M.; Walczak, J.; Malinska, D.; van Oppen, L.; Szczepanowska, J.; Koopman, W.J.H.; Wieckowski, M.R. Quantifying ROS levels using CM-H(2)DCFDA and HyPer. *Methods* **2016**, *109*, 3–11. [CrossRef]
31. Higgs, M.R.; Chouteau, P.; Lerat, H. 'Liver let die': Oxidative DNA damage and hepatotropic viruses. *J. Gen. Virol.* **2014**, *95 Pt 5*, 991–1004. [CrossRef]
32. Yoon, H.; Lee, H.K.; Jang, K.L. Hydrogen Peroxide Inhibits Hepatitis B Virus Replication by Downregulating HBx Levels via Siah-1-Mediated Proteasomal Degradation in Human Hepatoma Cells. *Int. J. Mol. Sci.* **2023**, *24*, 13354. [CrossRef] [PubMed]
33. Jiang, Y.; Han, Q.; Zhao, H.; Zhang, J. The Mechanisms of HBV-Induced Hepatocellular Carcinoma. *J. Hepatocell. Carcinoma* **2021**, *8*, 435–450. [CrossRef] [PubMed]
34. Yeom, S.; Jeong, H.; Kim, S.S.; Jang, K.L. Hepatitis B virus X protein activates proteasomal activator 28 gamma expression via upregulation of p53 levels to stimulate virus replication. *J. Gen. Virol.* **2018**, *99*, 655–666. [CrossRef] [PubMed]
35. Offutt, T.L.; Ieong, P.U.; Demir, Ö.; Amaro, R.E. Dynamics and Molecular Mechanisms of p53 Transcriptional Activation. *Biochemistry* **2018**, *57*, 6528–6537. [CrossRef] [PubMed]
36. Zhu, J.; Singh, M.; Selivanova, G.; Peuget, S. Pifithrin-α alters p53 post-translational modifications pattern and differentially inhibits p53 target genes. *Sci. Rep.* **2020**, *10*, 1049. [CrossRef] [PubMed]
37. Harris, S.L.; Levine, A.J. The p53 pathway: Positive and negative feedback loops. *Oncogene* **2005**, *24*, 2899–2908. [CrossRef]
38. Lee, S.G.; Rho, H.M. Transcriptional repression of the human p53 gene by hepatitis B viral X protein. *Oncogene* **2000**, *19*, 468–471. [CrossRef]
39. Zhao, J.; Wang, C.; Wang, J.; Yang, X.; Diao, N.; Li, Q.; Wang, W.; Xian, L.; Fang, Z.; Yu, L. E3 ubiquitin ligase Siah-1 facilitates poly-ubiquitylation and proteasomal degradation of the hepatitis B viral X protein. *FEBS Lett.* **2011**, *585*, 2943–2950. [CrossRef] [PubMed]
40. Belikov, A.V.; Schraven, B.; Simeoni, L. T cells and reactive oxygen species. *J. Biomed. Sci.* **2015**, *22*, 85. [CrossRef]
41. Tan, C.; Guo, H.; Zheng, M.; Chen, Y.; Huang, W. Involvement of mitochondrial permeability transition in hepatitis B virus replication. *Virus Res.* **2009**, *145*, 307–311. [CrossRef]
42. Kim, S.; Lee, H.-S.; Ji, J.-H.; Cho, M.-Y.; Yoo, Y.-S.; Park, Y.-Y.; Cha, H.-J.; Lee, Y.; Kim, Y.; Cho, H. Hepatitis B virus X protein activates the ATM–Chk2 pathway and delays cell cycle progression. *J. Gen. Virol.* **2015**, *96*, 2242–2251. [CrossRef] [PubMed]
43. Choi, Y.M.; Lee, S.Y.; Kim, B.J. Naturally Occurring Hepatitis B Virus Mutations Leading to Endoplasmic Reticulum Stress and Their Contribution to the Progression of Hepatocellular Carcinoma. *Int. J. Mol. Sci.* **2019**, *20*, 597. [CrossRef] [PubMed]
44. Liu, B.; Chen, Y.; St. Clair, D.K. ROS and p53: A versatile partnership. *Free Radic. Biol. Med.* **2008**, *44*, 1529–1535. [CrossRef] [PubMed]
45. Liochev, S.I.; Fridovich, I. The effects of superoxide dismutase on H_2O_2 formation. *Free Radic. Biol. Med.* **2007**, *42*, 1465–1469. [CrossRef]
46. Komarov, P.G.; Komarova, E.A.; Kondratov, R.V.; Christov-Tselkov, K.; Coon, J.S.; Chernov, M.V.; Gudkov, A.V. A chemical inhibitor of p53 that protects mice from the side effects of cancer therapy. *Science* **1999**, *285*, 1733–1737. [CrossRef] [PubMed]
47. Culmsee, C.; Zhu, X.; Yu, Q.S.; Chan, S.L.; Camandola, S.; Guo, Z.; Greig, N.H.; Mattson, M.P. A synthetic inhibitor of p53 protects neurons against death induced by ischemic and excitotoxic insults, and amyloid beta-peptide. *J. Neurochem.* **2001**, *77*, 220–228. [CrossRef] [PubMed]
48. Han, J.; Jang, K.L. All-trans Retinoic Acid Inhibits Hepatitis B Virus Replication by Downregulating HBx Levels via Siah-1-Mediated Proteasomal Degradation. *Viruses* **2023**, *15*, 1456. [CrossRef] [PubMed]
49. Zheng, Y.W.; Yen, T.S. Negative regulation of hepatitis B virus gene expression and replication by oxidative stress. *J. Biol. Chem.* **1994**, *269*, 8857–8862. [CrossRef] [PubMed]
50. Park, S.G.; Min, J.Y.; Chung, C.; Hsieh, A.; Jung, G. Tumor suppressor protein p53 induces degradation of the oncogenic protein HBx. *Cancer Lett.* **2009**, *282*, 229–237. [CrossRef]
51. Matsuzawa, S.-I.; Takayama, S.; Froesch, B.A.; Zapata, J.M.; Reed, J.C. p53-inducible human homologue of Drosophila seven in absentia (Siah) inhibits cell growth: Suppression by BAG-1. *EMBO J.* **1998**, *17*, 2736–2747. [CrossRef]
52. Quasdorff, M.; Protzer, U. Control of hepatitis B virus at the level of transcription. *J. Viral Hepat.* **2010**, *17*, 527–536. [CrossRef] [PubMed]
53. Tang, H.; Delgermaa, L.; Huang, F.; Oishi, N.; Liu, L.; He, F.; Zhao, L.; Murakami, S. The transcriptional transactivation function of HBx protein is important for its augmentation role in hepatitis B virus replication. *J. Virol.* **2005**, *79*, 5548–5556. [CrossRef] [PubMed]

54. Kwun, H.J.; Jang, K.L. Natural variants of hepatitis B virus X protein have differential effects on the expression of cyclin-dependent kinase inhibitor p21 gene. *Nucleic Acids Res.* **2004**, *32*, 2202–2213. [CrossRef] [PubMed]
55. Davarinejad, H. *Quantifications of Western Blots with ImageJ*; University of York: York, UK, 2015.

International Journal of
Molecular Sciences

MDPI

Article

T Cell Homeostasis Disturbances in a Cohort of Long-Term Elite Controllers of HIV Infection

José M. Benito [1,2,*], Daniel Jiménez-Carretero [3], Clara Restrepo [1,2], José M. Ligos [4], Jaime Valentín-Quiroga [5], Ignacio Mahillo [6], Alfonso Cabello [7], Eduardo López-Collazo [5], Fátima Sánchez-Cabo [3], Miguel Górgolas [7], Vicente Estrada [8] and Norma Rallón [1,2] on behalf of the ECRIS Integrated in the Spanish AIDS Research Network

1 HIV and Viral Hepatitis Research Laboratory, Instituto de Investigación Sanitaria Fundación Jiménez Díaz, Universidad Autónoma de Madrid (IIS-FJD, UAM), 28040 Madrid, Spain; clara.restrepo@hospitalreyjuancarlos.es (C.R.); norma.rallon@fjd.es (N.R.)
2 Hospital Universitario Rey Juan Carlos, 28933 Móstoles, Spain
3 Unidad de Bioinformática, Centro Nacional de Investigaciones Cardiovasculares (CNIC), 28029 Madrid, Spain; daniel.jimenez@cnic.es (D.J.-C.); fscabo@cnic.es (F.S.-C.)
4 Cytek Biosciences, Inc., Fremont, CA 94538, USA; jligos@cytekbio.com
5 Grupo de Respuesta Inmune Innata, IdiPAZ, Hospital Universitario La Paz, 28046 Madrid, Spain; jaimevquiroga@hotmail.com (J.V.-Q.); elopez.hulp@salud.madrid.org (E.L.-C.)
6 Department of Statistics, Instituto de Investigación Sanitaria Fundación Jiménez Díaz, Universidad Autónoma de Madrid (IIS-FJD, UAM), 28040 Madrid, Spain; imahillo@fjd.es
7 Hospital Universitario Fundación Jiménez Díaz, 28040 Madrid, Spain; acabello@fjd.es (A.C.); mgorgolas@fjd.es (M.G.)
8 Hospital Universitario Clínico San Carlos, 28040 Madrid, Spain; vicente.estrada@salud.madrid.org
* Correspondence: jose.benito@fjd.es or jbenito1@hotmail.com; Tel.: +34-91-544-37-20; Fax: +34-91-550-48-49

Abstract: Elite controllers (ECs) are people living with HIV (PLWH) able to control HIV replication without antiretroviral therapy and have been proposed as a model of a functional HIV cure. Much evidence suggests that this spontaneous control of HIV has a cost in terms of T cell homeostasis alterations. We performed a deep phenotypic study to obtain insight into T cell homeostasis disturbances in ECs maintaining long-term virologic and immunologic control of HIV (long-term elite controllers; LTECs). Forty-seven PLWH were included: 22 LTECs, 15 non-controllers under successful antiretroviral therapy (onART), and 10 non-controllers not receiving ART (offART). Twenty uninfected participants (UCs) were included as a reference. T cell homeostasis was analyzed by spectral flow cytometry and data were analyzed using dimensionality reduction and clustering using R software v3.3.2. Dimensionality reduction and clustering yielded 57 and 54 different CD4 and CD8 T cell clusters, respectively. The offART group showed the highest perturbation of T cell homeostasis, with 18 CD4 clusters and 15 CD8 clusters significantly different from those of UCs. Most of these alterations were reverted in the onART group. Interestingly, LTECs presented several disturbances of T cell homeostasis with 15 CD4 clusters and 13 CD8 clusters different from UC. Moreover, there was a specific profile of T cell homeostasis alterations associated with LTECs, characterized by increases in clusters of naïve T cells, increases in clusters of non-senescent effector CD8 cells, and increases in clusters of central memory CD4 cells. These results demonstrate that, compared to ART-mediated control of HIV, the spontaneous control of HIV is associated with several disturbances in CD4 and CD8 T cell homeostasis. These alterations could be related to the existence of a potent and efficient virus-specific T cell response, and to the ability to halt disease progression by maintaining an adequate pool of CD4 T cells.

Keywords: HIV control; long-term elite controllers (LTECs); T cell homeostasis; immunology

1. Introduction

HIV infection represents a paradigm of chronic viral infection without a model of spontaneous clearance. A subgroup of people living with HIV (PLWH) known as elite controllers (ECs) [1] are the closest model of this situation; they have the ability to limit viral replication to undetectable levels in the absence of treatment. These PLWH are thought to be the best example of a functional cure for HIV infection because of ECs' remarkable capacity to control HIV replication and disease progression. In this scenario, the virus is greatly suppressed but not completely eradicated, allowing the patient to maintain immunological and virological stability without needing to take life-long antiretroviral therapy (ART) [1,2].

Since ECs have this ability, numerous studies have been conducted to identify the immunological mechanisms underlying the spontaneous control of HIV disease [3]. This is necessary to identify surrogate immunological markers of protection against the progression of HIV disease, which is a necessary first step in developing immunotherapeutic strategies aimed at controlling the infection without the need for life-long ART. The immune response mediated by T cells, particularly CD8 T cells, has been one of the most extensively researched of the proposed mechanisms because of its significant function in controlling viral replication [4–8].

However, the fact that ECs are a diverse population of PLWH [9] has complicated efforts to identify correlates of HIV disease control using ECs as a model. A significant proportion of ECs are unable to sustain HIV control over extended periods of time, and a number of studies have found that EC cohorts experience varying rates of virological and/or immunological control loss [10,11]. Several factors have been associated with the loss of virological and/or immunological control in ECs, including a waning HIV-specific T cell response [12], a heightened inflammatory milieu [13], and alterations in several aspects of T cell homeostasis [14]. In fact, the percentage of ECs able to maintain long-term immunological and virological control—known as long-term elite controllers, or LTECs—may be as low as 0.15% of the entire PLWH population [11]. Therefore, research aiming at identifying the immunological traits associated with protection against HIV disease should concentrate on the population of LTECs, as they represent the closest model of an HIV functional cure.

Given the pivotal role of T cells in maintaining not only an adequate anti-viral immune response but also the homeostatic equilibrium necessary to halt immunological progression, the goal of the current study was to thoroughly characterize the T cell phenotypes in a well-defined cohort of LTEC patients in comparison to both non-controller patients whose viral replication was uncontrolled and non-controller patients whose viral replication was controlled by ART.

2. Results

2.1. Characteristics of the Study Groups

A total of 67 volunteers were included in the study, including 47 PLWH volunteers and 20 HIV-seronegative volunteers (UC group). Among the PLWH, 22 were LTECs, 15 were onART, and 10 were offART. Table 1 shows the main characteristics of the different study groups at inclusion. Age was similar in all study groups, whereas the distribution of sex was different, with a majority of males in the onART and offART groups, in contrast to a more balanced distribution of males and females in the LTEC and UC groups. CD4 counts were similar among the different PLWH groups. Length of HIV infection (since diagnosis) was lowest for the offART group. The length of EC status in the LTEC group was 13 years and the length of ART in the onART group was 6 years.

Table 1. Characteristics of groups included in this study.

Characteristic	LTEC (n = 22)	onART (n = 15)	offART (n = 10)	UC (n = 20)	*p*-Value
Age (years)	44 [35–49]	44 [42–49]	43 [36–49]	43 [36–49]	0.266
Sex (% of males)	59 *	80	100	50 *	**0.025**
Years since HIV diagnosis	15 * [6–20]	12 [5–15]	5 [3–9]	**NA**	**0.027**
Years as EC	13 [7–16]	NA	NA	NA	NA
Years on ART	NA	6 [3–9]	NA	NA	NA
Plasma HIV load (copies/mL)	50	50	57,937 [28,817–80,676]	NA	NA
CD4 count (cells/μL)	837 [603–1210]	820 [599–1127]	625 [518–920]	NA	0.200

Data are given as median [Q1–Q3], except sex that is expressed as a percentage. Statistical differences between the four study groups were tested by Kruskal-Wallis test for continuous variables and by Chi-square test for sex. Statistical significance was considered for *p*-values below 0.05 (in bold in the table). NA: not appicable. An asterisk denotes statistically significant differences ($p < 0.05$) with respect to the offART group.

2.2. Unsupervised Multidimensional Analysis of Flow Cytometry Data Yielded Several CD4 and CD8 T Cell Clusters That Differentiated the Study Groups

Spectral flow cytometry was used to examine eighteen CD4 and CD8 T cell markers in order to investigate various functional traits of these cells: maturation stage (CD45RA, CCR7, and CD28 markers); recent thymic emigrants (RTEs) (CD31 marker); activation (CD38 and HLADR markers); exhaustion (PD-1, Tim3, CTLA4, and TIGIT markers); senescence (CD57 and CD28 markers); apoptosis (CD95 and CD28 markers); homeostatic potential (CD127 marker); Tregulatory (Treg) cells (CD127, CD25, and CD39 markers); Thelper or Tcytotoxic 1 (Th1 or Tc1), Thelper or Tcytotoxic 2 (Th2 or Tc2), and Thelper or Tcytotoxic 17 (Th17 or Tc17) cells (CXCR3 and CCR6 markers); and peripheral follicular helper T cells (pTfh) (CXCR5 marker) (Supplementary Table S1). To create a two-dimensional map and reduce the dimensionality of the data, T-distributed stochastic neighbor embedding (tSNE) was used. The normalized density of events across this tSNE map was different in the four study groups, as well as the distribution of events from each group in the tSNE map (Supplementary Figure S1), suggesting the existence of differences in the phenotypes of CD4 and CD8 T cells between the study groups. Clustering analysis yielded 57 clusters of CD4 and 54 clusters of CD8 T cells (Supplementary Figure S2), differentiated by the expression levels of the markers analyzed (Supplementary Figure S3). The proportion of each cluster in the total population of CD4 or CD8 T cells was calculated for each study sample and a differential abundance analysis was performed, using the UC group as a reference.

Several clusters of CD4 (Figure 1 and Supplementary Table S2) and CD8 (Figure 2 and Supplementary Table S3) T cells were differentially expressed in the PLWH groups compared to the UC group, revealing a deep perturbation of T cell homeostasis in the context of HIV infection. Moreover, the heatmaps of fold-change of expression of each sample over the median expression in the UC group showed a clear separation between each PLWH group and the UC group (Supplementary Figures S4 and S5).

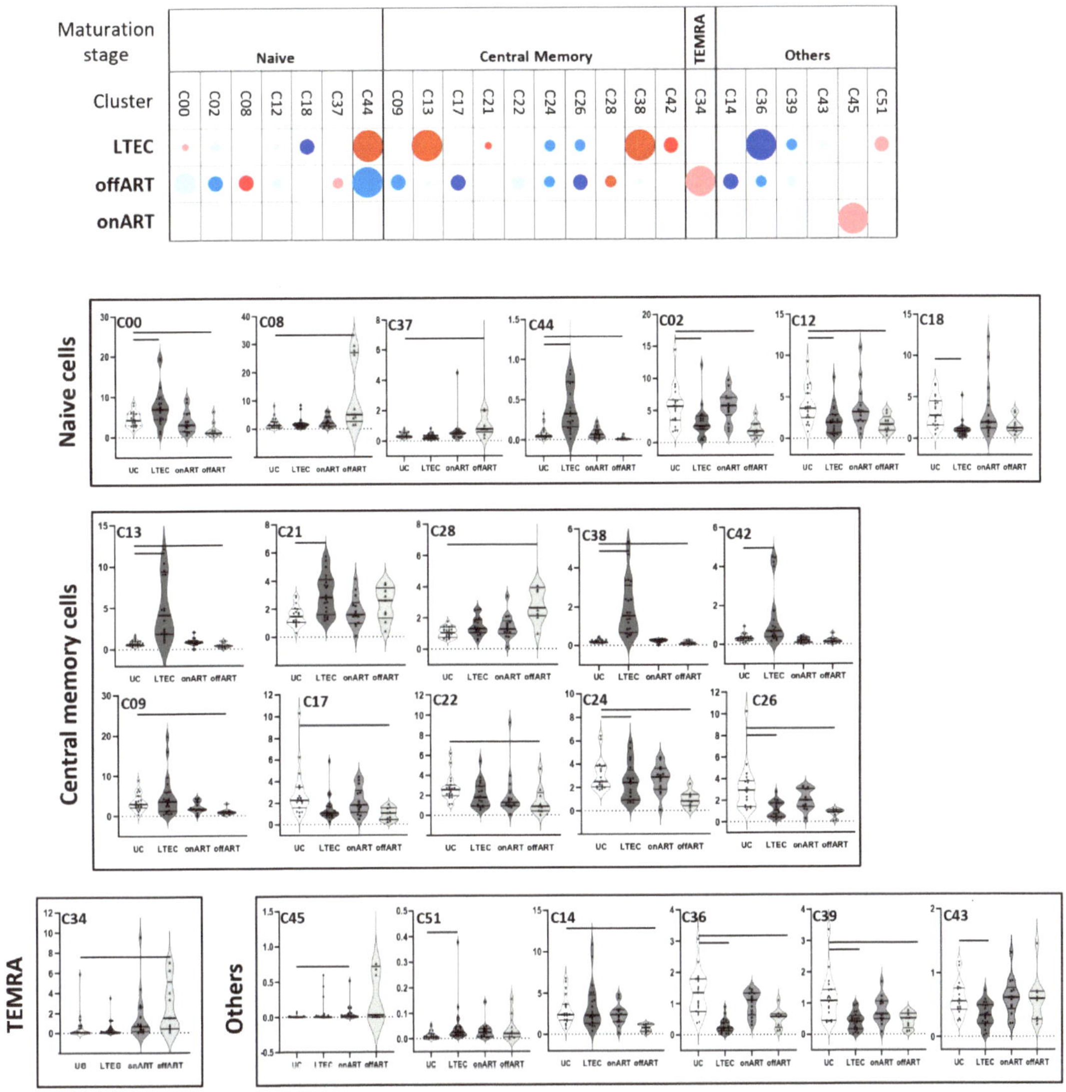

Figure 1. The upper graph of the figure shows a schematic representation (bubble diagram) of CD4 T cell clusters, with significant differences between the PLWH groups and the UC group. Clusters are grouped according to maturation stage (defined by CD45RA and CCR7 markers). Each bubble in the diagram represents a significant difference with respect to the UC group. The size of the bubble (from the smallest to the biggest size) indicates the degree of difference in fold-change: <2, 2–3, 3–4, 4–5, and >5. Red colors indicate an increase and blue colors indicate a decrease with respect to the UC group. The level of statistical significance (corrected p-value) is indicated by the color tone: 0.05–0.01, 0.01–0.001, and <0.001 for the light, medium, and dark tones, respectively. The lower graphs of the figure show violin plot graphs of the levels (expressed as a percentage of total CD4 T cells) of each cluster in the four study groups. As in the upper graph, clusters are grouped according to maturation stage. In each violin plot graph, bars inside the plots indicate those PLWH groups that showed a statistically significant difference with respect to the UC group.

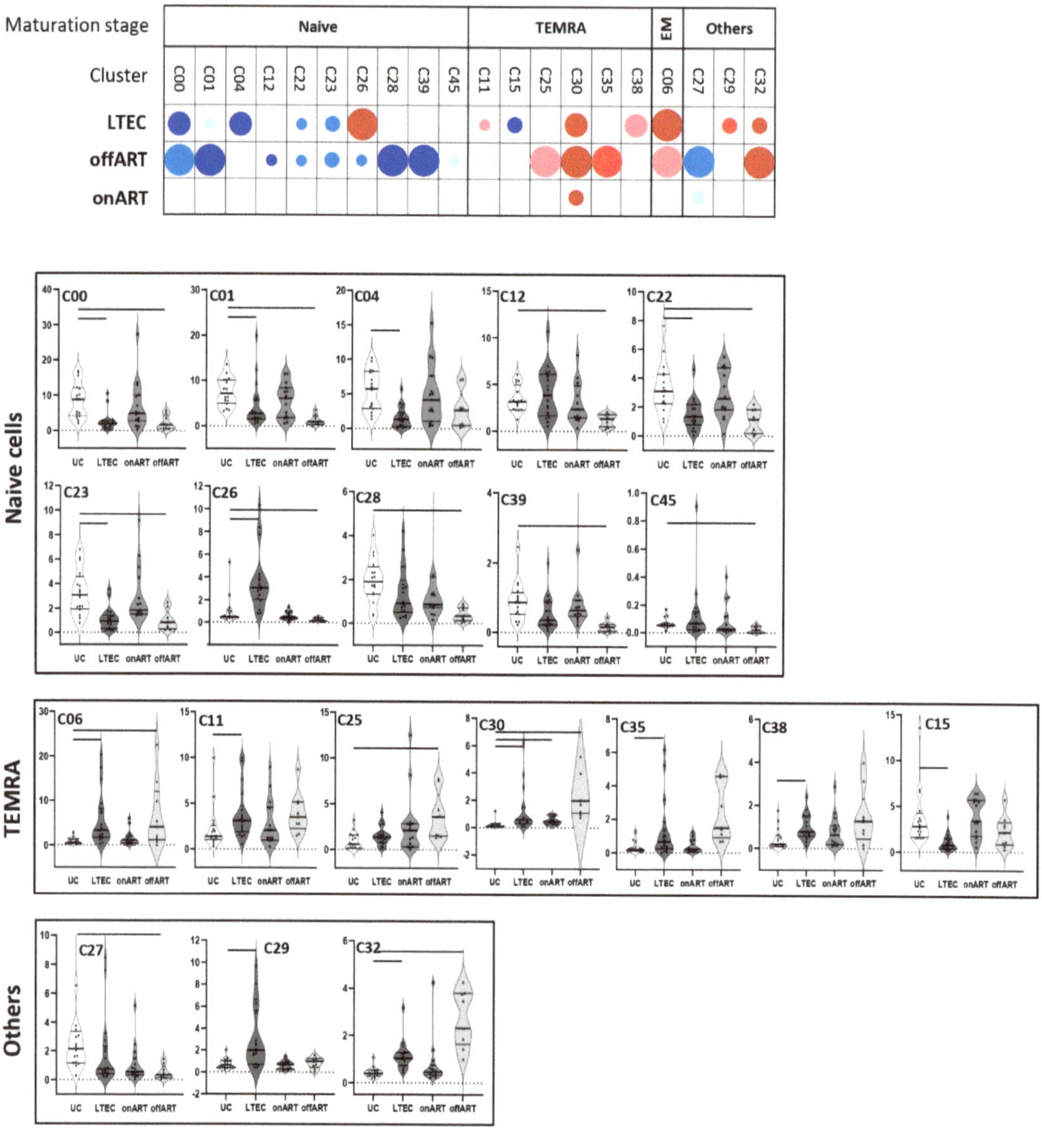

Figure 2. The upper graph of the figure shows a schematic representation (bubble diagram) of CD8 T cell clusters, with significant differences between the PLWH groups and the UC group. Clusters are grouped according to maturation stage (defined by CD45RA and CCR7 markers). Each bubble in the diagram represents a significant difference with respect to the UC group. The size of the bubble (from the smallest to the biggest size) indicates the degree of difference in fold-change: <2, 2–3, 3–4, 4–5, and >5. Red colors indicate an increase and blue colors indicate a decrease with respect to the UC group. The level of statistical significance (corrected p-value) is indicated by the color tone: 0.05–0.01, 0.01–0.001, and <0.001 for the light, medium, and dark tones, respectively. The lower graphs of the figure show violin plot graphs of the levels (expressed as a percentage of total CD8 T cells) of each cluster in the four study groups. As in the upper graph, clusters are grouped according to maturation stage. In each violin plot graph, bars inside the plots indicate those PLWH groups that showed a statistically significant difference with respect to the UC group.

2.3. Uncontrolled Viral Replication Is Associated with a Deep Perturbation of CD4 and CD8 T Cell Homeostasis

The highest perturbation of T cell homeostasis was observed in PLWH with uncontrolled viral replication (offART group). Regarding CD4 T cells, 18 different clusters were differentially expressed: most of them (14) decreased and only 4 increased with respect to UCs (Figure 1). Among the decreased clusters, four (C00, C02, C12, C44) were formed by naïve cells ($CD45RA^+CCR7^+CD28^+$) expressing CD127 and CD38 markers, with two of them also expressing CD31 (RTE cells). Another seven clusters (C09, C13, C17, C22, C24, C26, C38) were formed by central memory (CM) cells ($CD45RA^-CCR7^+CD28^+$): three with a Th1 phenotype ($CXCR3^+CCR6^-$), three with a Th2 phenotype ($CXCR3^-CCR6^-$), and another with a mixed Th1/Th17 phenotype ($CXCR3^+CCR6^+$). All seven clusters of central memory phenotypes expressed low levels of activation and exhaustion. Another two decreased clusters (C36 and C39) were formed by pTfh cells (Supplementary Figure S4). Among the increased CD4 clusters, one was formed by central memory cells with high expression of activation and exhaustion (C28) and another by TEMRA ($CD45RA^+CCR7^-CD28^-$) cells with high expression of activation, apoptosis, senescence, and exhaustion (C34) (Supplementary Figure S4).

Regarding CD8 T cells, 15 different clusters were differentially expressed: 10 decreased and 5 increased compared to the UC group (Figure 2). Among the decreased clusters, six (C00, C01, C12, C26, C28, C45) were formed by naïve cells expressing CD127, with some of them also expressing Tim3 or CD57. Another three clusters (C22, C23, C39) were also formed by naïve cells co-expressing CD127 and CD95 markers. Among the increased clusters, two of them (C06, C32) were formed by effector memory (EM) ($CD45RA^-CCR7^-CD28^-$) cells with high expression of activation and exhaustion markers. The other three clusters (C25, C30, C35) were formed by effector memory re-expressing CD45RA (TEMRA) cells with high expression of activation and senescence markers.

As another approach to analyze the distribution of CD4 and CD8 T cell clusters between the PLWH groups and the UC group as a reference, a random forest and hierarchical clustering were carried out using the levels of the different clusters (Supplementary Figure S6). The random forest model showed very good discrimination between the offART and UC groups. Moreover, a heatmap of the hierarchical clustering showed a clear difference in the profiles of CD4 and CD8 T cell cluster expression levels when comparing the offART and UC groups (Supplementary Figure S6).

2.4. The Majority of T Cell Homeostasis Disturbances Are Restored with Long-Term ART

PLWH with ART-mediated long-term suppression of viral replication (onART group) presented the lowest perturbation of T cell homeostasis, with a profile of CD4 and CD8 T cells approaching the profile of the UC group. However, some perturbation still remained, with one cluster of CD4 (C45) (Figure 1) and two clusters of CD8 (C27 and C30) (Figure 2) T cells differentially expressed in the onART compared to the UC group. Of note, two of the clusters (C30 and C45) were formed by TEMRA cells expressing high levels of activation, exhaustion, and senescence (Supplementary Figures S4 and S5).

2.5. PLWH with Spontaneous Control of HIV Replication Show a Deep Perturbation of CD4 and CD8 T Cell Clusters

The most interesting finding came from the LTEC group, in which a large number of CD4 and CD8 T cell clusters were differentially expressed compared to the UC group, even though plasma HIV load was undetectable, as in the onART group.

Regarding CD4 T cells, fifteen different clusters were differentially expressed in LTECs: eight decreased and seven increased (Figure 1). The decreased clusters included three clusters formed by Th2 naïve cells expressing CD127 and CD38 (C02, C18, and C12); two clusters formed by CM cells expressing CD95 and CD25 (C24 and C26); two clusters formed by pTfh (CXCR5+) cells (C36 and C39); and, lastly, one cluster formed by Treg (CD127-CD25+) cells (C43). The increased clusters included two clusters formed by Th2 naïve cells expressing CD127 and CD38 (C00 and C44); four clusters formed by CM cells with variable levels of activation and exhaustion (C13, C21, C38, and C42), three of them with a Th1 phenotype; and one cluster formed by TEMRA cells with high levels of activation, exhaustion, and senescence (C51) (Supplementary Figure S4). It is important to clarify that although LTECs exhibited both decreased (C02, C18, C12) and increased (C00, C44) clusters of Th2 naïve CD4 T cells, there were some important phenotypic differences between them. Thus, among the decreased clusters, two of them (C18, C12) expressed moderate levels of the activation marker CD25 and the other decreased cluster (C02) expressed moderate levels of the exhaustion marker CTLA4. In contrast, the two increased clusters (C00, C44) lacked the expression of CD25 or CTLA4. There were also both decreased (C24, C26) and increased (C13, C21, C38, C42) clusters of CM CD4 T cells in LTECs, but, again, there were some important phenotypic differences between the decreased and the increased central memory clusters, mainly in the expression profiles of activation and exhaustion markers (Supplementary Figure S3).

Regarding CD8 T cells, as many as thirteen clusters were differentially expressed in the LTEC group: seven increased and six decreased (Figure 2). Increased clusters were mainly formed by cells with TEMRA (C11, C30, C38) or EM (C06) phenotypes with high expression of activation and exhaustion markers. Among the decreased clusters, most of them were formed by cells with naïve phenotypes expressing CD127, with three of them (C00, C01, C04) also expressing Tim3 and two of them (C22, C23) also expressing CD95 (Supplementary Figure S5). As for the offART group, the random forest model showed very good discrimination between the LTEC and UC groups, and a heatmap of the hierarchical clustering showed a clear difference in the profiles of CD4 and CD8 T cell cluster expression levels when comparing the LTEC and UC groups (Supplementary Figure S6).

Notably, the profile of differentially expressed clusters (with respect to the UC group) partially overlapped between LTEC and offART groups (Supplementary Figure S7). Of the 15 CD4 T cell clusters altered in LTECs, 10 (67%) were in common with the clusters altered in the offART group. However, only six of these common clusters (C02, C12, C24, C26, C36, C39) were altered in the same direction in the LTEC and offART groups. All six clusters were decreased in both groups and consisted of naïve (C02, C12), CM (C24, C26), and pTfh (C36, C39) cells. In contrast, the other four clusters (C00, C13, C38, C44) were altered in opposite directions in the LTEC and offART groups (increased in LTECs and decreased in offART). As for CD8 T cell clusters, of the 13 clusters altered in LTECs, 7 (54%) were in common with the clusters altered in the offART group, and all of them altered in the same direction in the LTEC and offART groups. Of these common clusters, decreased clusters were formed by naïve cells (C00, C01, C22, C23), whereas increased clusters were formed by senescent and/or exhausted TEMRA cells (C30) or by EM cells (C06).

2.6. *A Specific Profile of CD4 and CD8 T Cell Clusters Associated with the Spontaneous Control of HIV Replication*

Although the profiles of T cell homeostasis alteration presented some similarities in the LTEC and offART groups, different findings demonstrated the existence of a specific profile of T cell homeostasis alteration associated with the LTEC group. First, there were some CD4 and CD8 T cell clusters that were significantly altered (with respect to the UC group) only in the LTEC group (Supplementary Figure S7). Five CD4 T cell clusters were significantly altered only in LTECs (C18, C21, C42, C43, C51), although C18 and C21 were altered in the same direction also in the offART group, without reaching statistical significance. Thus, only C42, C43, and C51 were distinctive for the LTEC group. Interestingly, C43 (decreased in LTECs) was formed by Treg cells with a mixed Th1/Th17 phenotype, whereas C42 (increased in LTECs) was formed by CM cells with a Th1 phenotype expressing high levels of activation (Supplementary Figure S3). As for CD8 T cell clusters, six clusters presented a significant alteration only in LTECs (C04, C11, C15, C26, C29, C38), although three of them (C04, C11, C38) were also altered in the same direction in the offART group, without statistical significance. Thus, only clusters C15, C26, and C29 were distinctive for the LTEC group. C15 (decreased in LTECs) was formed by senescent TEMRA cells, whereas C29 (increased in LTECs) was formed by cells with a mixed Tc1/Tc17 phenotype (CXCR3+CCR6+) expressing CD127, and C26 was formed by naïve cells with a Tc1 phenotype (Supplementary Figure S3).

Secondly, there were many T cell clusters differentially expressed between the LTEC and offART groups (Supplementary Figure S8), including twelve clusters of CD4 T cells and nine clusters of CD8 T cells. Among the CD4 T cell clusters, increased clusters in LTECs were formed by clusters of naïve cells (C00, C44), CM/Th1 cells (C14, C17, C38), CM/Th2 cells (C13), and CM/Th1-Th17 cells (C09, C42). Decreased clusters in LTECs were formed by clusters of naïve cells (C08, C37), one of them with high levels of exhaustion (C37); exhausted CM/Th1cells (C28); and senescent TEMRA cells (C34) (Supplementary Figure S3). Regarding CD8 T cell clusters, increased clusters in LTECs were formed by clusters of naïve cells (C01, C12, C26, C28) and cells with a mixed maturation stage expressing CD127 (C20, C29, C39). Decreased clusters in LTECs were formed by clusters of senescent/activated TEMRA cells (C30) and activated cells with a mixed maturation stage (C32) (Supplementary Figure S3).

Thirdly, several clusters were differentially expressed between the LTEC and onART groups, despite undetectable plasma HIV viremia in both groups (Supplementary Figure S9). With regard to CD4 clusters, eight clusters were down and four clusters were up for LTECs. The decreased clusters included naive (C02, C18, C37, C43), CM (C26), TEMRA (C35), and pTfh (C36, C39) clusters, while the increased clusters included naive (C00, C44) and CM (C13, C42) clusters. As for the CD8 clusters, LTECs showed a decrease in several naive clusters (C00, C04, C22, C23) and one TEMRA cluster (C15) and an increase in one naive cluster (C26), one EM cluster (C06), and two clusters composed of cells with a mixed maturation stage (C29, C32) (Supplementary Figure S9).

Lastly, there were a few T cell clusters differentially expressed in the LTEC group when compared with all the other study groups (UC, offART, and onART groups). Among CD4 cells, clusters C00 and C44 (formed by naïve cells) were greatly increased in LTECs compared to the UC, offART, and onART groups. Clusters C13, C38, and C42 (formed by CM cells) were also greatly increased in LTECs compared to the rest of the groups (Figure 1). Regarding CD8 cells, cluster C26 (formed by naïve cells) was greatly increased and cluster C15 (formed by senescent TEMRA cells) was decreased in LTECs compared to the rest of the groups (Figure 2).

2.7. Correlations between T Cell Clusters and Inmmuno-Virological Parameters

The potential associations of levels of CD4 and CD8 T cell clusters with virological (pVL) or immunological (CD4 counts) parameters were analyzed using non-parametric tests (Spearman rho correlation coefficient). In the offART group, the level of pVL was positively correlated with two clusters of naïve CD4 T cells (C80, C19) and two clusters of naïve CD8 T cells (C16, C26), whereas negative correlations were observed with CM clusters of CD4 T cells (C14, C21), pTfh clusters of CD4 T cells (C10, C36), and TEMRA clusters of CD8 T cells (C09, C19) (Figure 3). Regarding correlations with CD4 counts, the profile of correlations was different for the different PLWH groups (Figure 4). The offART and onART groups presented several positive correlations with naïve clusters of CD4 T cells; additionally, the offART group presented a negative correlation with pTfh and Treg clusters of CD4 T cells. In contrast, the LTEC group presented positive correlations between CD4 counts and several clusters of CD8 T cells, mainly of EM or TEMRA phenotypes, something that was also observed to a lesser extent in the offART group (Figure 4).

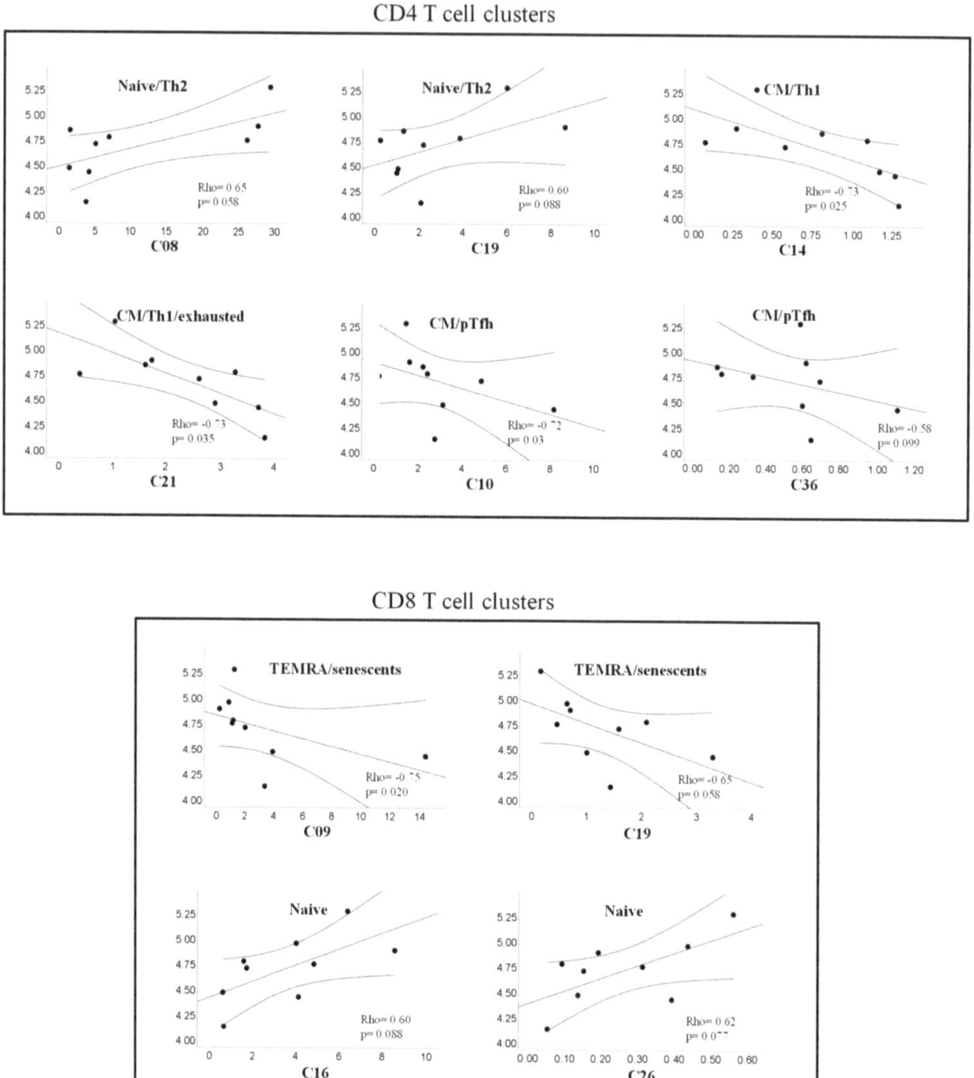

Figure 3. Scatter plots showing correlations between HIV plasma viral load and levels of different clusters of CD4 (upper plots) or CD8 (lower plots) T cells. Correlations with plasma HIV load were analyzed in the offART group. In all plots, the x-axis represents the level of each cluster expressed as a percentage of total CD4 or CD8 T cells and the y-axis represents the HIV plasma viral load expressed as copies/mL (log scale). The Spearman's Rho correlation coefficient and p-value are given inside each scatter plot and the legend over each plot describes the phenotype of cells forming that cluster.

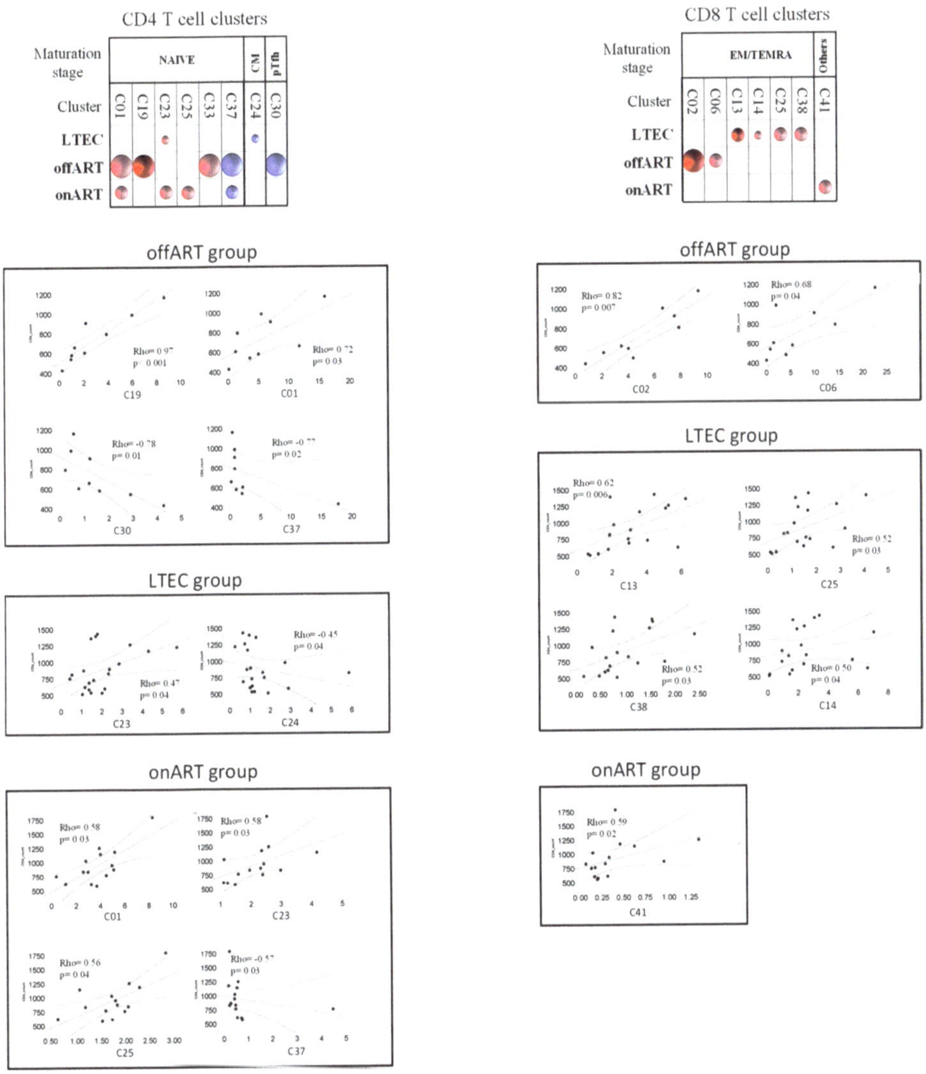

Figure 4. Schematic representation (bubble diagram) of bivariate correlations (Spearman rho coefficient) between CD4 (upper left diagram) or CD8 (upper right diagram) T cell clusters and CD4 counts in the different PLWH groups. Each bubble in the diagram represents a positive (red color) or negative (blue color) correlation. The size of the bubble (from the smallest to the biggest size) indicates the value of the Spearman rho coefficient: <0.5, 0.5–0.7, and >0.7. The level of statistical significance (p-value) is indicated by the color tone: 0.1–0.05, <0.05–0.01, and <0.01 for the light, medium, and dark tones, respectively. T cell clusters are grouped according to maturation stage. Below the diagrams, scatter plots of correlations between CD4 counts (y-axis on the plots) and the levels of different clusters of CD4 (left) and CD8 (right) T cells are shown. The Spesarman's rho coefficient and p-value are given inside each scatter plot.

3. Discussion

In order to identify the phenotypic traits of T cells linked to the long-term spontaneous control of viral replication, this study is the first, to our knowledge, to perform a thorough phenotypic analysis of T cell homeostasis in PLWH who are long-term elite controllers

(LTECs) compared with that found in non-controller PLWH with uncontrolled viremia and in non-controller PLWH with treatment-suppressed viremia. Our study's main conclusions are as follows: (a) the thorough phenotypic and multidimensional analysis approach allowed us to identify distinct variations in T cell homeostasis profiles between the various PLWH groups; (b) T cell homeostasis disturbances are most pronounced in PLWH with uncontrolled viral replication; (c) the majority of T cell homeostasis disturbances are restored by long-term ART-mediated suppression of viral replication; (d) interestingly, T cell homeostasis is significantly altered in LTECs compared to PLWH with treatment-suppressed viremia; (e) the T cell homeostasis profile that was specifically associated with the LTEC group was characterized by a decrease in Treg cell clusters; a decrease in cell clusters expressing high levels of activation, exhaustion, and senescence; and an increase in clusters of naïve cells and central memory cells.

Uncontrolled viral replication was linked to a significant disruption in the phenotypic profiles of T cells. In the offART group, alterations were found in 18 out of 57 (32%) CD4 T cell clusters and in 15 out of 54 (28%) CD8 T cell clusters when compared to the UC reference group. The most significant alterations in the offART group were characterized by decreases in several CD4 T cell clusters of naïve phenotype and of CM phenotype expressing low levels of activation and exhaustion as well as decreases in clusters of pTfh cells; there were also decreases in naïve clusters of CD8 T cells and increases in activated and exhausted EM and TEMRA clusters of CD8 T cells. Altogether, these results point to a significant impairment of T cell homeostasis in PLWH with uncontrolled HIV replication, in agreement with previous studies [15–17]. There were several correlations between T cell clusters and levels of HIV viremia, suggesting that many of these alterations may be the consequence of viral replication. Thus, levels of CM and pTfh CD4 T cells were negatively correlated with plasma viremia, in agreement with the diminished levels of these clusters in the offART group. It is interesting that, in spite of the decrease in several naïve clusters of CD4 T cells, some naïve clusters showed a positive correlation with HIV viremia (clusters C08 and C19), which supports the functional diversity of clusters with naïve phenotype. This was also observed for two naïve clusters of CD8 T cells (C16 and C26), which correlated positively with HIV viremia. Of note, the naïve clusters showing a positive correlation with HIV viremia also expressed CD31 (a marker of recent thymic emigrants) and lacked expression of the rest of the markers analyzed (activation, exhaustion, senescence), suggesting that these clusters may represent the thymus´s attempt to maintain the pool of naïve T cells [18,19]. Of note, some clusters of naïve CD4 T cells (C01 and C19) were positively correlated with CD4 counts, which supports a relevant role of these clusters in the maintenance of the CD4 T cell pool. A similar phenomenon was observed for TEMRA clusters of CD8 T cells, with some of them (C09 and C19) showing a negative correlation with HIV viremia, even though there was an increase in several TEMRA clusters in the offART group, which supports the functional heterogeneity of CD8 T cell clusters with TEMRA phenotype.

The majority of T cell homeostasis alterations seen in PLWH with uncontrolled viremia were reversed in PLWH with ART-mediated control of HIV replication, suggesting that these alterations may be the result of high viral replication. Interestingly, the few clusters that remained altered in the onART group were TEMRA clusters expressing high levels of activation, exhaustion, and senescence, which suggests the persistence of low viral replication driving the expansion of these highly differentiated T cell clusters [20,21]. Alternatively, the expansion of these highly differentiated subsets of T cells in the onART group may have been driven by the presence of cytomegalovirus (HCMV) infection [22]. We did not analyze either the serostatus of HCMV or the presence of HCMV reactivation and, thus, we could not test this hypothesis.

The most interesting findings of our study were those found in the LTEC group. In contrast to the onART group (in which the suppression of viremia was mediated by ART), the spontaneous control of HIV replication was associated with a deep perturbation of T cell homeostasis. Overall, our findings agree with those of previous studies addressing specific aspects of T cell homeostasis in different cohorts of PLWH who are ECs, such as the distribution of the maturation stage of T cells [23–25]; the activation, exhaustion, and senescence of T cells [26–28]; and Treg cell levels [29,30]. Our study confirms and extends these previous results by the analysis of a well-characterized cohort of elite controllers with long-term control of HIV infection, in which a comprehensive phenotypic analysis of T cells and a dimensionality reduction approach allowed us to perform an in-depth analysis of the homeostasis of this cell population. Moreover, this approach allowed us to find a specific profile of T cell homeostasis perturbation associated with the spontaneous control of HIV infection. Several findings from our study point to a profile of T cell homeostasis alteration specific to LTECs: (a) changes in different clusters of T cells that were only found in the LTEC group (when compared to the UC group as a reference); (b) differences in T cell cluster levels between the LTEC and onART groups, despite undetectable plasma viremia in both groups; (c) changes in several clusters of T cells that differentiated the LTEC group from the rest of the groups.

Of note, many of the perturbations observed in the LTEC group were similar to those observed in the context of uncontrolled viral replication. These perturbations could be driven by the existence of residual viral replication in LTECs at higher levels than in the ART-mediated control of HIV replication, a fact that has been previously reported [31–33]. In agreement with this hypothesis, previous studies have shown that antiretroviral treatment in elite controllers decreases T cell activation and exhaustion [34,35]. Alternatively, the existence of T cell homeostasis perturbations in LTECs may be a consequence of the existence of a potent and effective HIV-specific immune response. Interestingly, LTECs showed increased levels of non-senescent TEMRA clusters of CD8 T cells expressing low levels of activation and exhaustion (C11), something that was not observed in the offART group, in which the increased clusters of TEMRA CD8 T cells expressed high levels of activation, exhaustion, and senescence (C25, C30, C35). Taking into account the more preserved functionality of cluster C11 (non-senescent and non-exhausted), it is tempting to speculate that the expansion of this cluster in LTECs may be associated with the better anti-HIV CD8 T cell response reported in PLWH able to spontaneously suppress HIV replication [3]. Another perturbation of T cell homeostasis in the LTEC group that could be associated with the better HIV-specific immune response in this group is the existence of several expanded clusters of CM CD4 T cells in LTECs when compared not only to the UC group but also to the offART and onART PLWH groups. CM CD4 T cells are necessary to maintain a functional anti-viral CD8 T cell response [36]. Moreover, all these CM clusters expressed CD127, a surrogate marker of preserved homeostatic potential [37]. Lastly, some characteristics of T cell homeostasis specific of the LTEC group could be associated with the ability of this group of PLWH to maintain control of immunological progression (depletion of the CD4 T cell pool), among them the expansion of clusters of naïve CD4 (clusters C00, C44) and naïve CD8 (C26) T cells). C00 is a cluster of naïve cells expressing CD31 (recent thymic emigrants) and CD127 (proliferative potential), suggesting a high capacity of this cluster to repopulate the CD4 T cell pool. Cluster C44 has the same phenotype as C00, except that it lacks the expression of CD31, suggesting that this cluster may originate from the peripheral expansion of CD31+ naïve cells. Of note, a previous study of a cohort of elite controllers reported that a low level of naïve CD4 T cells is associated with HIV disease immunological progression [14].

In summary, the results of our study demonstrate the existence of a specific profile of T cell homeostasis in the population of PLWH that spontaneously control HIV disease in the long term (LTECs). This specific profile may be the consequence of different factors operating in this special population of PLWH: (a) the existence of low-level HIV replication; (b) the existence of a potent and efficient virus-specific T cell response; (c) the ability to halt disease progression by maintaining an adequate pool of CD4 T cells.

4. Materials and Methods

4.1. Study Design and Participants

This was a cross-sectional study including adult participants with chronic HIV infection (people living with HIV, PLWH) and HIV-seronegative healthy volunteers (uninfected controls, UC group). Three different groups of PLWH participants were included: PLWH with long-term spontaneous control of HIV infection (long-term elite controllers, LTEC group); PLWH non-controllers on antiretroviral treatment (ART) with an undetectable plasma HIV viral load (pVL) for at least five years at the time of inclusion in the study (onART group); and PLWH non-controllers off ART with detectable HIV pVL (offART group). The LTEC group was selected from the Spanish ECRIS database, a multicenter cohort of HIV elite controllers (ECs) launched in 2013 and described elsewhere [10]. The ECsRIS cohort includes patients from the long-term non-progressors (LTNPs) cohort, the AIDS research cohort (CoRIS), and several clinical centers in Spain (Supplementary Methods in Supplementary Materials). From this ECsRIS cohort, LTECs were defined as those ECs maintaining control (both virological and immunological) for a minimum of five years and throughout their whole follow-up period. Virological control was defined as having more than 90% of HIV pVL during the follow-up below the limit of detection. Immunological control was defined as having stable CD4 counts during follow-up. All study participants signed written informed consent. This study was approved by the Ethical Review Board of the Instituto de Investigación Sanitaria-Fundación Jiménez Díaz, Madrid, Spain (Approval ID: PIC097-19_FJD) and carried out in accordance with the Declaration of Helsinki.

4.2. Cell Samples

Blood samples were collected by venipuncture in tubes containing ethylenediaminetetraacetic acid (EDTA) as an anticoagulant, sent on the same day to the Spanish HIV HGM BioBank (http://hivhgmbiobank.com/?lang=en), immediately processed to obtain peripheral blood mononuclear cells (PBMCs) by gradient centrifugation over a ficoll-paque solution (Sigma-Aldrich, St. Louis, MA, USA), and kept frozen in liquid nitrogen. PBMC samples were kindly provided by the Spanish HIV HGM Biobank.

4.3. Multiparameter Flow Cytometry

One million PBMCs were stained with a panel of twenty-one different monoclonal antibodies (plus a viability dye) (Supplementary Table S4). A detailed staining protocol is given in the Supplementary Materials (Supplementary Methods). After staining, cells were acquired using an Aurora Spectral flow cytometer (Cytek Biosciences, Fremont, CA, USA), and a minimum of 200,000 events were acquired for further analysis. Data analysis was performed using FCS Express 7 v.7 (DeNovo software, Pasadena, CA, USA). Dead cells were excluded using the live/dead viability dye, and live lymphocytes were gated based on forward (FSC) and side (SSC) scatter. Lastly, single live lymphocytes were selected using FSC width vs. FSC height. Starting from the population of single live lymphocytes, a gate was placed to select T cells ($CD3^+$ lymphocytes) and then $CD3^+CD4^+CD8^-$ or $CD3^+CD8^+CD4^-$ T cells were selected for further analysis.

4.4. Preparation of Flow Cytometry Data for Unsupervised Multidimensional Analysis

Flow cytometry data from each study participant were manually gated using FCS Express 7 software as described above. Using the $CD3^+CD4^+CD8^-$ and $CD3^+CD8^+CD4^-$ gates, each fcs. file was split into two different files: one containing only the $CD3^+CD4^+CD8^-$ events and another containing only the $CD3^+CD8^+CD4^-$ events. Next, these new fcs. files were annotated using FCS Express 7 to include information about the positivity/negativity of each event contained in the file for each of the 18 different cell surface markers (lineage markers CD3, CD4, and CD8 were excluded from the annotation). These fcs. files were used for dimensionality reduction and clustering analysis.

4.5. Unsupervised Analysis of Flow Cytometry Data

All pre-gated data files containing $CD3^{+}CD4^{+}CD8^{-}$ and $CD3^{+}CD8^{+}CD4^{-}$ cells were included for unsupervised multidimensional analysis [38], including batch correction, dimensionality reduction, and clustering analysis. R software (version 4.1.1) was used for the implementation of the automated pipeline. The workflow for this analysis is shown in Supplementary Figure S10.

First, batch effects were corrected using the CytoNorm algorithm [39]. By taking advantage of the existence of the same control sample acquired in each batch or experiment along with the rest of the samples, non-biological variability was estimated and removed from the data before their unbiased analysis. Each file was subsampled to 25,000 cells to facilitate the proper processing of all samples at once and balance the contribution of each sample to the total number of cells analyzed. After transforming the data with the arcsinh function (cofactor = 6000), marker expression levels were jointly scaled via Zscore normalization. The whole panel of fluorescent markers (except the lineage markers CD3, CD4, and CD8) were used in the downstream analysis using the Seurat R package [40]. For the detection of cell subpopulations sharing a similar expression fingerprint, we used Louvain's algorithm [41], which identifies clusters by modularity optimization from a shared nearest neighbor graph (SNN). A level of cluster granularity corresponding to a resolution parameter value of 2.0 was used. To visualize single-cell measurements and clustering results, we used 2-dimension representations obtained with nonlinear dimensionality reduction techniques. T-distributed stochastic neighbor embedding (tSNE) maps were obtained using the FFT-accelerated Interpolation-based t-SNE (FIt-SNE) algorithm [42] (parameters: perplexity = 30, theta = 0.5, distance = Euclidean, rest default parameters). Cell frequencies in each cluster were computed and compared between each pair of groups using the Wilcoxon test. p-values were corrected using the Holm method. We also computed fold-change with respect to a reference group (typically the UC group) to represent the relative changes in cluster proportions between groups.

4.6. Statistical Analysis

The clinical and epidemiological characteristics of the different study groups were expressed as median and interquartile ranges, and differences were tested using non-parametric tests (Kruskall–Wallis test). Levels of the different clusters of CD4 and CD8 T cells were compared among the different study groups, as explained above. Moreover, using the levels (proportion) of the different CD4 and CD8 T cell clusters in each study volunteer, a random forest and hierarchical clustering analysis were performed as classification methods in order to ascertain the T cell clusters involved in discriminating between pairs of study groups. The Boruta algorithm was applied to select the most important variables for classification. Variables selected by Boruta were used to carry out the random forest and hierarchical clustering. The number of trees and variables for the random forest were optimized by minimizing the out-of-bag error. An internal validation, using the leave-one-out method, was conducted to assess the classification capabilities of the random forest. Hierarchical clustering was performed using the Euclidean distance and Ward's agglomerative method and graphically represented by a heat map with dendrograms.

Supplementary Materials: The supporting information can be downloaded at https://www.mdpi.com/article/10.3390/ijms25115937/s1. References [15,43–58] are cited in the supplementary materials.

Author Contributions: Data curation, A.C., M.G. and V.E.; funding acquisition, J.M.B. and N.R.; investigation, J.M.B. and N.R.; methodology, J.M.B., D.J.-C., C.R., J.V.-Q., E.L.-C., F.S.-C. and N.R.; project administration, J.M.B. and N.R.; software, D.J.-C., J.M.L., I.M. and F.S.-C.; supervision, J.M.B. and N.R.; writing—review and editing, J.M.B. and N.R. All authors have read and agreed to the published version of the manuscript.

Funding: This project was funded in part by grants RD16/0025/0013, PI16/01769, PI19/01237, and PI19/00973 and integrated into the State Plan for Scientific and Technical Research and Innovation; it was co-funded by the ISCIII Sub-Directorate General for Research Assessment and Promotion and the

European Regional Development Fund (ERDF). The HIV BioBank, integrated into the Spanish AIDS Research Network, is partially funded by the RD16/0025/0019 project as part of the Plan Nacional R+D+I and cofinanced by ISCIII Sub-Directorate General for Research Assessment and Promotion and the European Regional Development Fund (ERDF).

Institutional Review Board Statement: This study was approved by the Ethical Review Board of the Instituto de Investigación Sanitaria-Fundación Jiménez Díaz, Madrid, Spain (approval ID: PIC097-19_FJD) and carried out in accordance with the Declaration of Helsinki.

Informed Consent Statement: All study participants signed written informed consent.

Data Availability Statement: The raw data supporting the conclusions of this article will be made available by the authors upon request.

Acknowledgments: We want to acknowledge the patients in this study for their participation and the HIV BioBank integrated into the Spanish AIDS Research Network and collaborating Centers.

Conflicts of Interest: Author José M. Ligos was employed by Cytek Biosciences. The remaining authors declare that the research was conducted in the absence of any commercial or financial relationships that could be construed as a potential conflict of interest.

References

1. Autran, B.; Descours, B.; Avettand-Fenoel, V.; Rouzioux, C. Elite controllers as a model of functional cure. *Curr. Opin. HIV AIDS* **2011**, *6*, 181–187. [CrossRef] [PubMed]
2. Deeks, S.G.; Lewin, S.R.; Ross, A.L.; Ananworanich, J.; Benkirane, M.; Cannon, P.; Chomont, N.; Douek, D.; Lifson, J.D.; Lo, Y.R.; et al. International AIDS Society global scientific strategy: Towards an HIV cure 2016. *Nat. Med.* **2016**, *22*, 839–850. [CrossRef] [PubMed]
3. Kwaa, A.K.; Blankson, J.N. Immune responses in controllers of HIV. *Ann. Rev. Immunol.* **2024**, *42*, 21–33. [CrossRef] [PubMed]
4. Betts, M.R.; Nason, M.C.; West, S.M.; De Rosa, S.C.; Migueles, S.A.; Abraham, J.; Lederman, M.M.; Benito, J.M.; Goepfert, P.A.; Connors, M.; et al. HIV nonprogressors preferentially maintain highly functional HIV-specific CD8+ T cells. *Blood* **2006**, *107*, 4781–4789. [CrossRef] [PubMed]
5. Ndhlovu, Z.M.; Stampouloglou, E.; Cesa, K.; Mavrothalassitis, O.; Alvino, D.M.; Li, J.Z.; Wilton, S.; Karel, D.; Piechocka-Trocha, A.; Chen, H.; et al. The Breadth of Expandable Memory CD8+ T Cells Inversely Correlates with Residual Viral Loads in HIV Elite Controllers. *J. Virol.* **2015**, *89*, 10735–10747. [CrossRef] [PubMed]
6. Migueles, S.A.; Osborne, C.M.; Royce, C.; Compton, A.A.; Joshi, R.P.; Weeks, K.A.; Rood, J.E.; Berkley, A.M.; Sacha, J.B.; Cogliano-Shutta, N.A.; et al. Lytic granule loading of CD8+ T cells is required for HIV-infected cell elimination associated with immune control. *Immunity* **2008**, *29*, 1009–1021. [CrossRef] [PubMed]
7. Mothe, B.; Llano, A.; Ibarrondo, J.; Zamarreño, J.; Schiaulini, M.; Miranda, C.; Ruiz-Riol, M.; Berger, C.T.; Herrero, M.J.; Palou, E.; et al. CTL responses of high functional avidity and broad variant cross-reactivity are associated with HIV control. *PLoS ONE* **2012**, *7*, e29717. [CrossRef]
8. Kiepiela, P.; Ngumbela, K.; Thobakgale, C.; Ramduth, D.; Honeyborne, I.; Moodley, E.; Reddy, S.; de Pierres, C.; Mncube, Z.; Mkhwanazi, N.; et al. CD8+ T-cell responses to different HIV proteins have discordant associations with viral load. *Nat. Med.* **2007**, *13*, 46–53. [CrossRef]
9. Navarrete-Muñoz, M.A.; Restrepo, C.; Benito, J.M.; Rallón, N. Elite controllers: A heterogeneous group of HIV-infected patients. *Virulence* **2020**, *11*, 889–897. [CrossRef]
10. Leon, A.; Perez, I.; Ruiz-Mateos, E.; Benito, J.M.; Leal, M.; Lopez-Galindez, C.; Rallon, N.; Alcami, J.; Lopez-Aldeguer, J.; Viciana, P.; et al. Rate and predictors of progression in elite and viremic HIV-1 controllers. *AIDS* **2016**, *30*, 1209–1220. [CrossRef]
11. Grabar, S.; Selinger-Leneman, H.; Abgrall, S.; Pialoux, G.; Weiss, L.; Costagliola, D. Prevalence and comparative characteristics of long-term nonprogressors and HIV controller patients in the French Hospital Database on HIV. *AIDS* **2009**, *23*, 1163–1169. [CrossRef] [PubMed]
12. Pernas, M.; Tarancón-Diez, L.; Rodríguez-Gallego, E.; Gómez, J.; Prado, J.G.; Casado, C.; Dominguez-Molina, B.; Olivares, I.; Coiras, M.; León, A.; et al. Factors leading to the loss of natural elite control of HIV-1 infection. *J. Virol.* **2018**, *92*, e01805–e01817. [CrossRef]
13. Rodríguez-Gallego, E.; Tarancón-Diez, L.; García, F.; Del Romero, J.; Benito, J.; Alba, V.; Herrero, P.; Rull, A.; Dominguez-Molina, B.; Martinez-Madrid, O.; et al. Proteomic Profile Associated with Loss of Spontaneous HIV-1 Elite Control. *J. Infect. Dis.* **2019**, *291*, 867–876. [CrossRef] [PubMed]
14. Benito, J.M.; Ortiz, M.C.; León, A.; Sarabia, L.A.; Ligos, J.M.; Montoya, M.; Garcia, M.; Ruiz-Mateos, E.; Palacios, R.; Cabello, A.; et al. Class-modeling analysis reveals T-cell homeostasis disturbances involved in loss of immune control in elite controllers. *BMC Med.* **2018**, *16*, 30. [CrossRef]
15. Fenwick, C.; Joo, V.; Jacquier, P.; Noto, A.; Banga, R.; Perreau, M.; Pantaleo, G. T-cell exhaustion in HIV infection. *Immunol. Rev.* **2019**, *292*, 149–163. [CrossRef]

16. Douek, D.C. Disrupting T-cell homeostasis: How HIV-1 infection causes disease. *AIDS Rev.* **2003**, *5*, 172–177.
17. Soukas, C. Immunosenescence and aging in HIV. *Curr. Opin. HIV AIDS* **2014**, *9*, 398–404. [CrossRef] [PubMed]
18. Zakhour, R.; Tran, D.Q.; Heresi, G.P.; Degaffe, G.; Bell, C.S.; Donnachie, E.; Zhang, W.; Pérez, N.; Benjamins, L.J.; Del Bianco, G.; et al. CD31 Expression on CD4+ Cells: A Simple Method for Quantitation of Recent Thymus Emigrant CD4 Cells. *Am. J. Trop. Med. Hyg.* **2016**, *95*, 970–972. [CrossRef]
19. Hazra, R.; Mackall, C. Thymic function in HIV infection. *Curr. HIV/AIDS Rep.* **2005**, *2*, 24–28. [CrossRef]
20. Seth, N.; Kaufmann, D.; Lahey, T.; Rosenberg, E.S.; Wucherpfennig, K.W. Expansion and Contraction of HIV-Specific CD4 T Cells with Short Bursts of Viremia, but Physical Loss of the Majority of These Cells with Sustained Viral Replication. *J. Immunol.* **2005**, *175*, 6948–6958. [CrossRef]
21. Mollet, L.; Li, T.S.; Samri, A.; Tournay, C.; Tubiana, R.; Calvez, V.; Debré, P.; Katlama, C.; Autran, B.; RESTIM and COMET Study Groups. Dynamics of HIV-Specific CD81 T Lymphocytes with Changes in Viral Load. *J. Immunol.* **2000**, *165*, 1692–1704. [CrossRef] [PubMed]
22. Gianella, S.; Letendre, S. Cytomegalovirus and HIV: A Dangerous Pas de Deux. *J. Infect. Dis.* **2016**, *214*, S67–S74. [CrossRef] [PubMed]
23. George, A.F.; Luo, X.; Neidleman, J.; Hoh, R.; Vohra, P.; Thomas, R.; Shin, M.G.; Lee, M.J.; Blish, C.A.; Deeks, S.G.; et al. Deep Phenotypic Analysis of Blood and Lymphoid T and NK Cells From HIV+ Controllers and ART-Suppressed Individuals. *Front. Immunol.* **2022**, *13*, 803417. [CrossRef]
24. Ruiz-Mateos, E.; Ferrando-Martinez, S.; Machmach, K.; Viciana, P.; Pacheco, Y.M.; Nogales, N.; Genebat, M.; Leal, M. High Levels of CD57+ CD28− T-Cells, Low T-Cell Proliferation and Preferential Expansion of Terminally Differentiated CD4+ T-Cells in HIV Elite Controllers. *Curr. HIV Res.* **2010**, *8*, 471–481. [CrossRef] [PubMed]
25. Yang, Y.; Al-Mozainia, M.; Buzon, M.J.; Beamon, J.; Ferrando-Martinez, S.; Ruiz-Mateos, E.; Rosenberg, E.S.; Pereyra, F.; Yu, X.G.; Lichterfeld, M. CD4 T-cell regeneration in HIV-1 elite controllers. *AIDS* **2012**, *26*, 701–706. [CrossRef] [PubMed]
26. Bansal, A.; Sterrett, S.; Erdmann, N.; Westfall, A.O.; Dionne-Odom, J.; Overton, E.T.; Goepfert, P.A. Normal T-cell activation in elite controllers with preserved CD4+ T-cell counts. *AIDS* **2015**, *29*, 2245–2254. [CrossRef] [PubMed]
27. López, M.; Soriano, V.; Peris-Pertusa, A.; Rallón, N.; Restrepo, C.; Benito, J.M. Elite controllers display higher activation on central memory CD8 T cells than HIV patients successfully on HAART. *AIDS Res. Hum. Retroviruses* **2011**, *27*, 157–165. [CrossRef] [PubMed]
28. Kamya, P.; Tsoukas, C.M.; Boulet, S.; Routy, J.P.; Thomas, R.; Côté, P.; Boulassel, M.R.; Lessard, B.; Kaul, R.; Ostrowski, M.; et al. T cell Activation does not drive CD4 decline in longitudinally followed HIV-infected Elite Controllers. *AIDS Res. Ther.* **2011**, *8*, 20. [CrossRef] [PubMed]
29. Caetano, D.G.; de Paula, H.H.S.; Bello, G.; Hoagland, B.; Villela, L.M.; Grinsztejn, B.; Veloso, V.G.; Morgado, M.G.; Guimarães, M.L.; Côrtes, F.H. HIV-1 elite controllers present a high frequency of activated regulatory T and Th17 cells. *PLoS ONE* **2020**, *15*, e0228745. [CrossRef]
30. Brandt, L.; Benfield, T.; Mens, H.; Clausen, L.N.; Katzenstein, T.L.; Fomsgaard, A.; Karlsson, I. Low Level of Regulatory T Cells and Maintenance of Balance Between Regulatory T Cells and TH17 Cells in HIV-1–Infected Elite Controllers. *J. Acquir. Immun. Defic. Syndr.* **2011**, *57*, 101–108. [CrossRef]
31. Landay, A.; Golub, E.T.; Desai, S.; Zhang, J.; Winkelman, V.; Anastos, K.; Durkin, H.; Young, M.; Villacres, M.C.; Greenblatt, R.M.; et al. HIV RNA levels in plasma and cervical-vaginal lavage fluid in elite controllers and HAART recipients. *AIDS* **2014**, *28*, 739–743. [CrossRef] [PubMed]
32. Dinoso, J.B.; Kim, S.Y.; Siliciano, R.F.; Blankson, J.N. A Comparison of Viral Loads between HIV-1–Infected Elite Suppressors and Individuals Who Receive Suppressive Highly Active Antiretroviral Therapy. *Clin. Infect. Dis.* **2008**, *47*, 102–104. [CrossRef] [PubMed]
33. Hatano, H.; Delwart, E.L.; Norris, P.J.; Lee, T.H.; Dunn-Williams, J.; Hunt, P.W.; Hoh, R.; Stramer, S.L.; Linnen, J.M.; McCune, J.M.; et al. Evidence for Persistent Low-Level Viremia in Individuals Who Control Human Immunodeficiency Virus in the Absence of Antiretroviral Therapy. *J. Virol.* **2009**, *83*, 329–335. [CrossRef] [PubMed]
34. Sedaghat, A.R.; Rastegar, D.A.; O'Connell, K.A.; Dinoso, J.B.; Wilke, C.O.; Blankson, J.N. T Cell Dynamics and the Response to HAART in a Cohort of HIV-1–Infected Elite Suppressors. *Clin. Infect. Dis.* **2009**, *49*, 1763–1766. [CrossRef]
35. Li, J.Z.; Segal, F.P.; Bosch, R.J.; Lalama, C.M.; Roberts-Toler, C.; Delagreverie, H.; Getz, R.; Garcia-Broncano, P.; Kinslow, J.; Tressler, R.; et al. Antiretroviral Therapy Reduces T-cell Activation and Immune Exhaustion Markers in Human Immunodeficiency Virus Controllers. *Clin. Infect. Dis.* **2020**, *70*, 1636–1642. [CrossRef] [PubMed]
36. Castellino, F.; Germain, R.N. Cooperation between CD4+ and CD8+ T cells: When, where, and how. *Ann. Rev. Immunol.* **2006**, *24*, 519–540. [CrossRef] [PubMed]
37. Fry, T.J.; Mackall, C.L. The many faces of IL-7, from thymopoiesis to peripheral T cell maintenance. *J. Immunol.* **2005**, *174*, 6571–6576. [CrossRef] [PubMed]
38. Jimenez-Carretero, D.; Ligos, J.M.; Martínez-López, M.; Sancho, D.; Montoya, M.C. Flow Cytometry Data Preparation Guidelines for Improved Automated Phenotypic Analysis. *J. Immunol.* **2018**, *200*, 3319–3331. [CrossRef] [PubMed]
39. Van Gassen, S.; Gaudilliere, B.; Angst, M.S.; Saeys, Y.; Aghaeepour, N. CytoNorm: A Normalization Algorithm for Cytometry Data. *Cytom. A* **2020**, *97*, 268–278. [CrossRef]

40. Hao, Y.; Hao, S.; Andersen-Nissen, E.; Mauck, W.M., III; Zheng, S.; Butler, A.; Satija, R. Integrated analysis of multimodal single-cell data. *Cell* **2021**, *184*, 3573–3587. [CrossRef]
41. Waltman, L.; Van Eck, N.J. A smart local moving algorithm for large-scale modularity-based community detection. *Eur. Phys. J. B* **2013**, *86*, 471. [CrossRef]
42. Linderman, G.C.; Rachh, M.; Hoskins, J.G.; Steinerberger, S.; Kluger, Y. Fast interpolation-based t-SNE for improved visualization of single-cell RNA-seq data. *Nat. Methods* **2019**, *16*, 243–245. [CrossRef] [PubMed]
43. Emu, B.; Moretto, W.J.; Hoh, R.; Krone, M.; Martin, J.N.; Nixon, D.F.; Deeks, S.G.; McCune, J.M. Composition and function of T cell subpopulations are slow to change despite effective antiretroviral treatment of HIV disease. *PLoS ONE* **2014**, *21*, e85613. [CrossRef] [PubMed]
44. Mahnke, Y.D.; Brodie, T.M.; Sallusto, F.; Roederer, M.; Lugli, E. The who's who of T-cell differentiation: Human memory T-cell subsets. *Eur. J. Immunol.* **2013**, *43*, 2797–2809. [CrossRef] [PubMed]
45. Kohler, S.; Thiel, A. Life after the thymus: CD31+ and CD31- human naive CD4+ T-cell subsets. *Blood* **2009**, *113*, 769–774. [CrossRef] [PubMed]
46. Liu, Z.; Cumberland, W.G.; Hultin, L.E.; Prince, H.E.; Detels, R.; Giorgi, J.V. Elevated CD38 antigen expression on CD8+ T cells is a stronger marker for the risk of chronic HIV disease progression to AIDS and death in the Multicenter AIDS Cohort Study than CD4+ cell count, soluble immune activation markers, or combinations of HLA-DR and CD38 expression. *J. Acquir. Immune. Defic. Syndr. Hum. Retrovirol.* **1997**, *16*, 83–92. [PubMed]
47. Paiardini, M.; Müller-Trutwin, M. HIV-associated chronic immune activation. *Immunol. Rev.* **2013**, *254*, 78–101. [CrossRef] [PubMed]
48. Pangrazzi, P.; Reidla, J.; Carmona-Arana, J.A.; Naismith, E.; Miggitsch, C.; Meryk, A.; Keller, M.; Krause, A.A.N.; Melzer, F.L.; Trieb, K.; et al. CD28 and CD57 define four populations with distinct phenotypic properties within human CD8+ T cells. *Eur. J. Immunol.* **2020**, *50*, 363–379. [CrossRef]
49. Strioga, M.; Pasukoniene, V.; Characiejus, D. CD8+ CD28- and CD8+ CD57+ T cells and their role in health and disease. *Immunology* **2011**, *134*, 17–32. [CrossRef]
50. Hove-Skovsgaard, M.; Zhao, Z.; Tingstedt, J.L.; Hartling, H.J.; Thudium, R.F.; Benfield, T.; Afzal, S.; Nordestgaard, B.; Ullum, H.; Gerstoft, J.; et al. Impact of Age and HIV Status on Immune Activation, Senescence and Apoptosis. *Front. Immunol.* **2020**, *11*, 583569. [CrossRef]
51. Crawley, A.M.; Angel, J.B. The influence of HIV on CD127 expression and its potential implications for IL-7 therapy. *Semin. Immunol.* **2012**, *24*, 231–240. [CrossRef] [PubMed]
52. Dunham, R.M.; Cervasi, B.; Brenchley, J.M.; Albrecht, H.; Weintrob, A.; Sumpter, B.; Engram, J.; Gordon, S.; Klatt, N.R.; Frank, I.; et al. CD127 and CD25 expression defines CD4+ T cell subsets that are differentially depleted during HIV infection. *J. Immunol.* **2008**, *180*, 5582–5592. [CrossRef] [PubMed]
53. Rouas, R.; Merimi, M.; Najar, M.; El Zein, N.; Fayyad-Kazan, M.; Berehab, M.; Agha, D.; Bron, D.; Burny, A.; Rachidi, W.; et al. Human CD8+ CD25 + CD127 low regulatory T cells: microRNA signature and impact on TGF-β and IL-10 expression. *J. Cell. Physiol.* **2019**, *234*, 17459–17472. [CrossRef] [PubMed]
54. Takenaka, M.V.; Robson, S.; Quintana, F.J. Regulation of the T Cell Response by CD39. *Trends Immunol.* **2016**, *37*, 427–439. [CrossRef]
55. Groom, J.R.; Luster, A.D. CXCR3 in T cell function. *Exp. Cell. Res.* **2011**, *317*, 620–631. [CrossRef] [PubMed]
56. Silveira-Mattos, S.M.; Narendran, G.; Akrami, K.; Fukutani, K.F.; Anbalagan, S.; Nayak, K.; Subramanyam, S.; Subramani, R.; Vinhaes, C.L.; Oliveira-de Souza, D.; et al. Differential expression of CXCR3 and CCR6 on CD4+ T-lymphocytes with distinct memory phenotypes characterizes tuberculosis-associated immune reconstitution inflammatory síndrome. *Sci. Rep.* **2019**, *9*, 1502. [CrossRef] [PubMed]
57. Lee, A.Y.S.; Körner, H. CCR6/CCL20 chemokine axis in human immunodeficiency virus immunity and pathogenesis. *J. Gen. Virol.* **2017**, *98*, 338–344. [CrossRef]
58. Vinuesa, C.G.; Linterman, M.A.; Yu, D.; MacLennan, I.C.M. Follicular Helper T Cells. *Annu. Rev. Immunol.* **2016**, *34*, 335–368. [CrossRef]

International Journal of
Molecular Sciences

MDPI

Article

The Diverse Nature of the Molecular Interactions That Govern the COV-2 Variants' Cell Receptor Affinity Ranking and Its Experimental Variability

Fredy Sussman * and Daniel S. Villaverde

Department of Organic Chemistry, Faculty of Chemistry, Universidad de Santiago de Compostela, 15784 Santiago de Compostela, Spain; daniel.sussman@raii.usc.es
* Correspondence: fredy.sussman@usc.es

Abstract: A critical determinant of infectivity and virulence of the most infectious and or lethal variants of concern (VOCs): Wild Type, Delta and Omicron is related to the binding interactions between the receptor-binding domain of the spike and its host receptor, the initial step in cell infection. It is of the utmost importance to understand how mutations of a viral strain, especially those that are in the viral spike, affect the resulting infectivity of the emerging VOC, knowledge that could help us understand the variant virulence and inform the therapies applied or the vaccines developed. For this sake, we have applied a battery of computational protocols of increasing complexity to the calculation of the spike binding affinity for three variants of concern to the ACE2 cell receptor. The results clearly illustrate that the attachment of the spikes of the Delta and Omicron variants to the receptor originates through different molecular interaction mechanisms. All our protocols unanimously predict that the Delta variant has the highest receptor-binding affinity, while the Omicron variant displays a substantial variability in the binding affinity of the spike that relates to the structural plasticity of the Omicron spike–receptor complex. We suggest that the latter result could explain (at least in part) the variability of the in vitro binding results for this VOC and has led us to suggest a reason for the lower virulence of the Omicron variant as compared to earlier strains. Several hypotheses have been developed around this subject.

Keywords: COV-2; variants of concern; variant infectivity; spike–receptor affinities; molecular mechanism; molecular mechanics calculations

1. Introduction

Severe acute respiratory syndrome coronavirus 2 (SARS-CoV-2) is the causative agent of COVID-19, which was first reported in Wuhan (China) at the end of 2019 and which escalated into a global pandemic in 2020, producing, to date, hundreds of millions of infections and more than 6,000,000 deaths besides the countless human, social and economic disruptions, according to the World Health Organization (WHO). SARS-CoV-2 continuously undergoes mutations due to changes in the genetic code that usually occur during the replication of its genome. The new strains go unchecked because the virus lacks proofreading machinery. In the case of SARS-CoV-2, the mutants that have spread more widely and produced a larger number of deaths have been named variants of concern (VOCs) by the WHO. The ones that have proven to have the most lasting effects are the Delta [1] and Omicron [2] VOCs and their variants.

The SARS-CoV-2 viral membrane has a transmembrane glycoprotein, the spike (S) glycoprotein, and the envelope protein, and it surrounds the flexible helical nucleocapsid. The spike glycoprotein engages human angiotensin-converting enzyme (ACE2) through the spikes that protrude from the virion and give it its characteristic crown shape. The interaction between the virus receptor and the cell allows for viral genetic material to be delivered to the host cell cytoplasm for replication [3].

The segment of the spike that enters in direct contact with the ACE2 cell receptor is called the receptor-binding domain (RBD) and it contains many of the mutations that distinguish one VOC from others. Figure 1 displays the Delta and Omicron RBDs with the residue variants with respect to the native strain.

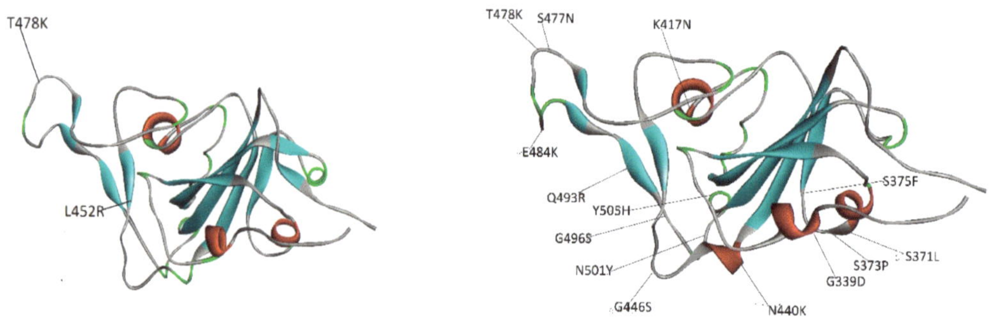

Figure 1. Placement of the mutations in the RBD spike of the Delta variant (**left-hand side**) and the Omicron variant (**right-hand side**). The Delta RBD has only 2 mutations whereas the Omicron RBD has 15 mutations. In the case of Delta, one mutation is in a loop (T478K) and the other is in a sheet (L452R). For its part, in Omicron, most of the mutations are in loops and sheets.

1.1. Hypothesis for Infectivity and Virulence of the VOCs

It is of the utmost importance to understand how mutations of a viral strain, especially those that are in the viral spike, affect the resulting infectivity of the emerging VOC, knowledge that could help us understand the variant virulence and inform the therapies applied or the vaccines developed. Several hypotheses have been developed around this subject.

1.1.1. Spike–ACE2 Receptor Interactions

Since the initial step in cell infection is the viral cell entry that starts with the interaction of the spike trimer or monomer with the cell ACE2 receptor, it has been thought that this could be a rate-limiting step for viral entry into the cell [4–7] and hence an increase in the spike–ACE2 receptor affinity for the different VOCs could translate into enhanced infectivity. It has been observed that a greater infectivity does not always lead to enhanced virulence and lethality. Even though the Omicron variant has shown to be very infectious, it causes milder symptoms than previous VOCs, like the Wild Type and Delta, since it infects mostly the respiratory tracts rather than affecting the lungs.

The experimental spike–receptor affinities for a variety of mutant strains have been obtained independently by various research groups through Surface Plasmon Resonance (SPR) experiments [4,7,8] and microscale thermophoresis [9]. The results are shown in Table 1.

Table 1. Spike–receptor affinities from various laboratories.

Source	Wild Type	Delta	Omicron
Mannar et al. (kd, nM) [4]	3.0 ± 0.3	2.0 ± 0.2	2.1 ± 0.3
Kim et al. (Kd, nM) [9]	27.5 ± 4.8	21.5 ± 2.9	5.5 ± 1.4
Han et al. (Kd, nM) [7]	24.63 ± 5.00	25.07 ± 6.70	31.40 ± 11.62
Zhang et al. (Kd, nM) [8]	------	2.9	8.9

As seen from this table, the results from different labs are dissimilar in many aspects. While the results of Han et al. indicate that the spike–receptor affinities for the Wild Type and Delta variants are very similar [7], the results of other labs show an increase in affinity for the latter strain (Delta). Moreover, only the results of the Im lab indicate that the larger number of mutations in the Omicron variant produces an increase in affinity [9]. Understanding the origin of the variability seen in these experiments is a very important step in determining the importance of the spike–receptor interaction (the first step in the virus cell entrance) in these VOCs' infectivity.

1.1.2. Spike–Antibody Interactions

An alternative source of increased infectivity upon the emergence of a new VOC is the ability of the new strain to reduce the affinity of the viral spikes for the antibodies produced by the organism and hence evade the immune system, as proposed for the omicron variant [4]. This knowledge should inform the design of antibody-based vaccines [10–13].

There has been an intensive effort aimed at understanding the spike–cell receptor and spike–antibody interactions using computational chemistry and especially molecular mechanics/molecular dynamics (MM/MD) tools [14–27]. The study of protein associations by MM/MD calculations represents a grand challenge in computational biochemistry since they involve the study of protein–protein interactions (PPIs), a difficult task given the size of the systems involved and the fact that protein–protein interactions (like antibody–antigen) are mediated by residue residents in loops, the most flexible secondary structure [27,28]. As seen from Figure 1, the spike interface with the receptor is predominantly made up of loop segments. Hence, most of the mutations of the Delta and Omicron variants reside in these segments. Given the flexibility, an exponential explosion of conformations should be expected, leading to sizable standard deviations in the calculation of binding affinities from MD simulations, which may exceed the average value of this property. Some of the authors have sidestepped this problem by calculating the binding affinities from the average over those frames that have the lowest binding energy [24]. In most of the calculations, the spike RBD–receptor complexes for the latter strains (Delta, Omicron) were modeled from the structure obtained by mutating the Wild Type strain (PDB entry 6m0j [29]) in silico, rather than starting from the actual experimental structures for the spike–receptor structures, a feature of the calculations that may have influenced the results. Since then, cryo-EM-based structures of the VOC spike RBD with the receptor are now available in the case of the Delta variant (PDB entry 7tew [30]) and in more than one case for the Omicron mutant (PDB entries 7u0n, 7wpb, 7wbp, 7t9l) [4,6,30–32]. The present work uses these experimentally determined structures, a route that will allow us to avoid the uncertainty of starting from modeled structures.

Most previous work aiming at ranking the spike–receptor affinities of the VOCs studied in this work resulted in the following order (see, for instance, refs. [20,21,24]).

Native < Delta < Omicron.

Sometimes, this was with sizable differences favoring the Omicron variant [20].

In this work, we propose a few protocols (some of them MD-based) aimed at predicting the affinities of any COV-2 RBD spike variant for the cell receptor. The methodology used here was designed to substantially reduce computer expense and hence can be applied to a wider range of larger systems describing the viral cell entry, without having to delve into coarse-grained approaches used previously [17]. The binding affinity prediction protocols have been organized as a battery of techniques of increasing complexity to evaluate the spike–receptor affinity. This strategy is expected to eliminate the bias from a single protocol and allow us to look for possible consensus amongst the predictions provided by these methods. As shown below, it has allowed us to determine that the Delta and Omicron variants' RBD spikes enhance their ACE2 receptor affinity through very different molecular interactions. Our results have also allowed us to put forward an explanation for the variability observed in different SPR binding affinity experiments shown in Table 1.

2. Results and Discussion

2.1. Ranking Affinity Prediction by a Suite of Protocols of Increasing Complexity

The most basic protocol for evaluating protein–protein affinities is related to the calculation of the surface area that is buried upon protein association, called Buried Surface

Area (BSA) (For a review, see ref. [33]). We present the BSA-based affinity values from the six known PDB structures (see Table 2) as well as the average values generated by the spike–receptor association, obtained from the MD trajectories (see the Section 3 for details) for the three VOCs studied here (see Table 3). As shown from both types of calculations, the Delta variant is predicted to increase the binding affinity with respect to the original (Wild Type) strain, in line with the experiments of Mannar et al. [4] and Kim et al. [9]. On the other hand, this protocol predicts that the large number of mutations present in Omicron does not translate into a rise in affinity above the native variant.

Table 2. Spike–receptor association energies from Buried Surface Area calculations.

VOC (PDB Entry)	BSA [(a)]	Binding Energy [(b)]
Wild Type (6m0j)	−1719.2	−43.0
Omicron (7u0n)	−1720.6	−43.0
Omicron (7wbp)	−1733.1	−43.3
Omicron (7wpb)	−1710.7	−42.8
Omicron (7t9l)	−1678.2	−42.0
Delta (7tew)	−1807.6	−45.2

[(a)] BSA values in $Å^2$ [(b)] energies in Kcal/mol.

Table 3. Spike–receptor association energies from MD frames on a BSA approach.

VOC/PDB	BSA [(c)]	Binding Energy [(a)]
Wild Type (6M0J)	−1750.9 (32.0) [(b)]	−43.7 (0.8) [(b)]
Delta (7TEW)	−1849.4 (19.6)	−46.2 (0.5)
Omicron (7T9L)	−1708.0 (48.0)	−42.7 (1.2)
Omicron (7U0N)	−1682.5 (50.9)	−42.1(1.3)

[(a)] These are the average binding energies from the MD trajectories in Kcal/mol. [(b)] The number in parentheses is the standard deviation from the average of the property. [(c)] BSA values in $Å^2$.

The above protein–protein interaction energy BSA-based protocol neglects many components of the binding free energy or at best only includes them indirectly. That is the case of the electrostatic interaction energy, which comprises both charge–charge interactions and the process leading to the desolvation of the charged groups during the binding process, as well as the contribution of entropy to the binding affinity.

There have been many efforts in developing protocols that include all the relevant components of the protein–protein binding free energy. One of them is based on chemical thermodynamics principles and used for the automatic detection of macromolecular assemblies in the Protein Data Bank (PDB) entries in the PISA server [34]. Here, we have applied this method to all known spike RBD–receptor complexes. They include the Wild Type and Delta variants and the four Omicron structures (PDB entries 7t9l, 7u0n, 7wpb and 7wbp). The evaluation of these latter structures will allow us to determine how the structural variability of the loops where mutations reside (see Figure 1) affects the binding energy results. The results are shown in Table 4.

Perusal of this table indicates that (once again) the spike RBD from the Delta variant is the one that is predicted to have the highest affinity for the ACE2 receptor. For some of the Omicron structures, their association energies are comparable to the ones obtained by Delta (as in the case of the 7t9l and 7wbp), but other Omicron structures like 7wpb and 7u0n present binding energies lower than even the native strain.

Table 4. Results for the binding of the RBD spike segment to cell receptor using PISA.

VOC	PDB Entry [b]	Binding Energy (Pisa) [a]
Wild Type	6m0j	−5.1
Delta	7tew	−6.3
Omicron	7t9l	−6.1
Omicron	7wpb	−3.4
Omicron	7u0n	−4.4
Omicron	7wbp	−6.2

[a] Results in Kcal/mol, [b] calculations were performed directly from PDB structures.

To determine the possibility of interconversion of one structure into the others, we performed MD studies using the protocol described in the Section 3 for Omicron structures 7u0n and 7t9l as well as the Delta (7tew) and Wuhan (6m0j) structures. The binding energies were obtained using the molecular mechanics energy differences supplemented by the corresponding changes in implicit generalized Born solvation energy upon binding, performed on frames that resulted from extensive MD simulations (see the Section 3 for details). The binding affinities that resulted from these calculations are listed in Table 5.

Table 5. Binding energies from MD/GBSW protocol on MD trajectories.

VOC	PDB Entry	Binding Energy [a]
Native	6m0j	−76.7(1.4) [b]
Delta	7tew	−87.3 (4.3)
Omicron	7u0n	−67.0 (2.0)
Omicron	7t9l	−83.9 (1.0)

[a] Energies in Kcal/mol. [b] Numbers in parentheses are standard deviations.

As seen from Table 5, the Delta variant is (once again) the mutant with the highest affinity, in agreement with all previous protocols employed here (see Tables 2–4). Moreover, this variant has the largest standard deviation (4.3 Kcal/mol), an outcome that has implications for the ranking of the Omicron spike binding whose evaluation started from the 7t9l structure. Although the spike mean binding energy from the latter structure is more positive than the average from Delta, their combined standard deviations make the Omicron binding energy results indistinguishable from those for the Delta strain.

The results displayed in Tables 1–5 support some of the conclusions reached by Mannar et al. [4] rather than the ones by Han et al. [7] or Kim et al. [9]. The mutations that give rise to the Delta variant (T478K and L452R) are definitely the ones that enhance the binding affinity of the viral spike for the receptor. The Omicron variant shares a mutation (T478K) with the Delta variant, but that feature is not enough to increase the Omicron spike–receptor affinity above the one found for the Delta variant, and in some cases not even above the native variant (see Tables 2–4). This outcome would imply that the couple of mutations present in the RBD of the Delta variant (T478K and L452R) indeed increase the affinity of the spike for the receptor. The surplus of mutations present in the Omicron variant (over the Delta one) does not seem to increase the binding affinity, a result obtained previously within the PISA prediction approach. This outcome is probably due to the fact that the contributions of single mutations seem to cancel each other out, a result that is backed up by high-throughput single-mutation assays [4,7]. Hence, it would seem that the large number of mutations present in the RBD of the Omicron spike is not designed to increase its affinity for the receptor but rather fulfill some other tasks like reducing the affinity of the antibodies for them, a feature that sets the Omicron mutant apart from other VOCs [4].

2.2. *Affinity Ranking and Structural Variability*

The variability in the ranking of the binding affinities for the native, Delta and Omicron (see Table 1) variants has been attributed by other authors to the differences in the experimental setup of these labs [9]. Our multi-protocol approach for the evaluation of the binding affinity of the spike RBD for its cognate receptor allowed us to reach the consensus conclusion that the Delta variant RBD consistently has the highest affinity amongst all VOCs, in all evaluations. The only outstanding issue relates to the affinity of the spike of the Omicron variant. The variability of the calculated spike–receptor binding energies (that have, as a starting point, structures obtained in different labs) may be puzzling at first sight. The superimpositions of any two Omicron structures indicate that their maximum RMS value hovers around 1.0 Å, a rather small value. This issue draws attention to the structural plasticity of the PPIs, especially the most variable regions. The rationale for this variability in the binding affinities results in vitro and in silico could be related to the sizable extent of the protein–protein interface. This feature may lead to different predicted binding receptor affinities even for the same spike strain for different starting structures. It may even lead to different binding poses (for the spike RBD). In this study, we have observed this trend for the Omicron variant. The protein–protein interface areas for the original PDB Omicron structures display different values (see Table 2).

Still, the binding energy predictions for these two structures are quite different when a full MM/GBSA protocol is applied. To understand the structural underpinnings of this difference, we have superimposed the 7t9l and 7u0n structures (see Figure 2). In spite of the small RMS between these two structures (1.1 Å), there are some noticeable differences between these two structures, more evident in the spike component, which has a preponderance of loops and beta sheets, whose conformation differs among these structures.

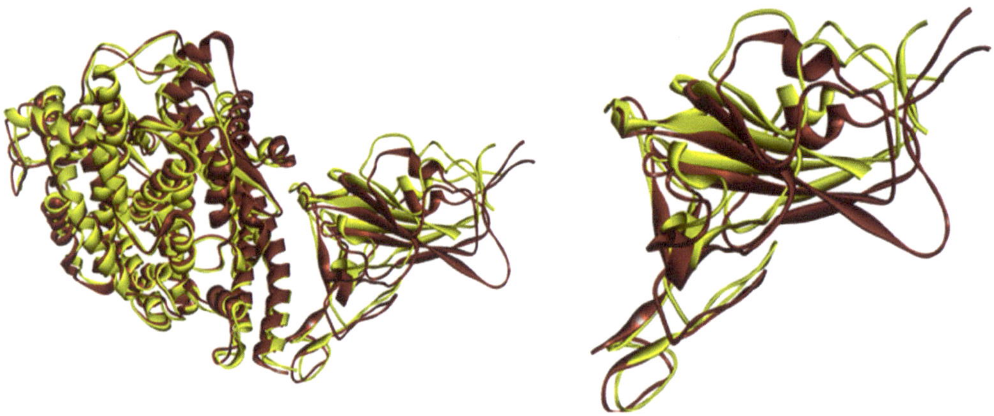

Figure 2. Overlap of the PDB structures 7u0n (yellow) and 7t9l (red). On the left panel is the full complex and on the right are the spike RBDs for both molecules. Notice the change in structure especially in the spike regions that are spanned by loops and sheets.

One feature that sets the Omicron mutant apart from the earlier variants (Delta and native) is the very strong difference in the cell type tropism of the former type. While it is known that the native and Delta strains have a broad cell tropism that directly impacts into substantial viral replication rates in many tissues, the Omicron variant shows more restricted cell preferences. Its replication rates are similar for Delta in human nasal epithelial cultures and kidney cells [35] although it decreases substantially in gut cells and more importantly for lung cells [36].

The differences in replication were mapped to the entry efficiency of the virus on the basis of spike-pseudotyped virus assays [35]. Various rationales have been put forward to explain these differences including the lower efficiency of the S1/S2 cleavage of the Omicron variant by the TMPRSS2 protease [37].

In the present work, our results indicate that the first step in cell entry (that is, the spike–cell receptor interaction) is governed by a different molecular interaction mechanism for the Omicron mutant than for the native and Delta strains. Possibly, as a result, our calculations indicate a broad range of affinities of the Omicron spike for the receptor that varies from below the one found for the native strain to values that are close to the Delta mutant. It has occurred to us that these results could have a direct bearing on the Omicron cell type tropism. It could happen that in the lung tissue environment, the spike–receptor affinities have lower affinities than in some in vitro experiments, possibly lower than the ones for the native variant. This outcome could explain the low virulence of the Omicron viral type.

The results obtained here have allowed us to obtain insights into the molecular interactions that underpin the VOC binding affinities for the receptor and hence their binding ranking. As seen from Tables 2 and 3, the results from the BSA-based affinity predictors clearly single out the Delta spike variant as the one with the higher affinity (over the native strain). Since the contact area calculated from the BSA should be based primarily on Van der Waals interactions, we expect that these are paramount for the increased affinity of the Delta strain. Only when electrostatic interactions and the corresponding desolvation terms are included in the calculations does the Omicron spike affinity for the receptor reach values close to that of the Delta spike (see Tables 4 and 5). This outcome indicates that diverse VOCs' spikes may increase their affinity over the native variant through different molecular interactions.

One outstanding issue is how the mutations found in the Delta variant modify the BSA due to spike–receptor interactions. Saville et al. argued that these residue modifications increase electrostatic complementarity between both proteins [30]. Perusal of the structure indicates indeed that new hydrogen bonds in the region of residues 500–505 of the spike are created in the Delta variant [30]. The mutants that define the Delta variant (T478K and L452R) lie far away from the above segment, so a question arises as to how they could produce this effect.

Overlap of the protein–protein complex for the native and Delta structures (see Figure 3) sheds some light. As expected, there are some changes in the segment that spans from 500 to 505 in the RBD. Nevertheless, there are wider differences in the backbone of the two loops (see residues 471–490) that are closer to Lys 478, one of two mutated residues. A possible explanation is that the latter residue exerts a pull effect that changes the location of the adjacent loops, resulting in an enhanced buried surface and binding affinity. The ACE2 cell receptor has many acidic residues and hence a highly negative charge (−23) that could produce an electrostatic attraction on Lys 478 that results in the loop drag mentioned above.

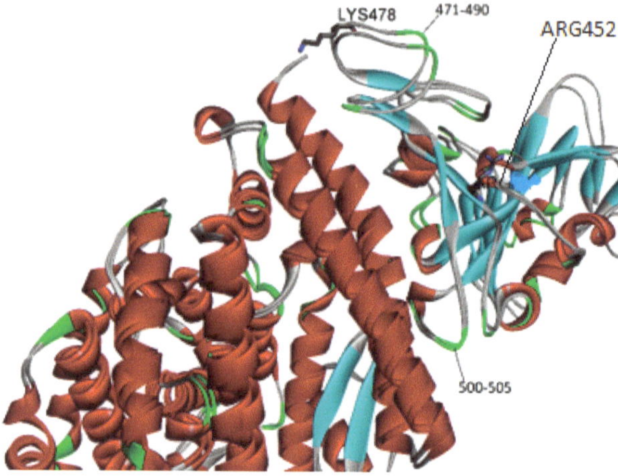

Figure 3. Close-up of the spike–receptor interface for the native (PDB entry 6m0j) and the Delta strain (PDB entry 7tew). Notice that the mutations (T478K and L452R) are not close to the receptor.

3. Materials and Methods

All the structures used in this work were downloaded from the Protein Data Bank (PDB). The structures used in this study were the native (6m0j), the Delta (7tew) and the four Omicron structures obtained independently (7u0n, 7t9l, 7twbp and 7wpb. For the molecular mechanics-based studies, we only used the 7u0n and 7t9l structures for the Omicron studies. The input files for the molecular dynamics simulations were generated by the CHARMM-GUI Web interface [38]. All water and sugar molecules were discarded. The number of residues in each of the PDB structures differs slightly at the N and C terminals, so we took a template of the spike RBD and the ACE2 receptor, the one found in the native structure, and modified the other structures accordingly. The force field used in all simulations was CHARMM_36 [39]. To avoid the use of all atom solvents, we used the generalized Born with switching (GBSW) implicit solvation function [40]. In order to mimic the friction provided by the solvent, we used Langevin dynamics [41] with a friction factor (fbeta) of 20. In all cases, the time step was 2 femtoseconds, and a cutoff of 18 Å was used for non-bonded interactions. The dynamics protocol was divided into two stages: In the first one, a dynamic trajectory of 2.5 ns was performed with initial heating, equilibration and preliminary production steps that added up to 0.5 ns, followed by ten steps, each one 0.2 ns. The structures saved after each of these steps were cooled down to 50K in two stages. The final frame of each trajectory (a total of 10) was saved for analysis of the spike–receptor interaction energies, which was evaluated using the same MM/GBSW protocol as the dynamics runs.

Two other alternative evaluation protocols were applied. The simplest is based on the calculation of the Buried Surface Area (BSA) generated by the protein–protein association. This quantity has been shown to be a primary descriptor of the binding affinity or interaction energy of two macromolecular entities. A first-order approximation assumes that the BSAs could be proportional to the binding energies with a proportionality constant of 0.025 Kcal/mol per Å^2 of surface protein removed from contact with water [33]. We applied this protocol to both the original spike–receptor structures found in the PDB, as well as to the frames that resulted from the last stage of our MD protocol, using the tools present in the graphical interface from Discovery Studio visualizer [42]. To find the accessible surface, we used a sphere the size of a water molecule (1.4 Å). The last and more elaborated protocol used here is based on a chemical thermodynamics approach which includes explicit entropic contributions to the protein–protein affinity [34].

Associated Method Information

The input files for the MM/MD simulations (which included the parameter, connectivity files, coordinate files, etc.) were obtained through the input generator option of the CHARMM-GUI interface https://www.charmm-gui.org/. Last accessed on the 10 December 2023.

The MM/MD calculations were carried out by CHARMM version 46. The software can be found at https://www.charmm.org/archive/charmm/showcase/news/free-charmm/ and the documentation at https://www.charmm.org/archive/charmm/documentation. Last accessed 12 December 2023.

The parameter files and connectivity library can be found at https://github.com/fsussman/TOPPAR1.

The binding free energy calculations based on the Buried Surface Area (BSA) approach were performed on the PDB structures on the PISA-PDB server https://www.ebi.ac.uk/pdbe/pisa/. Documentation is available at the site. Last accessed on 5 November 2023.

The binding free energy calculations based on the BSA approach for the frames resulting from the MD trajectories were performed with the free version of Discovery Studio V3.5, which can be found at https://discover.3ds.com/discovery-studio-visualizer-download?gclid=CjwKCAjwu4WoBhBkEiwAojNdXpgnprVjkxLBhnSIdL1mRc6-7_twnRb26LNEP0y1-2G68t65nrpqSRoCH2wQAvD_BwE, Last accessed on 1 January 2023. Documentation is available for every task in the graphic interface.

4. Conclusions

The initial step in the COV-2 cell infection process (and possibly the rate-limiting step) is the interaction of the viral spike with its cognate receptor, the ACE2 enzyme.

In this work, we have applied a battery of protocols of increasing complexity to the calculation of the spike binding affinity for three variants of concern to the ACE2 cell receptor. The results clearly illustrate that binding to the receptor (of the Delta and Omicron VOC spikes) occurs through different kinds of molecular interactions. While the Delta variant enhances its binding affinity above the one found for the native by an increased contact area between the proteins involved and hence through augmented Van der Waal contacts, the Omicron variant reaches the highest affinity by increasing electrostatic and entropic contributions.

The existence of various structures of Omicron complexes obtained by cryo-EM has allowed us to determine the effect of structural variability on the results. The results obtained indicate that there is a high variability in the binding affinity of the Omicron spike depending on the starting cryo-EM structures used. The highest binding affinity reaches at most the affinity value of the Delta variant, indicating that the surplus of mutations in the Omicron RBD segment of the spike exists to fulfill a different role possibly related to antibody escape [4]. Other cryo-EM structures produce binding affinities that are even lower than those calculated for the native strain (see Table 5), indicating that under some conditions, the Omicron spike may be a weaker binder than the native variant. The experimental results used here (see Table 1) were obtained in vitro. We have thought that the results obtained here could give some clue about the in vivo infectivity and lethality. It is known that the Omicron strain binds to the respiratory tract rather than to the lungs, a result that explains the high infectivity but low virulence of this variant. Our results indicate that the Omicron spike structure is endowed with a high variability and that some of the possible structures may bind to lung ACE2 receptors with an affinity predicted to be lower than the one calculated for the native variant spike, a result that could explain the lower virulence of this strain.

The protocols presented here for the evaluation of the spike–receptor affinities involve substantial computer time savings within all atom molecular mechanics approaches for protein–protein interactions and will be especially useful for research in this field that requires intensive computer resources. One of the research lines we are involved in is the antibody evasion by some of the COV-2 VOCs. This is a major computational endeavor given the number of known antibodies, each one targeting different epitopes within the spike [43].

Author Contributions: Conceptualization, supervision and resources: F.S.; Methodology: F.S. and D.S.V.; Calculations: F.S. and D.S.V.; Formal analysis: F.S. and D.S.V.; Writing first draft: F.S.; review and writing of additional drafts: F.S. and D.S.V. All authors have read and agreed to the published version of the manuscript.

Funding: This research received no external funding.

Institutional Review Board Statement: Not applicable.

Informed Consent Statement: Not applicable.

Data Availability Statement: Data is contained within the article.

Acknowledgments: The calculations were performed at the Centro de Super Computación de Galicia (CESGA). We would like to thank Ramón Estevez and Mercedes Torneiro for useful comments.

Conflicts of Interest: The authors declare no conflict of interest.

References

1. Mlcochova, P.; Kemp, S.; Dhar, M.S.; Papa, G.; Meng, B.; Mishra, S.; Whittaker, C.; Mellan, T.; Ferreira, I.; Datir, R.; et al. SARS-CoV-2 B.1.617.2 Delta variant replication, sensitivity to neutralising antibodies and vaccine breakthrough. *Nature* **2021**, *599*, 114–119. [CrossRef]
2. Viana, R.; Moyo, S.; Amoako, D.G.; Tegally, H.; Scheepers, C.; Althaus, C.L.; Anyaneji, U.J.; Bester, P.A.; Boni, M.F.; Chand, M.; et al. Rapid Epidemic Expansion of the SARS-CoV-2 Omicron Variant in Southern Africa. *Nature* **2022**, *603*, 679–686. [CrossRef]
3. Jackson, C.B.; Farzan, M.; Chen, B.; Choe, H. Mechanisms of SARS-CoV-2 entry into cells. *Nat. Rev. Mol. Cell Biol.* **2022**, *23*, 3–20. [CrossRef]
4. Mannar, D.; Saville, J.W.; Zhu, X.; Srivastava, S.S.; Berezuk, A.M.; Tuttle, K.S.; Marquez, A.C.; Sekirov, I.; Subramaniam, S. SARS-CoV-2 Omicron variant: Antibody evasion and cryo-EM structure of spike protein-ACE2 complex. *Science* **2022**, *375*, 760–764. [CrossRef]
5. Wrapp, D.; Wang, N.; Corbett, K.S.; Goldsmith, J.A.; Hsieh, C.-L.; Abiona, O.; Graham, B.S.; McLellan, J.S. Cryo-EM structure of the 2019-nCoV spike in the prefusion conformation. *Science* **2020**, *367*, 1260–1263. [CrossRef]
6. Han, P.; Li, L.; Liu, S.; Wang, Q.; Zhang, D.; Xu, Z.; Han, P.; Li, X.; Peng, Q.; Su, C.; et al. Receptor binding and complex structures of human ACE2 to spike RBD from omicron and delta SARS-CoV-2. *Cell* **2022**, *185*, 630–640. [CrossRef] [PubMed]
7. Starr, T.N.; Greaney, A.J.; Hilton, S.K.; Ellis, D.; Crawford, K.H.; Dingens, A.S.; Navarro, M.J.; Bowen, J.E.; Tortorici, M.A.; Walls, A.C.; et al. Deep Mutational Scanning of SARS-CoV-2 Receptor Binding Domain Reveals Constraints on Folding and ACE2 Binding. *Cell* **2020**, *182*, 1295–1310. [CrossRef] [PubMed]
8. Zhang, X.; Wu, S.; Wu, B.; Yang, Q.; Chen, A.; Li, Y.; Zhang, Y.; Pan, T.; Zhang, H.; He, X. SARS-CoV-2 Omicron strain exhibits potent capabilities for immune evasion and viral entrance. *Signal Transduct. Target. Ther.* **2021**, *6*, 430. [CrossRef]
9. Kim, S.; Liu, Y.; Ziarnik, M.; Cao, Y.; Zhang, X.F.; Im, W. Binding of human ACE2 and RBD of omicron enhanced by unique interaction patterns among SARS-CoV-2 variants of concern. *J. Comp. Chem.* **2023**, *44*, 594–601. [CrossRef]
10. Jangra, S.; Ye, C.; Rathnasinghe, R.; Stadlbauer, D.; Krammer, F.; Simon, V.; Martinez-Sobrido, L.; Garcia-Sastre, A.; Schotsaert, M. The E484K mutation in the SARS-CoV-2 spike protein reduces but does not abolish neutralizing activity of human convalescent and post-vaccination sera. *MedRxiv* **2021**. *preprint*. [CrossRef]
11. Greaney, A.J.; Starr, T.N.; Gilchuk, P.; Zost, S.J.; Binshtein, E.; Loes, A.N.; Hilton, S.K.; Huddleston, J.; Eguia, R.; Crawford, K.H.D.; et al. Complete Mapping of Mutations to the SARS-CoV-2 Spike Receptor-Binding Domain that Escape Antibody Recognition. *Cell Host Microbe* **2021**, *29*, 44–57. [CrossRef]
12. Liu, Z.; VanBlargan, L.A.; Bloyet, L.-M.; Rothlauf, P.W.; Chen, R.E.; Stumpf, S.; Zhao, H.; Errico, J.M.; Theel, E.S.; Liebeskind, M.J.; et al. Identification of SARS-CoV-2 spike mutations that attenuate monoclonal and serum antibody neutralization. *Cell Host Microbe* **2021**, *29*, 477–488. [CrossRef]
13. Thomson, E.C.; Rosen, L.E.; Shepherd, J.G.; Spreafico, R.; da Silva Filipe, A.; Wojcechowskyj, J.A.; Davis, C.; Piccoli, L.; Pascall, D.J.; Dillen, J.; et al. Circulating SARS-CoV-2 spike N439K variants maintain fitness while evading antibody-mediated immunity. *Cell* **2021**, *184*, 1171–1187. [CrossRef]
14. Ali, A.; Vijayan, R. Dynamics of the ACE2–SARS-CoV-2/SARS-CoV spike protein interface reveal unique mechanisms. *Sci. Rep.* **2020**, *10*, 14214. [CrossRef] [PubMed]
15. Bai, C.; Warshel, A. Critical Differences between the Binding Features of the Spike proteins of SARS-CoV-2 and SARS-CoV. *J. Phys. Chem. B* **2020**, *124*, 5907–5912. [CrossRef]
16. Laurini, E.; Marson, D.; Aulic, S.; Fermeglia, M.; Pricl, S. Computational Alanine Scanning and Structural Analysis of the SARS-CoV-2 Spike Protein/Angiotensin-Converting Enzyme 2 Complex. *ACS Nano* **2020**, *14*, 11821–11830. [CrossRef]

17. Bai, C.; Wang, J.; Chen, G.; Zhang, H.; An, K.; Xu, P.; Du, Y.; Ye, R.D.; Saha, A.; Zhang, A.; et al. Predicting Mutational Effects on Receptor Binding of the Spike Protein of SARS-CoV-2 Variants. *J. Am. Chem. Soc.* **2021**, *143*, 17646–17654. [CrossRef]
18. Verkhivker, G.; Agajanian, S.; Kassab, R.; Krishnan, K. Computer Simulations and Network-Based Profiling of Binding and Allosteric Interactions of SARS-CoV-2 Spike Variant Complexes and the Host Receptor: Dissecting the Mechanistic Effects of the Delta and Omicron Mutations. *Int. J. Mol. Sci.* **2022**, *23*, 4376. [CrossRef] [PubMed]
19. Pitsillou, E.; Liang, J.J.; Beh, R.C.; Hung, A.; Karagiannis, T.C. Molecular dynamics simulations highlight the altered binding landscape at the spike-ACE2 interface between the Delta and Omicron variants compared to the SARS-CoV-2 original strain. *Comput. Biol. Med.* **2022**, *149*, 106035. [CrossRef]
20. Khan, A.; Khan, S.A.; Zia, K.; Altowyan, M.S.; Barakat, A.; Ul-Haq, Z. Deciphering the Impact of Mutations on the Binding Efficacy of SARS-CoV-2 Omicron and Delta Variants With Human ACE2 Receptor. *Front. Chem.* **2022**, *10*, 892093. [CrossRef] [PubMed]
21. An, K.; Yang, X.; Luo, M.; Yan, J.; Xu, P.; Zhang, H.; Li, Y.; Wu, S.; Warshel, A.; Bai, C. Mechanistic study of the transmission pattern of the SARS-CoV-2 omicron variant. *Proteins* **2024**, 1–15. [CrossRef]
22. Kumar, R.; Murugan, N.A.; Srivastava, V. Improved Binding Affinity of Omicron's Spike Protein for the Human Angiotensin-Converting Enzyme 2 Receptor Is the Key behind Its Increased Virulence. *Int. J. Mol. Sci.* **2022**, *23*, 3409. [CrossRef]
23. Verkhivker, G.M.; Di Paola, L. Integrated Biophysical Modeling of the SARS-CoV-2 Spike Protein Binding and Allosteric Interactions with Antibodies. *J. Phys. Chem. B* **2021**, *125*, 4596–4619. [CrossRef]
24. da Costa, C.H.; de Freitas, C.A.; Alves, C.N.; Lameira, J. Assessment of mutations on RBD in the Spike protein of SARS-CoV-2 Alpha, Delta and Omicron variants. *Sci. Rep.* **2022**, *12*, 8540. [CrossRef] [PubMed]
25. Jawad, B.; Adhikari, P.; Podgornik, R.; Ching, W.Y. Key Interacting Residues between RBD of SARS-CoV2 and ACE2 Receptor: Combination of Molecular Dynamics Simulation and Density Functional Calculation. *J. Chem. Inf. Model.* **2021**, *61*, 4425–4441. [CrossRef] [PubMed]
26. Jawad, B.; Adhikari, P.; Podgornik, R.; Ching, W.Y. Binding Interactions between Receptor-Binding Domain of Spike Protein and Human Angiotensin Converting Enzyme-2 in Omicron Variant. *J. Phys. Chem. Lett.* **2022**, *13*, 3915–3921. [CrossRef] [PubMed]
27. Gumbart, J.C.; Roux, B.; Chipot, C. Efficient determination of protein-protein standard binding free energies from first principles. *J. Chem. Theory Comput.* **2013**, *9*, 3789–3798. [CrossRef] [PubMed]
28. Sun, Z.; Yan, Y.N.; Yang, M.; Zhang, J.Z. Interaction entropy for protein-protein binding. *J. Chem. Phys.* **2017**, *146*, 124124. [CrossRef]
29. Lan, J.; Ge, J.; Yu, J.; Shan, S.; Zhou, H.; Fan, S.; Zhang, Q.; Shi, X.; Wang, Q.; Zhang, L.; et al. Structure of the SARS-CoV-2 spike receptor-binding domain bound to the ACE2 receptor. *Nature* **2020**, *581*, 215–220. [CrossRef] [PubMed]
30. Saville, J.W.; Mannar, D.; Zhu, X.; Srivastava, S.S.; Berezuk, A.M.; Demers, J.P.; Zhou, S.; Tuttle, K.S.; Sekirov, I.; Kim, A.; et al. Structural and biochemical rationale for enhanced spike protein fitness in delta and kappa SARS-CoV-2 variants. *Nat. Commun.* **2022**, *13*, 742. [CrossRef] [PubMed]
31. Geng, Q.; Shi, K.; Ye, G.; Zhang, W.; Aihara, H.; Li, F. Structural Basis for Human Receptor Recognition by SARS-CoV-2 Omicron. Variant BA.1. *J. Virol.* **2022**, *96*, e00249-22. [CrossRef]
32. Yin, W.; Xu, Y.; Xu, P.; Cao, X.; Wu, C.; Gu, C.; He, X.; Wang, X.; Huang, S.; Yuan, Q.; et al. Structures of the Omicron spike trimer with ACE2 and an anti-Omicron antibody. *Science* **2022**, *375*, 1048–1053. [CrossRef]
33. Kastritis, P.L.; Bonvin, A.M.J.J. On the binding affinity of macromolecular interactions: Daring to ask why proteins interact. *J. R. Soc. Interface* **2013**, *10*, 20120835. [CrossRef]
34. Krissinel, E.; Henrick, K. Inference of Macromolecular Assemblies from the Crystalline State. *J. Mol. Biol.* **2007**, *372*, 774–797. [CrossRef] [PubMed]
35. He, W.; Liu, X.; Hu, B.; Li, D.; Chen, L.; Li, Y.; Tu, Y.; Xiong, S.; Wang, G.; Deng, J.; et al. Mechanisms of SARS-CoV-2 Infection-Induced Kidney Injury: A Literature Review. *Front. Cell. Infect. Microbiol.* **2022**, *12*, 838213. [CrossRef] [PubMed]
36. Meng, B.; Abdullahi, A.; Ferreira, I.A.; Goonawardane, N.; Saito, A.; Kimura, I.; Yamasoba, D.; Gerber, P.P.; Fatihi, S.; Rathore, S.; et al. Altered TMPRSS2 usage by SARS-CoV-2 Omicron impacts infectivity and fusogenicity. *Nature* **2022**, *603*, 706–714. [CrossRef] [PubMed]
37. Padmanabhan, P.; Dixit, N.M. Modelling how increased Cathepsin B/L and decreased TMPRSS2 usage for cell entry by the SARS-CoV-2 Omicron variant may affect the efficacy and synergy of TMPRSS2 and Cathepsin B/L inhibitors. *J. Theo. Biol.* **2023**, *572*, 111568. [CrossRef]
38. Jo, S.; Kim, T.; Iyer, V.G.; Im, W. CHARMM-GUI: A web-based graphical user interface for CHARMM. *J. Comput. Chem.* **2008**, *29*, 1859–1865. [CrossRef] [PubMed]
39. Brooks, B.R.; Brooks, C.L., 3rd; Mackerell, A.D., Jr.; Nilsson, L.; Petrella, R.J.; Roux, B.; Won, Y.; Archontis, G.; Bartels, C.; Boresch, S.; et al. CHARMM: The biomolecular simulation program. *J. Comput. Chem.* **2009**, *30*, 1545–1614. [CrossRef] [PubMed]
40. Im, W.; Lee, M.S.; Brooks, C.L., III. Generalized Born Model with a Simple Smoothing Function. *J. Comput. Chem.* **2003**, *24*, 1691–1702. [CrossRef]
41. Pastor, R.W. Techniques and Applications of Langevin Dynamics Simulations. In *The Molecular Dynamics of Liquid Crystals*; Luckhurst, G.R., Veracini, C.A., Eds.; NATO ASI Series; Springer Netherlands: Dordrecht, The Netherlands, 1994; Volume 431, pp. 85–138.

42. BIOVIA Dassault Systèmes. *Discovery Studio Visualizer*, Version 3.5; Dassault Systèmes: San Diego, CA, USA, 2018.
43. Chen, Y.; Zhao, X.; Zhou, H.; Zhu, H.; Jiang, S.; Wang, P. Broadly neutralizing antibodies to SARS-CoV-2 and other human coronaviruses. *Nat. Rev. Immunol.* **2023**, *23*, 189–199. [CrossRef] [PubMed]

International Journal of
Molecular Sciences

MDPI

Article

The African Swine Fever Virus Virulence Determinant DP96R Suppresses Type I IFN Production Targeting IRF3

Niranjan Dodantenna [1], Ji-Won Cha [1], Kiramage Chathuranga [1], W. A. Gayan Chathuranga [1], Asela Weerawardhana [1], Lakmal Ranathunga [1], Yongkwan Kim [2], Weonhwa Jheong [2] and Jong-Soo Lee [1,*]

1 College of Veterinary Medicine, Chungnam National University, Daejeon 34134, Republic of Korea; niranjan3k@gmail.com (N.D.); siniloon@naver.com (J.-W.C.); chathurangakiramage@gmail.com (K.C.); gayachathu123@gmail.com (W.A.G.C.); aselasampath2009@gmail.com (A.W.); lakmalranathunga13@gmail.com (L.R.)

2 Wildlife Disease Response Team, National Institute of Wildlife Disease Control and Prevention, Gwangju 62407, Republic of Korea; kyk5388@korea.kr (Y.K.); purify@korea.kr (W.J.)

* Correspondence: jongsool@cnu.ac.kr

Abstract: DP96R of African swine fever virus (ASFV), also known as uridine kinase (*UK*), encodes a virulence-associated protein. Previous studies have examined *DP96R* along with other genes in an effort to create live attenuated vaccines. While experiments in pigs have explored the impact of DP96R on the pathogenicity of ASFV, the precise molecular mechanism underlying this phenomenon remains unknown. Here, we describe a novel molecular mechanism by which DP96R suppresses interferon regulator factor-3 (IRF3)-mediated antiviral immune responses. DP96R interacts with a crucial karyopherin (KPNA) binding site within IRF3, disrupting the KPNA-IRF3 interaction and consequently impeding the translocation of IRF3 to the nucleus. Under this mechanistic basis, the ectopic expression of DP96R enhances the replication of DNA and RNA viruses by inhibiting the production of IFNs, whereas DP96R knock-down resulted in higher IFNs and IFN-stimulated gene (ISG) transcription during ASFV infection. Collectively, these findings underscore the pivotal role of DP96R in inhibiting IFN responses and increase our understanding of the relationship between DP96R and the virulence of ASFV.

Keywords: African swine fever virus; DP96R (*UK* gene); IRF3; KPNA

1. Introduction

African swine fever, a highly contagious and fatal viral disease caused by the African swine fever virus (ASFV), is extremely harmful to the swine industry. This viral disease, linked to significant fatality rates in domestic pigs, poses a risk to worldwide pork production and food safety [1,2]. ASFV, the only member of the family *Asfarviridae*, is an icosahedral, enveloped, double-stranded DNA arbovirus (genus *Asfivirus*) with a genome of 170–193 kb [3–5]. The virus replicates within the cytoplasm of mononuclear phagocytic cells, and its genome encodes 150–167 proteins that are responsible for viral replication, interaction with host molecules, and regulation of the host innate immune response [6,7]. However, many of these viral proteins have unresolved functions [8,9]; this lack of knowledge impedes our understanding of disease pathophysiology and immune evasion, making vaccine development a challenge. Despite significant research to date, there is no safe and efficient commercial vaccine against ASFV.

Type I IFNs and pro-inflammatory cytokines play a vital role in a host's defense against invading viruses [10]. The sensor cGAS [cyclic GMP-AMP (2′3′cGAMP) synthase] is primarily responsible for recognizing cytosolic viral DNA [11]. Upon activation, cGAS undergoes a reconfiguration of its catalytic pocket before binding its substrates, adenosine triphosphate (ATP) and guanosine triphosphate (GTP), which are then used to generate the mammalian second messenger 2′3′cGAMP [12,13]. The synthesized 2′3′cGAMP molecule attaches itself to the endoplasmic reticulum (ER) membrane adaptor STING, thereby triggering the structural modifications that are required for its activation [14]. Once activated, STING relocates to the ER–Golgi intermediate compartment, where it recruits and activates TANK-binding kinase 1 (TBK1), which in turn triggers the phosphorylation of interferon regulatory factor 3 (IRF3). Finally, cytoplasmic KPNA molecules recruit phosphorylated and dimerized IRF3 for nuclear translocation; translocated IRF3 then induces the transcription of type I IFNs [15,16]. As a result, to facilitate effective replication within the host, viruses have developed various antagonistic tactics to avoid host type I IFN and inflammatory responses [17–20]. Previous studies suggest that the virulent ASFV isolate Armenia/07 inhibits generation of IFN-β via the cGAS-STING pathway [21], and that diverse ASFV proteins play various roles in blocking type I IFNs and pro-inflammatory cytokines to facilitate innate immune evasion, which is necessary for the successful replication of ASFV inside macrophages [6,22–34].

ASFV DP96R, the *UK* gene, encodes a 9.6 kDa protein with no significant homology to other known genes in databases. In the late 1990s, a study demonstrated that deleting this highly conserved DP96R from a pathogenic ASFV strain E70 reduced its virulence and protected immunized pigs from challenge with homologous ASFV [35]. Thus, DP96R has been utilized, in combination with other genes, to develop live attenuated vaccines [36–38]. While the impact of DP96R on ASFV pathogenicity has been explored in vivo, the underlying molecular mechanism is unknown. Here, we describe a novel mechanism by which DP96R suppresses host type I IFNs and pro-inflammatory reactions by interacting directly with IRF3. Thus, DP96R plays a crucial role in ASFV virulence and could serve as a promising target for the development of live attenuated vaccines.

2. Results

2.1. DP96R Targets IRF3

Immune evasion is a critical virulence factor. ASFV uses a wide range of immune evasion strategies to inhibit the generation of IFNs and pro-inflammatory cytokines to enable efficient reproduction within the host. To examine whether ASFV proteins play a role in modulating type I IFN signaling or promoting virus replication, we subjected 158 ASFV viral proteins to individual testing using the Adenoviral replication system. PK-15 cells were transiently transfected with plasmids encoding discrete ASFV genes, or with a control vector (as a negative control). Twenty-four hours post-transfection (hpt), cells were exposed for two hours to a green fluorescent protein (GFP)-expressing adenovirus (ADV-GFP) at a multiplicity of infection (MOI) of 1. Cells infected with the virus were collected 24 h after infection (24 hpi), and the harvested cell pellets were collected to quantify the amount of replicated virus using a fluorescence modulator. As shown in Figure S1, we selected a number of ASFV proteins that increased viral replication. Among them, we identified DP96R. Previous studies have used DP96R in combination with other genes to develop live attenuated vaccines, and the effect of DP96R on the pathogenicity of ASFV has been studied in pigs [35–38]. However, the exact molecular mechanism underlying virulence remains unknown. Based on our screening results and on previous information, we selected the ASFV *DP96R* gene and used it to determine the precise molecular mechanism that is involved in immune evasion. To explore the crucial role and precise target of ASFV DP96R during type I IFN signaling, we conducted a luciferase promoter assay. This assay involved co-expression of the *DP96R* gene alongside various adapters that are associated with the cGAS-STING pathway. The findings revealed that DP96R inhibited poly(dA:dT)-, cGAS-, 2′3′cGAMP-, STING-, TBK1-, IKKε-, IRF3-, and IRF3-5D (a constitutively active form of IRF3)-mediated activation of the IFN-β promoter in a dose-dependent manner (Figure 1A). This suggests that ASFV DP96R suppresses the cGAS-STING pathway and might target IRF3 to inhibit its dimerization/nuclear translocation or nuclear translocation alone. To confirm the results of the luciferase promoter activity assay, we conducted a series of co-immunoprecipitation (IP) assays to examine the association between ASFV DP96R and IRF3 or several IRF3 deletion mutants (Figure 1B–D,G,H). First, using molecules from the cGAS-STING pathway, we performed an IP assay to determine whether DP96R interacts specifically with IRF3. As shown in Figure 1B, DP96R interacted specifically with IRF3. Furthermore, the DP96R-IRF3 interaction was confirmed by an overexpression IP assay in HEK293T cells (Figure S2A,B). Subsequently, we validated the endogenous interaction between DP96R and IRF3 in HEK293T cells (Figure 1C) and porcine alveolar macrophages (PAMs) (Figure 1D). A Flag-tagged IRF3 and Strep-tagged DP96R overexpression immunofluorescence colocalization assay in PK-15 cells with a significant correlation of 0.603 verified this interaction. It is worth highlighting that ASFV DP96R was localized primarily within the cytoplasm and was not translocated to the nucleus (Figure 1E).

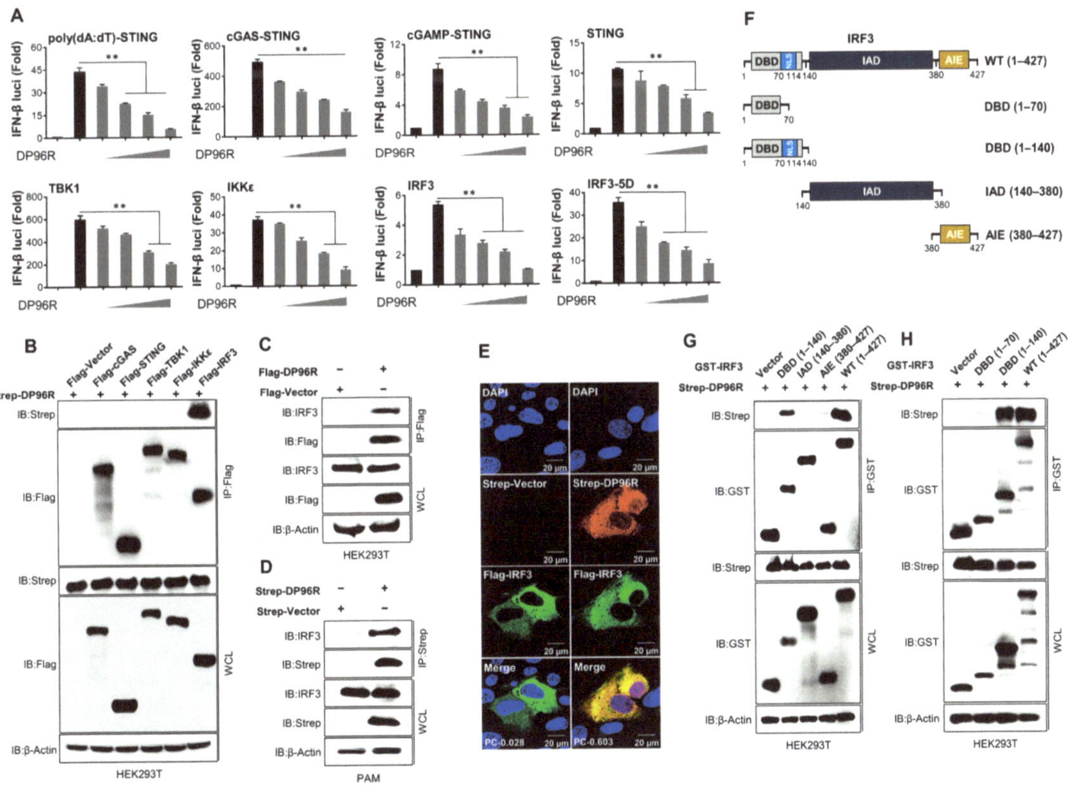

Figure 1. DP96R targets the IRF3 nuclear import domain. (**A**) IFN-β luciferase assay. HEK293T cells were used for STING-, TBK1-, IKKε-, IRF3-, and IRF3-5D-induced IFN-β luciferase assays. HEK293T cells were transfected with Flag-DP96R plasmid, as indicated with relevant stimulants, and firefly luciferase reporter plasmid encoding the IFN-β promoter plus TK-renilla plasmid as transfection control to normalize firefly luciferase activity. Expression plasmids of STING, TBK1, IKKε, IRF3, and IRF3-5D were used as stimulants of the cGAS-STING pathway. STING overexpressing 293-Dual™ hSTING-A162 cells were used for poly(dA:dT)-, cGAS-, and cGAMP-induced luciferase assays. 293-Dual™ hSTING-A162 cells were transfected with Flag-DP96R plasmid as indicated, along with 3×Flag-cGAS expression plasmid, or transfected with poly(dA:dT) or 2′3′cGAMP for 12 h. Thirty six hpt, the luciferase activity of each sample was measured. Results are expressed relative to those renilla luciferases alone. The first and second black bars represent negative and positive controls, respectively. Grey bars represent DP96R dose-dependent transfection. (**B**) HEK293T cells were co-transfected with Strep-DP96R with Flag-tagged cGAS, STING, TBK1, IKKε, IRF3, and Flag

control plasmids. Cell lysates were subjected to immunoprecipitation (IP) by Flag antibody, followed by immunoblotting with an anti-Strep and anti-Flag antibodies. (**C**) HEK293T and (**D**) PAMs were transfected with Flag-DP96R and Strep-DP96R and their control plasmids, respectively. Cell lysates were subjected to IP by Flag antibody and Strep beads, followed by immunoblotting with anti-IRF3, anti-Flag or anti-Strep antibodies. (**E**) Colocalization of DP96R and IRF3. PK-15 cells were transfected with Strep-tagged DP96R plasmid and its control plasmid with Flag-tagged IRF3 plasmid, followed by confocal microscopy assay with anti-Flag and anti-Strep primary and anti-mouse (red) and anti-rabbit (green) secondary antibodies. Nuclei were stained with DAPI (blue). (**F**) IRF3 domain constructs. GST-tagged IRF3 wild-type (WT) and amino acid 1–140 (DBD), 140–380 (IAD), 380–427 (AIE), and control vector (GST) were co-transfected to HEK293T cells together with Strep-tagged ASFV DP96R plasmid (**G**). GST-tagged IRF3 WT and amino acid 1–140 (DBD), amino acid 1–70 (DBD excluding NLS), and control vector (GST) were co-transfected to HEK293T cells together with Strep-tagged ASFV DP96R plasmid (**H**). Cell lysates were subjected to GST-PD and immunoblotted with anti-Strep antibodies after immunoblotting the WCL with anti-Strep and anti-GST antibodies. Luciferase data represent three independent experiments, each with similar results, and all the values are expressed as mean ± SD of two biological replicates. All the immunoblot and confocal data represent at least two independent experiments, each with similar results. The scale bar represents 20 μM. Student's *t*-test: **, $p < 0.01$.

Next, we constructed expression plasmids containing GST-tagged DBD (a DNA-binding domain spanning aa 1–140), IAD (an IRF association domain spanning aa 140–380), and AIE (an autoinhibition element spanning aa 380–427) domains to identify which of them interact with DP96R (Figure 1F). We found that DP96R precipitated with the DBD of IRF3, but not with the IAD domain (Figure 1G). Thus, we hypothesized that DP96R prevents IRF3 from translocating to the nucleus by inhibiting the IRF3 nuclear localization signal (NLS). The NLS region between IRF3 aa 70–114 comprises two clusters of basic amino acids (K77, R78, and R86, K87) that are embedded in the DBD [39]. Therefore, we removed aa 70–140 from the IRF3-DBD, which includes the NLS region (Figure 1F), and investigated the interaction with DP96R. Exclusion of the NLS impaired the DP96R-IRF3-DBD interaction, confirming that IRF3-NLS is required for the DP96R interaction (Figure 1H). These data strongly suggest that ASFV DP96R hinders type I IFN signaling by interacting with IRF3 and its NLS.

2.2. DP96R Impairs the Nuclear Translocation of IRF3

Subsequently, to ascertain how ASFV DP96R impacts the activation of IRF3, we examined the phosphorylation, dimerization, and nuclear translocation of IRF3. Different cells express IRF3 constitutively, where it resides within the cytoplasm in an inactive form. Phosphorylation of the C-terminal domain of IRF3 by virus-activated kinases triggers dimerization of IRF3 and its translocation to the nucleus. This event, in turn, leads to the activation of IFN genes [40–42]. First, we examined phosphorylation of IRF3 induced by the overexpression of TBK1 in response to escalating quantities of ASFV DP96R protein. Figure 2A shows that increasing doses of ASFV DP96R protein did not affect IRF3 phosphorylation. Furthermore, we examined the interaction between IRF3-5D and DP96R and compared it with the activity of wild-type (WT) IRF3. Concomitant with the results of the IRF3-5D-IFNβ-luciferase assay, and similar to the DP96R-IRF3-WT interaction, DP96R interacted with IRF3-5D (Figure 2B), confirming that DP96R does not interfere with the five amino acid residues of IRF3 that are necessary for its phosphorylation. Second, we performed an IRF3 dimerization assay using IRF3 plasmids tagged with Strep and GST and exposed them to increasing doses of Flag-tagged DP96R. Again, DP96R did not affect IRF3 dimerization (Figure 2C). Third, we performed an ADV-GFP-induced IRF3 nuclear translocation assay using DP96R and a control vector stably expressing PAMs. Cytoplasmic and nuclear fractions were obtained from harvested PAMs, and immunoblotting was carried out to identify specific proteins. The nuclear uptake of IRF3 by DP96R-expressing PAMs was diminished significantly (Figure 2D,E). At 16 hpi, nuclear translocation of IRF3 in DP96R-expressing PAMs was impaired by nearly ~80% compared with that in control PAMs. Even though DP96R does not target IRF3 phosphorylation, DP96R-expressing PAMs showed suppression of phosphorylated IRF3 in the cytoplasmic fraction only at 16 h, which might be due to the inhibition of the IFN signaling pathway by nearby bystander cells. Suppression of type I IFN reduces the transcription of antiviral genes that affect the activation of the cGAS-STING axis. To further confirm the suppressive impact of ASFV DP96R on type I IFN signaling, we carried out an immunofluorescence assay in PK-15 cells overexpressing DP96R or control plasmids after infection with Sendai virus (Sev) at specified time intervals. Upon viral infection, the quantity of nuclear-translocated IRF3 in control cells at 16 hpi increased to nearly ~80% of the total cellular level. By contrast, the amount of IRF3 within the nucleus of PK-15 cells expressing ASFV DP96R was nearly ~30% of the total cellular level (Figure 2F,G). At each time point, DP96R colocalized with endogenous IRF3, and the interaction was strongest at 16 hpi. These findings indicate that ASFV DP96R disrupts the nuclear translocation of IRF3, leading to a reduction in type I IFN production.

2.3. DP96R Disrupts Type I IFN Signaling and Subsequent Transcription of Antiviral Genes

To conduct a more thorough assessment of the effects of DP96R on the virus-induced type I IFN signaling pathway, we examined the phosphorylation status of TBK1, IRF3, IKKα/β, IκBα, P65, and STAT1 in PAMs stably expressing DP96R, as well as in PK-15 cells that were transfected with DP96R. ADV-GFP-infected cells were harvested at the specified time points. Specific antibodies were used to assess the phosphorylation status of signaling molecules involved in the cGAS-STING signaling axis, the NF-κB signaling axis, and the type I IFN pathway that is associated with virus-related processes. The levels of TBK1, IRF3, IKKα/β, IκBα, P65, and STAT1 phosphorylation were notably reduced in PAMs expressing DP96R (Figure 3A) and in PK-15 cells (Figure 3B) compared with control cells. Subsequently, to clarify the impact on IFN signaling and the IFN-antagonistic function of ASFV DP96R, we examined the transcription of genes related to IFN (*IFN-β* and *IFN-γ*), a pro-inflammatory cytokine (*IL-6*), and IFN-stimulated genes (ISGs) *IFIT1*, *ISG15*, *OASL*, *MX-1*, and *PKR*. For this experiment, PAMs stably overexpressing DP96R and PK-15 cells transiently overexpressing DP96R, both of which had been transiently transfected, were subsequently infected with ADV-GFP, followed by quantitative real-time PCR (qRT-PCR) using specific primers for the genes of interest (Table S1). Compared with control cells, DP96R-overexpressing PAMs (Figure 3C) and PK-15 cells (Figure 3D) showed a marked reduction in their expression of mRNA encoding *IFN-β*, *IFN-γ*, *IL-6*, and other antiviral genes.

These findings indicate that the interaction of DP96R with IRF3 has a detrimental impact on host type I IFN signaling pathways, thereby suppressing the transcription of antiviral genes.

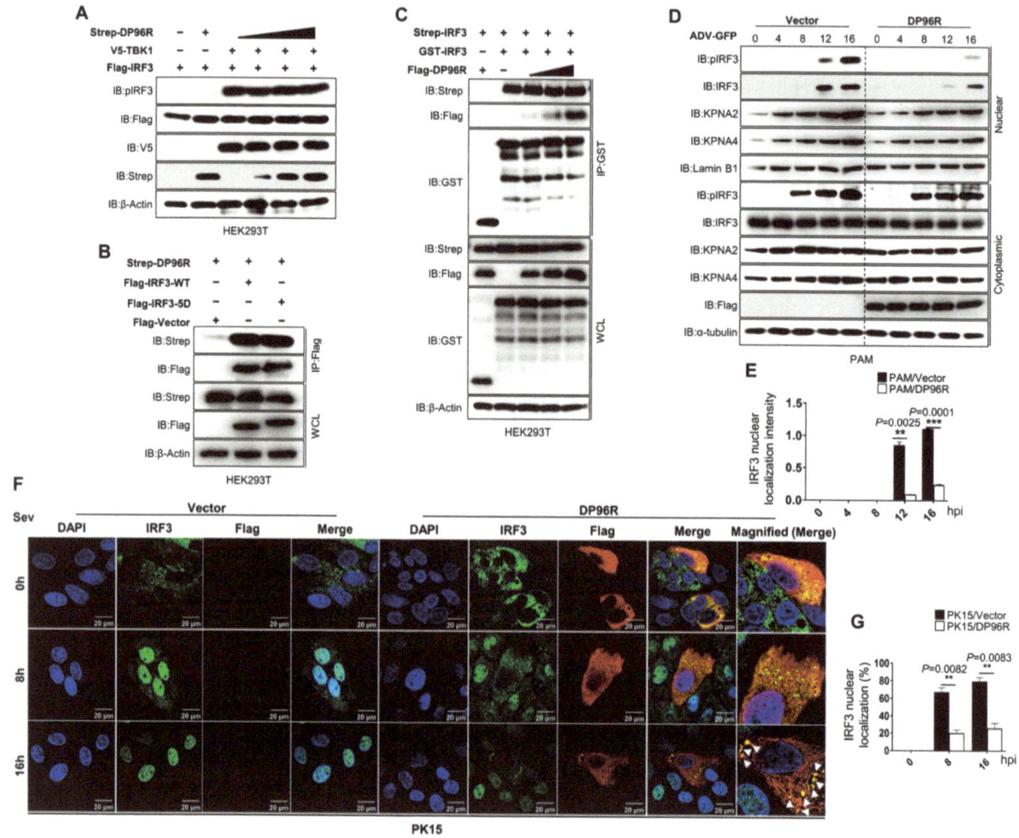

Figure 2. DP96R inhibits the nuclear localization of IRF3. (**A**) IRF3 phosphorylation inhibition assay. HEK293T cells were transfected with Flag-IRF3, and V5-TBK1 with Strep-DP96R dose-dependently. Cell lysates were immunoblotted with anti-pIRF3, -Flag, -V5, and -Strep antibodies. (**B**) HEK293T cells were transfected with Flag-IRF3-WT and Flag-IRF3-5D, and its control vector with Strep-DP96R. Cell lysates were subjected to Flag IP and immunoblotted with anti-Flag and anti-Strep antibodies. (**C**) IRF3 dimerization inhibition assay. HEK293T cells were transfected with Strep and GST-tagged IRF3 with Flag-DP96R dose-dependently. Cell lysates were subjected to GST PD and immunoblotted with anti-Flag, -GST, and -Strep antibodies. (**D**) Cellular fractionation assay. PAMs expressing Flag-DP96R, and its control plasmid were infected with ADV-GFP (1MOI) and harvested at indicated time points. Cytoplasmic and nuclear extracts were then subjected to immunoblot with anti-pIRF3, -IRF3, KPNA2, and KPNA4 antibodies. Lamin B1 and α-tubulin were used to confirm equal loading of proteins of nuclear and cytoplasmic fractions, respectively. (**E**) The histogram represents the relative quantification of the protein levels of the Western blot. Nuclear localization intensity of IRF3 quantified by IRF3 band intensity in nuclear fraction adjusted to nuclear Lamin B1 fraction. (**F**) PK-15 cells were transfected with the Flag control or Flag DP96R plasmid, followed by the Sev (1MOI) infection. Cells were fixed at indicated time points, followed by confocal microscopy assay with anti-IRF3, anti-Flag primary and anti-rabbit (green) and ant-mouse (red) secondary anti-bodies with DAPI (blue) to stain the nuclei. (**G**) By dividing the number of cells expressing nuclear expression of IRF3 by the total number of IRF3-positive cells, the percentages of cells demonstrating nuclear translocation of IRF3 were computed. All the immunoblot and confocal data represent at least two independent experiments, each with similar results. The scale bar represents 20 μM. Student's t-test: **, $p < 0.01$; ***, $p < 0.001$.

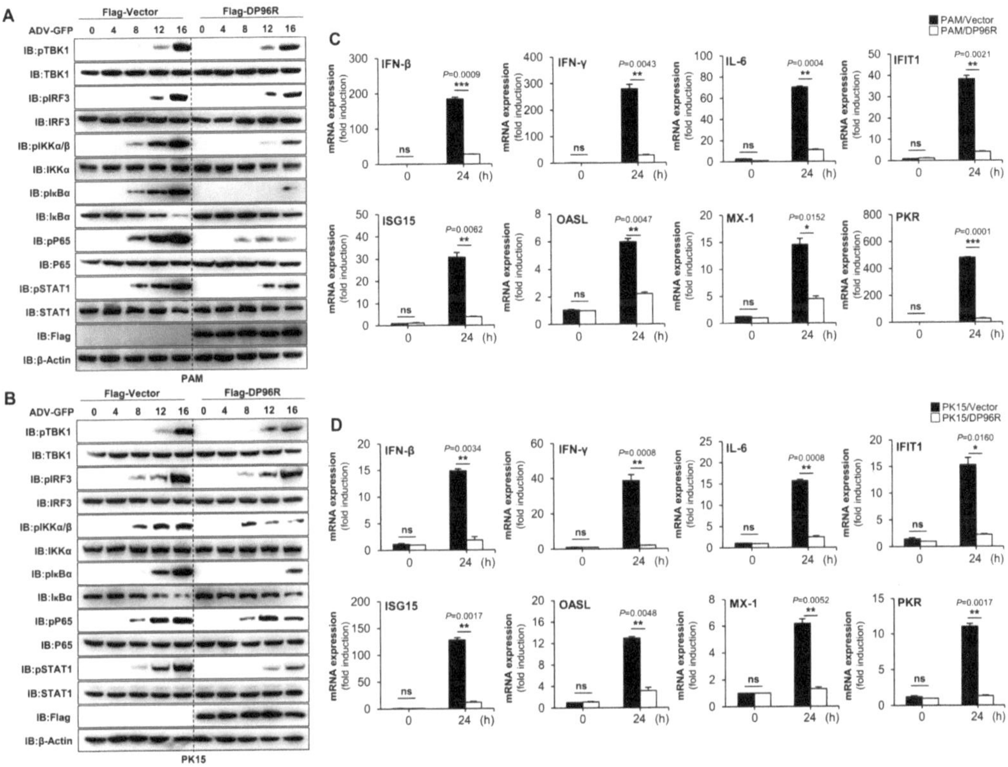

Figure 3. DP96R impairs cGAS-STING signaling and transcription of antiviral genes. Flag-tagged DP96R stably expressing PAMs (**A**) and transfected PK-15 cells (**B**) were infected with ADV-GFP (1MOI), and cells were harvested at indicated time points. DP96R protein expression level and total and phosphorylated TBK1, IRF3, IKKα (pIKKα/β), IκBα, P65, and STAT1 were measured by immunoblotting. β-actin was used as a loading control indicator. PAMs (**C**) and PK-15 cells (**D**), expressing Flag-DP96R with Flag control, were mock-infected and infected with Adenovirus GFP (1MOI), and following 0 hpi and 24 hpi, total RNA was extracted at indicated time points. Quantitative RT-PCR analyzed the mRNA transcripts of the indicated genes relative to internal control of porcine GAPDH. All qRT-PCR data represent at least two independent experiments, each with similar results, and the values are expressed as the mean ± SD of two biological replicates. All the immunoblot data represent at least two independent experiments, each with similar results. Student's *t*-test: *, $p < 0.05$; **, $p < 0.01$; ***, $p < 0.001$; ns, not significant.

2.4. DP96R Negatively Regulates Innate Immune Responses against Viral Infection

So far, the data imply that ASFV DP96R inhibits type I IFN signaling and transcription of antiviral genes by targeting the nuclear translocation of IRF3. To investigate the role of DP96R during DNA virus replication in vitro, we infected PAMs stably expressing Flag-tagged DP96R or control vector, porcine immortalized bone marrow-derived macrophages (PIBs), and monkey kidney epithelial cells (MA-104) with three GFP proteins expressing DNA viruses. Additionally, we transfected PK-15 cells with a Flag-tagged DP96R or control vector, followed by infection with a DNA virus. Cells were infected with three IFN-sensitive surrogate DNA viruses expressing GFP protein (ADV-GFP, herpes simplex virus (HSV-GFP), and vaccinia virus (VACV-GFP)) instead of ASFV [43–45]. The efficacy of transient transfection with the plasmids and the expression of DP96R protein in each cell line was validated by immunoblotting (Figure S3A–D). Remarkably, virus replication was more robust in PAMs, porcine epithelial cells, and the monkey cell line overexpressing DP96R than in control cells (Figure 4A–D, Figures S4A–D and S5A–D). Next, we conducted ELISAs to determine the amounts of IFN-β and IL-6 that were secreted by each virus-infected cell line. In alignment with the outcome of the viral replication test, we observed that DP96R-overexpressing cells secreted lower levels of cytokines than control cells (Figure 4E–H, Figures S4E–H and S5E–H). In line with the earlier findings, both the virus replication and ELISA assays confirmed that DP96R inhibited production of type I IFNs and pro-inflammatory cytokines. As a result, the replication of viral DNA in porcine macrophages, epithelial cells, and monkey epithelial cells is increased.

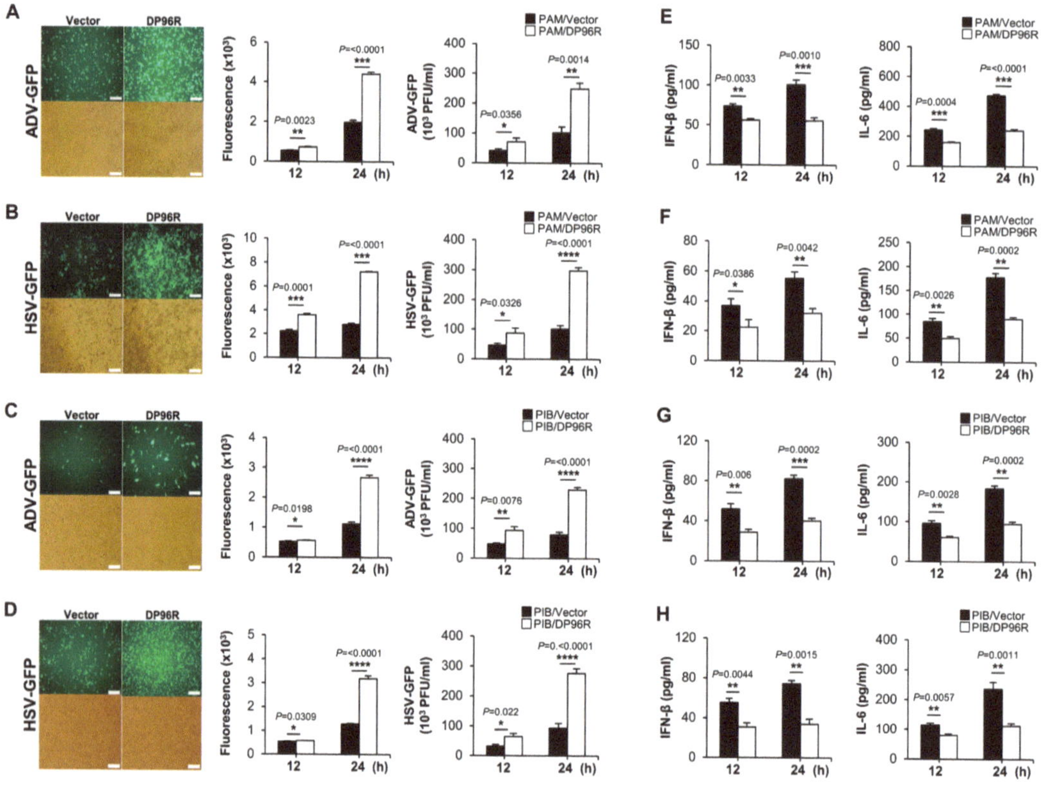

Figure 4. DP96R impairs virus-induced innate immune responses. Stably expressing Flag-DP96R protein in PAMs and PIBs with Flag-control cells were infected with ADV-GFP (1MOI) (**A**,**C**) and HSV-GFP (1MOI) (**B**,**D**). Viral replication was determined at 24 hpi by GFP expression levels by fluorescence microscopy and quantified at 12 hpi and 24 hpi by a fluorescence modulator. The virus titers of each sample were determined by plaque assay in A549 cells and Vero cells. Porcine IFN-β and IL-6 secretion in cell culture supernatant at 12 hpi and 24 hpi were determined by ELISA (**E**–**H**). Data represent at least two independent experiments, each with similar results, and the values are expressed as mean ± SD of three biological replicates. The scale bar represents 50 μM. Student's t-test: *, $p < 0.05$; **, $p < 0.01$; ***, $p < 0.001$; ****, $p < 0.0001$.

IRF3 plays a crucial role in multiple innate immune signaling pathways, including RIG-I-like Receptor (RLR) signaling, cGAS/STING signaling, and TLR3 signaling [46]. Viral-derived RNAs trigger RLR signaling, resulting in the activation of IRF3 [47,48]. Thus, we used two RNA viruses, the GFP-expressing Newcastle disease virus (NDV-GFP) and the PR8 strain of H_1N_1 virus (PR8-GFP), to infect PAMs and MA104 cells stably expressing

DP96R protein. As expected, both cell lines expressing ASFV DP96R protein showed a significant increase in GFP expression and virus replication relative to control cells (Figure S6A–D). Subsequently, we quantified the amounts of IFN-β and IL-6 that was secreted by each virus-infected cell line. Consistent with the virus replication results, the ELISAs showed that cells stably expressing DP96R secreted lower amounts of IFNs and cytokines than control cells (Figure S6E–H). These findings reinforce the idea that DP96R promotes viral replication by interacting with IRF3, thereby exerting a negative regulatory effect on the production of type I IFN.

2.5. DP96R Inhibits Interaction between IRF3 and KPNA

To explore the mechanism by which DP96R hinders the nuclear translocation of IRF3, we examined the interaction between karyopherin α (KPNA) and IRF3 in the presence of DP96R. Proteins that have an NLS are shuttled from the cellular cytoplasm to the cellular nucleoplasm by KPNA, which recognizes the NLS of the cargo protein and recruits the KPNB subunit for nuclear localization [49,50]. Many virus proteins block the cargo-NLS-KPNA interaction to inhibit the movement of transcription factors into the cell nucleus, thereby impairing secretion of type I IFNs and pro-inflammatory cytokines [29,51,52]. Previous studies demonstrated that KPNA3 and KPNA4 interact with IRF3 [53–55], whereas a recent study revealed that of the six KPNAs, only KPNA2 interacts with IRF3 [56]. Therefore, we next examined the interaction between IRF3 and KPNA1-KPNA6. Interestingly, we found that KPNA1, KPNA2, KPNA3, and KPNA4 interacted strongly with IRF3 (Figure 5A). Next, we checked the interaction between KPNA and the DBD of IRF3, including the NLS region. As shown in Figure 5B, KPNA1, KPNA2, KPNA3, and KPNA4 interacted strongly with the IRF3-DBD. Accordingly, we hypothesized that, despite the slight differences in sequence homology, KPNA1, KPNA2, KPNA3, and KPNA4 interact with a similar motif within the IRF3-NLS region. Next, we performed a competition binding assay in which we overexpressed KPNA1–KPNA4, IRF3, and DP96R in HEK293T cells in a dose-dependent manner. Then, we examined the IRF3-DP96R interaction and the KPNA-IRF3 interaction. An increase in the interaction between DP96R and IRF3 coincided with a reduction in the interaction between KPNA and IRF3 (Figure 5C–F). These results suggest that the DP96R and IRF3 interaction interferes with the KPNA-IRF3 interaction in a dose-dependent manner. The inhibition of the IRF3-KPNA interaction in this overexpression system was verified under virus-induced conditions by performing a competition assay in PAMs stably expressing DP96R. In the first step, we determined the optimal time point of the association between cellular IRF3 and KPNA2 and KPNA4 in PAMs. PAMs were exposed to ADV-GFP (MOI = 1) and then collected at specified intervals. Cellular IRF3 was pulled down and immunoblotted with anti-KPNA2 and anti-KPNA4 antibodies. Figure 5G shows that the ADV-GFP infection-induced interaction between KPNA2 and KPNA4 with IRF3 was strong at 16 hpi, at which point the phosphorylation of IRF3 was augmented (later time points were not tested). Next, we used an endogenous competition binding assay using PAMs that were transiently transfected with DP96R or empty vector, followed by infection with ADV-GFP. As the interaction between DP96R and IRF3 increased, the interaction between IRF3 and KPNA2 or KPNA4 decreased (Figure 5H).

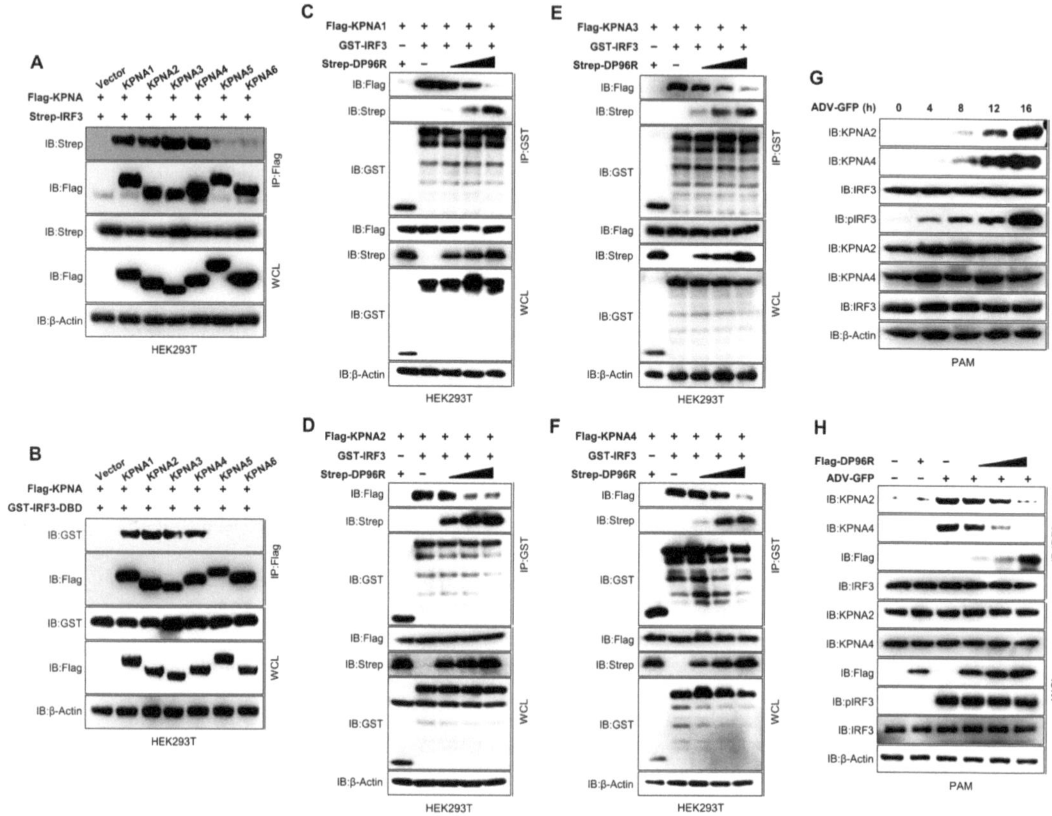

Figure 5. DP96R impairs IRF3-KPNA interaction. (**A**) HEK293T cells were transfected with Flag-KPNA1-KPNA6 and its control plasmid along with the Strep IRF3 plasmid. Cell lysates were subjected to Flag IP and immunoblotted with anti-Strep and anti-Flag antibodies. (**B**) HEK293T cells were transfected with Flag-KPNA1-KPNA6, its control plasmid, and GST-IRF3-DBD mutant plasmid. Cell lysates were subjected to Flag IP and immunoblotted with anti-Flag and anti-GST antibodies. (**C**–**F**) KPNA-DP96R competition assay. HEK293T cells were transfected with (**C**) Flag-KPNA1, (**D**) KPNA2, (**E**) KPNA3, (**F**) KPNA4, GST IRF3, its control, and Strep DP96R plasmids in a dose-dependent manner. Cell lysates were subjected to GST PD and immunoblotted with anti-Flag, -Strep, and -GST antibodies. (**G**) PAMs were infected with ADV-GFP (1MOI), and cells were harvested at indicated time points. Cell lysates were subjected to IRF3 IP and immunoblotted with anti-KPNA2, -KPNA4, -pIRF3, and -IRF3 antibodies. (**H**) PAMs were transfected with Flag-DP96R dose-dependently and infected with ADV-GFP (1MOI), and cells were harvested at indicated time points. Cell lysates were subjected to IP with IRF3 antibodies and immunoblotted with anti-KPNA2, -KPNA4, -pIRF3, -IRF3, and -Flag antibodies. Data represents at least two independent experiments, each with similar results.

2.6. DP96R Interacts with the Major KPNA-Binding Site within IRF3

Previously, Shun Li et al. reported that dephosphorylation of serine 97 (S97) within IRF3 is critical for the nuclear translocation of IRF3 [57]. Upon virus infection, nuclear translocation of IRF3 was inhibited by the S97D substitution in IRF3 (which imitates the phosphorylated condition) in IRF3- and IRF7-knockout cells. Zeng Cai et al. also showed that the S97D substitution disrupts the KPNA2-IRF3 interaction, whereas IRF3 S97A (which mimics the dephosphorylated state) triggers the interaction. These findings imply that when IRF3 is dephosphorylated at S97, it strengthens the interaction with KPNA2, thereby facilitating the nuclear translocation of IRF3 [56]. To determine how DP96R interferes with the IRF3-KPNA interaction, we performed a series of binding assays using several mutant IRF3 plasmids. First, we asked whether DP96R prevents the dephosphorylation of IRF3 at position S97 to inhibit nuclear translocation. To this end, we confirmed that the S97D and S97A substitutions in IRF3 inhibit and enhance the interaction with KPNA4, respectively (Figure S7A,B). Consequently, we found that both S97A and S97D IRF3 mutant proteins interacted with DP96R at a similar level to WT IRF3, suggesting that DP96R does not interfere with the S97 of IRF3 (Figure 6A,B). Next, we conducted a competitive binding assay using IRF3 S97A and KPNA4 and added DP96R incrementally. Figure 6C shows that the interaction between IRF3 S97A and KPNA4 was inhibited by DP96R, suggesting that DP96R disrupts the KPNA-IRF3 interaction without interfering with the dephosphorylation of IRF3 S97.

To examine the IRF3-KPNA and IRF3-DP96R interactions at the amino acid level, we performed several immunoprecipitation assays using two mutant IRF3 plasmids. Previous studies suggested that the basic amino acid clusters K77/R78 and R86/K87 within IRF3, which occupy the minor and major binding sites within IRF3, respectively, are important for the nuclear translocation of IRF3 and may interact with KPNA [39,53]. The NLS region of IRF3, which contains these amino acid clusters, is highly conserved in most mammalian species (Figure S7C). Armed with this information, we performed a binding assay using K77N/R78G and R86L/K87Q double-mutant IRF3 plasmids, along with Flag-KPNA4 and Strep-DP96R. First, we checked the IRF3-KPNA4 and IRF3-DP96R interactions using the K77N/R78G IRF3 double-mutant plasmid. The mutant IRF3 protein still interacted with KPNA4 and DP96R (Figure 6D,E), indicating that DP96R does not interact with the KPNA4 minor binding site within IRF3. Second, we performed a similar interaction analysis using the R86L/K87Q double-mutant IRF3 plasmid. By contrast, we found that the mutant IRF3 protein interacted neither with KPNA4 (Figure 6F) nor with DP96R (Figure 6G), suggesting that DP96R targets the KPNA major interacting site within IRF3 exclusively to inhibit the IRF3-KPNA interaction.

IRF7, which belongs to the IRF family of proteins, is activated by the cytosolic DNA-sensing pathway to generate IFNs. To test whether the interaction between DP96R and IRF3 is specific, we investigated the interaction between DP96R and IRF7. As shown in Figure S7D, IRF7 did not precipitate DP96R, suggesting that the interaction between endogenous IRF3 and DP96R is unique. Finally, we compared the NLS sequences of IRF3 and IRF7 and discovered that the KPNA-binding amino acids in IRF3 differ from those in IRF7 (Figure S7E). These findings strongly suggest that DP96R interacts with IRF3 via the IRF3 major binding site of KPNA, which then inhibits the IRF3-KPNA interaction and impairs the nuclear translocation of IRF3.

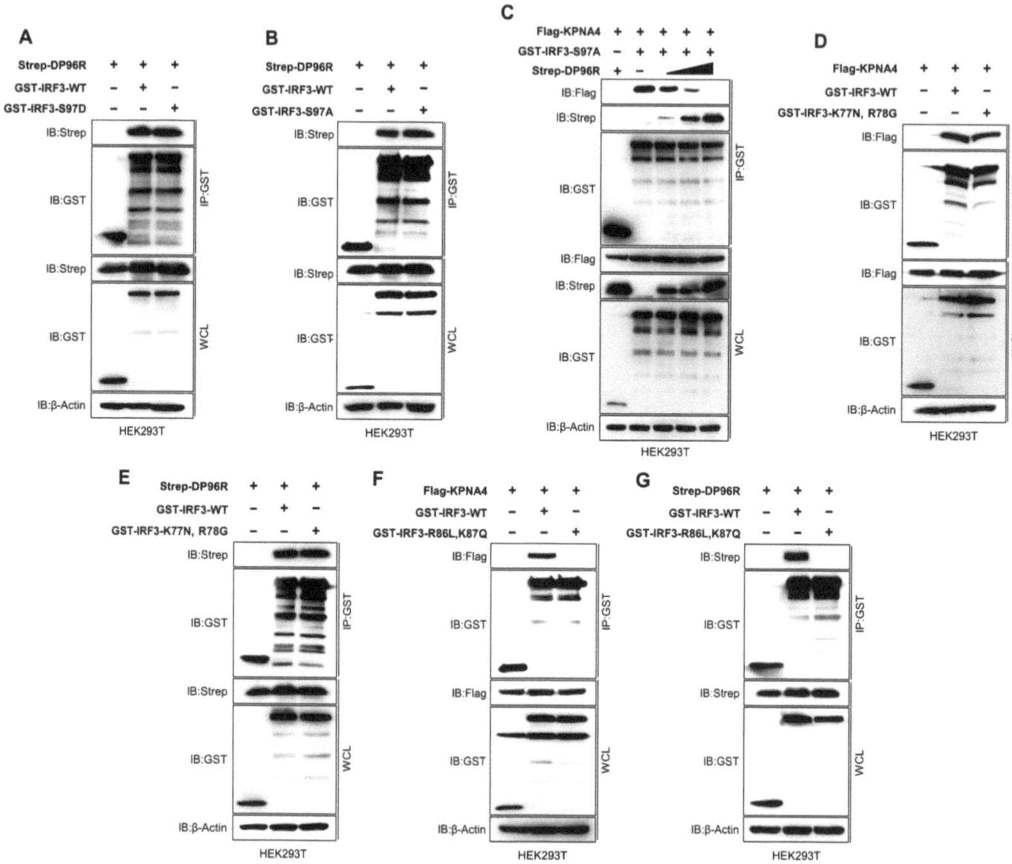

Figure 6. DP96R interacts with the KPNA major binding site of IRF3. HEK293T cells were transfected with GST-IRF3-WT, GST-IRF3 (S97D) (**A**) or GST-IRF3 (97A) (**B**), and Strep-DP96R plasmid. Cell lysates were subjected to GST PD and immunoblotted with anti-Strep, and -GST antibodies. (**C**) KPNA4-DP96R competition assay. HEK293T cells were transfected with Flag-KPNA4, GST-IRF3 (S97A), its control, and Strep DP96R plasmid in a dose-dependent manner. Cell lysates were subjected to GST PD and immunoblotted with anti-Flag, -Strep, and -GST antibodies. (**D**,**E**) HEK293T cells were transfected with GST-IRF3-WT, GST-IRF3 (K77N, R78G), its control plasmid, and Flag-KPNA4 (**D**) or Strep-DP96R (**E**) plasmid. Cell lysates were subjected to GST PD and immunoblotted with anti-Flag, -Strep and -GST antibodies. (**F**,**G**) HEK293T cells were transfected with GST-IRF3-WT, GST-IRF3 (R86L, K87Q) and its control plasmid along with Flag-KPNA4 (**F**) or Strep-DP96R (**G**) plasmid. Cell lysates were subjected to GST PD and immunoblotted with anti-Flag, -Strep, and -GST antibodies. Data represents at least two independent experiments, each with similar results.

2.7. DP96R Is an Early Transcribed Protein Involved in the Antagonism of IFN and ISG Transcription

To assess the transcription kinetics of DP96R, total RNA was isolated from primary porcine alveolar macrophages (primary PAMs) infected with an ASFV (Korea/wild boar/Hwacheon/2020-2287) strain at a MOI of 0.5. The mRNA levels of the viral gene, relative to the cellular beta-actin (β-actin), were determined as outlined in previous protocols [58]. As depicted in Figure 7A, the transcription pattern of DP96R closely resembled that of CP302L, an early-transcribed gene, but differed from that of B646L, a late-transcribed gene. This observation supports the conclusion that DP96R is an early-transcribed gene of

ASFV. To investigate the immunomodulatory functions of DP96R during ASFV infection, we knocked down DP96R mRNA that was induced by ASFV infection using DP96R-specific siRNA (siDP96R) in primary PAMs. Subsequently, these cells were infected with ASFV at a MOI of 0.5 for 12 h. DP96R mRNA was knocked down with almost 51% efficiency at 12 hpi (Figure 7B). In accordance with that, upon DP96R knock-down in ASFV-infected primary PAMs, the transcriptions of IFNs and ISGs were upregulated (Figure 7C). These data demonstrate that DP96R exerts inhibitory effects on IFN transcription, leading to reduced ISGs and thus inhibiting viral replication.

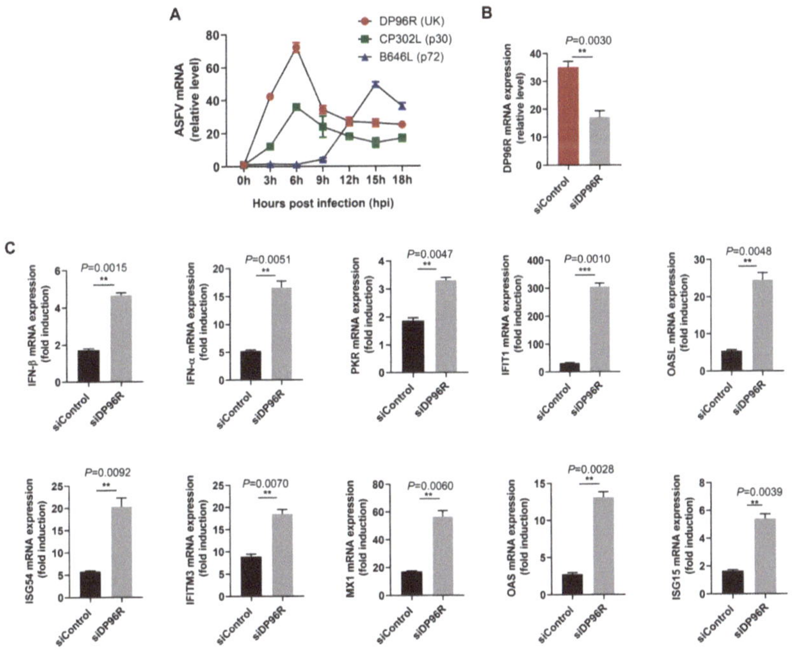

Figure 7. DP96R protein is transcribed early, inhibiting the transcription of IFNs and ISGs. Primary PAMs were infected with ASFV at 0.5 MOI and harvested at indicated time points. DP96R, CP302L, and B646L transcription expression levels during ASFV infection were determined by qRT-PCR. The p30 and p72 are shown as indicators of early and late genes, respectively (**A**). Primary PAMs in 24-well plates were transfected with the siRNAs against the DP96R or control siRNA for 6 h and then infected with the ASFV at an MOI of 0.5. The transcription levels of the indicated genes in the ASFV-infected primary PAMs were examined at 12 hpi (**B**,**C**). Data represents two independent experiments, each with similar results, and all the values are expressed as mean ± SD of two biological replicates. Student's t-test: **, $p < 0.01$; ***, $p < 0.001$.

2.8. The Central Region of DP96R Regulates Immune Evasion

The results described so far suggest that ASFV DP96R inhibits the nuclear translocation of IRF3. Next, we aimed to pinpoint the precise region within DP96R that is responsible for inhibiting IFNs, cytokines, and chemokines. To do this, we created two plasmids that express truncated mutants of DP96R spanning aa 1–29 (D1) and 30–96 (D2) (Figure S8A); the construction of these mutants is described elsewhere [59]. Moreover, we further narrowed down DP96R, generating 30–60 (D3) and 61–96 (D4) fragments. Subsequently, we conducted IFN-β luciferase assays, mediated by TBK1 and IRF3-5D, and IRF3 interaction assays using truncated mutants of ASFV DP96R. The luciferase assays indicated that the D3 fragment of DP96R led to a decrease in luciferase activity, whereas the D1 and D4 fragments did not exhibit such an effect (Figure S8B,C). Next, we examined the interaction between DP96R fragments and IRF3. We found that the DP96R-D3 fragment is vital for this interaction (Figure S8D,E). Taken together, our findings suggest that the central region of DP96R, specifically amino acids 30–60, plays a pivotal role in suppressing the production of IFNs, pro-inflammatory cytokines, and chemokines by interacting with IRF3.

3. Discussion

Immune evasion mechanisms are important virulence factors. ASFV uses various immune evasion strategies to inhibit the production of IFNs and pro-inflammatory cytokines. These strategies involve manipulating and controlling multiple components within the IFN and NF-κB signaling pathways. Prior studies suggest that specific ASFV genes suppress host IFN or pro-inflammatory cytokine responses by targeting elements within the cGAS-STING-IRF3 and IKK-NF-κB signaling pathways. A recent study examined the idea that ASF viruses lacking these specific genes show reduced virulence, potentially serving as the basis for developing live attenuated vaccine strains [60]. The recombinant ASFV virus Benin ΔMGF, which lacks MGF family genes, shows impaired inhibition of IFN signaling, leading to reduced pathogenicity and protection of pigs against a subsequent challenge [61]. Infection with ASFV ΔMGF505-7R virus, which lacks the *MGF505-7R* gene, resulting in the induction of type I IFN production and NF-κB activation, also led to increased secretion of IFN-β in porcine serum and a reduction in viral virulence [6,23]. Additionally, ASFV CD2v, which encodes EP402R, and ASFV I267L regulate type I IFN signaling. Both ASFV ΔCD2v and ASFV ΔI267L show attenuated virulence in pigs [62,63]. Hence, investigating ASFV genes that are involved in evading IFN, or inflammatory responses may serve as a foundation for developing live attenuated ASFV vaccines.

ASFV *DP96R* encodes a protein that was previously demonstrated to be an inhibitor of cGAS-STING signaling [59], is transcribed towards the right end of the genome and does not exhibit significant similarity to other known genes in existing databases. Analysis of the *DP96R* gene sequence of various pathogenic ASFVs from Europe, the Caribbean, and Africa shows that this gene is remarkably conserved among pathogenic isolates, including those from ticks and pigs. In the late 1990s, a study demonstrated that the ASFV ΔDP96R virus lacking DP96R was not virulent in inoculated pigs and protected against challenge with homologous ASFV-E70 [35]. Therefore, a double-gene-deleted ASFV-G-Δ9GL/ΔDP96R virus [36], a seven-gene-deleted ASFV-HLJ/-18-7GD (including ASFV *UK*) virus [38], and a double-gene-deleted ASFV-SY18-ΔCD2v/ΔDP96R virus [37] were constructed and showed attenuated virulence and protected immunized pigs from challenge with parental ASFV. Although the relationship between DP96R and ASFV virulence has been studied in vivo, the molecular mechanisms, including immune invasion related to DP96R, remain unknown.

Here, we identified a novel molecular mechanism by which ASFV DP96R inhibits IRF3-mediated antiviral immune responses. First, DP96R interacts with a crucial karyopherin (KPNA) binding site within IRF3. Second, in vitro mutagenesis of IRF3 showed that DP96R prevents interaction between activated IRF3 and KPNA and consequently impedes the translocation of IRF3 to the nucleus. Third, overexpression of DP96R boosts the replication of both DNA and RNA viruses by impeding the cGAS-STING signaling pathway and subsequent transcription of antiviral genes. Fourth, DP96R knock-down using DP96R-specific siRNA resulted in higher IFNs and ISG transcription during ASFV infection.

The IRF family in mammals includes IRF1, IRF2, IRF3, IRF4, IRF5, IRF6, IRF7, IRF8, and IRF9 [41,64,65]. Indeed, among IRF family members, IRF3 is a key transcription factor, crucial for initiating the production of type I IFNs. Expressing numerous genes that are involved in innate immune responses is essential and highlights their significance in antiviral defense mechanisms and overall immune regulation [66]. Upon infection by RNA or DNA viruses, activated TBK1 and IKKε/i phosphorylate IRF3 at specific serine residues [67,68]; and IRF3 then undergoes nuclear translocation to initiate the transcription of type I IFN by binding to specific regulatory elements known as IFN-stimulated response elements (ISREs). This process is crucial for activating the innate immune response against viral infections [69].

IRF3 comprises three key domains: AIE for phosphorylation, IAD for dimerization, and an NLS-associated DBD for nuclear translocation and binding to promoter DNA. The DBD and AIE of IRF3, but not the IAD, are highly conserved in humans and pigs. Given the central role of IRF3 in virus-induced innate immunity, many RNA or DNA viruses, including ASFV, have developed mechanisms to avoid or counteract IRF3-mediated responses within host cells [18,70]. Indeed, studies demonstrate that several viral proteins interact specifically with IRF3, which effectively hinders its nuclear translocation. For instance, the V protein of simian virus 5 [71]; the NS1 protein of influenza B virus [72]; the L protein of Theiler's murine encephalomyelitis virus [73]; the ICP0 protein of HSV-1 [74]; and the ORF6 [75], NSP12 [76], and NSP5 [77] proteins of SARS-CoV-2 all inhibit the nuclear translocation of IRF3.

Reports suggest that the inhibition of the nuclear translocation of phosphorylated IRF3 in the virulent isolate ASFV Armenia/07 is greater than that of IRF3 in the non-virulent ASFV isolate NH/P68 [21], suggesting that there are specific ASFV proteins that impair the type I IFN signaling cascade by inhibiting the translocation of IRF3 to the nucleus. A recent study showed that MGF 505-7R of ASFV-HLJ/18 associates with IRF3 to inhibit the nuclear translocation of IRF3, although the mechanism is unclear [6]. Here, we show that ASFV DP96R interacts with a critical binding site in IRF3, which is essential for the interaction with KPNA. This interaction between DP96R and IRF3 effectively hinders the translocation of IRF3 to the nucleus by disrupting the IRF3-KPNA interaction. KPNA molecules play a pivotal role in the nuclear translocation of IRF3 by acting as an adaptor protein that links NLS-containing proteins (e.g., IRF3, IRF7, P65, and STAT1) with KPNB [78]. The KPNA family comprises KPNA1–KPNA6. Previously, Zeng Cai et al. showed that IRF3 only interacts with KPNA2, although this result may be due to the disproportionate expression of KPNA molecules in the experiments [56]. In the current study, we show that IRF3 interacts with KPNA1, KPNA2, KPNA3, and KPNA4 (Figure 5A,B) and that DP96R impedes interaction between IRF3 and all of these molecules (Figure 5C–H).

Shun Li et al. demonstrated that S97 of IRF3 must be dephosphorylated for nuclear translocation of IRF3 [57], and Zeng Cai et al. showed that KPNA2 interacts with IRF3 when S97 is dephosphorylated [56]. Therefore, we asked whether DP96R inhibits the dephosphorylation of IRF3 S97 or blocks the KPNA-S97 dephosphorylated IRF3 interaction. When we used an S97A mutant of IRF3, DP96R was not involved in the dephosphorylation of IRF3 S97, and the interaction between dephosphorylated IRF3 with KPNA4 was still disrupted by ASFV DP96R, suggesting that DP96R blocks the KPNA-S97-dephosphorylated IRF3 (Figure 6C).

The IRF3 protein possesses a bipartite NLS that is defined by two clusters of basic amino acids: one formed by lysine 77 (K77) and arginine 78 (R78), located at the minor binding site, and the other comprising arginine 86 (R86) and lysine 87 (K87), located at the major binding site of KPNA. Notably, prior to our study, there was no evidence of an interaction between these components [39]. Here, we demonstrate that R86 and K87 of IRF3 are necessary for the interaction with KPNA, whereas K77 and R78 of IRF3 are not (Figure 6D,F). Specifically, DP96R also targets R86 and K87 of IRF3 (Figure 6G). These results suggest that DP96R competes with KPNA molecules for binding to IRF3 without affecting the dephosphorylation of S97.

Previously, Wang et al. also described a mechanism of cGAS-STING pathway inhibition by DP96R degrading TBK1 [59]. However, we did not observe evidence of endogenous or overexpressed TBK1 degradation by DP96R in a dose-dependent manner in our condition. Moreover, our luciferase data showed that DP96R dose-dependently reduces IRF3 and IRF3-5D that are mediated IFN-β luciferase activity. During our study, we depended on the dosage-dependent effects of DP96R, which could potentially be a determining factor in the varied outcomes obtained. In summary, we report a novel molecular mechanism by which DP96R acts as a negative modulator of type I IFNs and pro-inflammatory responses by impairing the nuclear translocation of IRF3. These data increase our understanding of the connection between DP96R and the virulence of ASFV and reinforce the idea that DP96R represents a crucial target for the development of live attenuated vaccines.

4. Materials and Methods

4.1. Cells and Antibodies

HEK293T cells (ATCC® CRL11268™), 293-Dual™ hSTING-A162 cells (Invivogen, San Diego, CA, USA, 293d-a162), PK-15 cells (ATCC® CCL-33), A549 cells (ATCC CCL-185), Vero cells (ATCC® CCL-81™), and MA104 cells were cultured in Dulbecco's Modified Eagle Medium (DMEM) (Cytiva, Marlborough, MA, USA) media and PAMs (ATCC® CRL2843™) in Roswell Park Memorial Institute Medium (RPMI) (Cytiva) media. For the culture of PIBs, RPMI media were utilized, supplemented with 5% Glutamax (Gibco, Waltham, MA, USA) along with 300 μL of normacin (Invivogen). Each cell type was provided with 10% fetal bovine serum (Gibco) and 1% antibiotic/antimycotic (AA) (Gibco). Subsequently, they were placed in a humidified incubator with 5% CO_2 at 37 °C. Primary PAM (Optipharm Inc., Cheongju, Republic of Korea) was cultured in 10% fetal bovine serum (FBS) (Gibco™) and 1% Penicillin-Streptomycin (Gibco™)-added RPMI-1640 medium (Hyclone™, Cytiva) in an incubator at 37 °C and 5% CO_2 atmosphere. The antibodies employed for both immunoblot and immunoprecipitation analyses are as follows: Flag-mouse (Cell Signaling, Danvers, MA, USA, 8146), Flag-Rabbit (Sigma, St. Louis, MO, USA, F7425), GST (Santa Cruz, sc-138), Strep (QIAGEN, Venlo, The Netherlands, 34850), IRF3 (Cell Signaling 4302s), phospho IRF3 (Ser396) (Cell Signaling, 4947), IKKα (Santa Cruz, Dallas, TX, USA, sc-7606), phospho IKKα/β (Cell Signaling, 2697S), p65 (Cell Signaling, 4764S), phospho p65 (Cell Signaling, 3031S), TBK1 (Cell Signaling, 3504S), phospho TBK1 (Cell Signaling, 5483S), phospho IκBα (Cell Signaling, 2859S), IκBα (Cell Signaling, 9242S), STAT1 (Cell Signaling, 9172S), pSTAT1 (Cell Signaling, 9167S), KPNA2 (Santa Cruz, sc-55538), KPNA4 (Santa Cruz, 390535), α-tubulin (Santa Cruz, sc-8035), Lamin B1(Cell Signaling, 12586), and β-actin (Santa Cruz, SC 47778). For the immunofluorescence assay, cy3-conjugated donkey mouse IgG (The Jackson Laboratory, Bar Harbor, ME, USA, 715-165-150) and Alexa 488 goat rabbit IgG (Invivogen, A11034) were used. These antibodies were utilized following the manufacturer's provided instructions.

4.2. Porcine Bone Marrow-Derived Macrophage Isolation

Cells derived from porcine bone marrow were isolated following the subsequent protocols [79–81]: Five ribs from the rear portion were extracted from each side of the animal, and the bone surfaces were cleansed with alcohol. Both ends were incised, and the bone marrow was flushed from both sides using a 20 mL syringe and an 18 G needle, with RPMI 1640 (Cytiva) supplemented with 5 mM EDTA (Gibco) to prevent clotting. The cells were filtered, subjected to centrifugation, and then suspended in red blood cell lysis buffer (ACK lysing buffer, Gibco) for 2 min. After the cells were lysed, they underwent centrifugation, and the resulting pellet was washed first with PBS and then with RPMI 1640. Lastly, the bone marrow cells were suspended in a freezing medium consisting of 90% heat-inactivated FBS and 10% DMSO. They were then frozen overnight in a "Mr. Frosty" isopropanol box (Nalgene, Rochester, NY, USA) at −80 °C, enabling gradual temperature reduction for controlled freezing. On the subsequent day, the cells were transferred to a freezer set at a temperature of −196 °C for extended, long-term storage. To retrieve the cells from the freezer, they were rapidly thawed in a 37 °C water bath and then cautiously diluted by gradually adding complete medium drop by drop over 2–3 min. This gradual dilution was performed to prevent any abrupt changes in osmolarity due to DMSO. The cells were slowly diluted by adding 40 mL of warm PBS drop by drop over a 2–3-min period, preventing the sudden dilution shock from DMSO. Following the DMSO removal through washing, the cells were cultured in a complete medium consisting of RPMI 1640, 10% heat-inactivated FBS (Gibco), AA (Gibco), and GlutaMAX-I supplement (Gibco). Porcine BMDMs were generated by culturing bone marrow cells on sterile 100 mm^2 petri dishes for 5–7 days in the presence of pGM-CSF-1 (20 ng/mL; R&D Systems, Minneapolis, MN, USA,

711-pg-010). The obtained macrophages were detached by forcefully splashing them with a medium using a syringe and an 18 G needle. Afterward, they were washed, counted, and then seeded into tissue culture plates at a concentration of 1 million cells per milliliter (10^6 cells/mL) in a medium containing pGM-CSF-1.

4.3. Porcine Bone Marrow-Derived Macrophage Immortalization

Before the lentiviral transduction, a G418 dose selection assay was performed, and the minimum G418 concentration for the onset of pBMDM cell death was determined. First, pBMDMs were allowed to differentiate for 5–7 days. Then, the pBMDMs were split into a 24-well cell culture plate. For G410, 100µg/mL to 400 µg/mL doses were tested, and cell death was shown at 400 µg/mL. Subsequently, primary porcine BMDMs were divided and placed into individual wells of a 24-well cell culture plate. Each well contained 0.5 mL of RPMI1640 supplemented with 10% FBS, GlutaMax-I, and AA, with a cell density of 1 million cells (1×10^6 cells) per well. The plates were then incubated at 37 °C in a 5% CO_2 environment overnight.

At the time of transduction, cell confluence was 50–70%. Preformed lentiviral particles expressing SV40 large T antigen under the CMV promoter and containing neomycin markers were purchased (AMSBIO, Abingdon, UK). The calculated virus particle and polybrene levels were 5 µL and 0.6 µg/µL, respectively. A determined amount of retrovirus was added into 1.5 mL microcentrifuge tubes, bringing the volume up to 0.5 mL using complete media. Polybrene was introduced into the mixture to achieve a final concentration of 8 µg/mL. The medium was removed from the cell culture plate, and 500 µL of the prepared mixture of complete medium, including lentivirus and polybrene, was added without disturbing the cells. The cells were cultured with the virus for a period of 48 to 72 h in a 37 °C incubator under 5% CO_2 conditions.

Three days after transduction, cells were treated with complete media containing 800 µg/mL G418 to select resistant cells. The integration of hTERT and SV40LT was verified using PCR. For the template for PCR, the genomic DNA was isolated using an RNA/DNA mini kit. PCR primers for hTERT were as follows: forward primer- 5′-GCCGAGACCAAGCACTTCCTCTACT-3′, reverse primer-5′-GCAACTTGCTCCAGACA CTCTTCCG-3′ and SV40LT; forward primer- 5′-GATGGCTGGAGTTGCTTGGCTACAC-3′, reverse primer-5′GCCTGAAATGAGCCTTG GGACTGTG-3′. The PCR protocol included an initial denaturation cycle of 5 min at 95 °C, followed by 35 cycles consisting of 30 s at 94 °C, 30 s at 63 °C, and 45 s at 72 °C. Finally, there was a concluding extension step of 5 min at 72 °C.

4.4. Continuous ASFV DP96R Protein-Expressing Cell Generation

DP96R continuously expressing PAMs, PIBs, and MA104 cells were generated by transient transfection of the DP96R-pIRES-Flag plasmid with Lipofectamine 2000 (Invitrogen, Carlsbad, CA, USA). Positive colonies were selected in 10% FBS-containing DMEM treated with 2 µg/mL, 0.3 µg/mL, or 4 µg/mL puromycin (Thermo Fisher Scientific, Waltham, MA, USA) for each cell line for at least two weeks. The presence of DP96R protein in the cells was validated through immunoblotting using anti-Flag antibodies.

4.5. Plasmids

The complete form of the ASFV DP96R protein-expressing sequence (GenBank, Bethesda, MD, USA: FR682468.1) was cloned into Flag-tagged pIRES, Strep-tagged pEXPR, and GST-tagged pEGB vectors. Truncated DP96R gene fragments were cloned into a pEGB vector with a GST tag. To generate IRF3 different constructs, IRF3 was amplified from template DNA using PCR and cloned into pIRES-Flag, pEXPR-Strep, or pEBG-GST vectors. IRF3 domains were subcloned into the pEGB vector. The mutation cloning kit (Thermo Fisher, 00940669) generated the IRF3 point mutated sequences. KPNA (Importin α) 1 to 6 were cloned into a pIRES vector with a Flag tag. Generation procedures of the IFN-β and NF-κB promoter and luciferase reporter plasmids have been described elsewhere [82]. cGAS 3×

Flag-tagged plasmid was kindly donated by Dr Jae U. Jung (Lerner Research Institute, 9500 Euclid Avenue, Cleveland, OH, USA). Flag-tagged STING, TBK1, IKKε, IRF3, P65, and V5-tagged TBK1 were generated by amplifying template DNA and cloning it into the pIRES vector. The integrity of all sequences was verified by sequencing analysis. Plasmids utilized in the studies encode human proteins.

4.6. Virus Infection and Plasmid Transfection

ADV-GFP was amplified in PK-15 cells, while HSV-GFP and VACV-GFP were amplified in Vero cells and quantified using a plaque assay. The Sendai virus Cantell strain, GFP-tagged H_1N_1 virus-PR8 strain (PR8-GFP), and NDV-GFP were amplified in specific pathogen-free (SPF) eggs. Before virus infection, the culture medium was replaced with DMEM containing 1% FBS, and the virus was introduced into the target cells at a specific multiplicity of infection (MOI). After a 2 h incubation at 37 °C, the extracellular viruses were eliminated, and the medium was substituted with DMEM containing 10% FBS. At the specified time points, the cells were detached from the culture plates along with the supernatants, and the mixture was subjected to centrifugation at 3000 rpm for 3 min. The supernatant of each sample was separated from the cell pellet for ELISA. The cell pellet was reconstituted in 300 μL of phosphate-buffered saline (PBS), and the fluorescence of each sample was assessed using a fluorometer, specifically, the Glomax detection system from Promega. The plasmids were introduced into PK-15, PAMs, PIBs, and MA104 cells using Lipofectamine 2000 (Invitrogen), and for HEK293T and 293-Dual™ hSTING-A162 cells, Polyethylenimine (PEI; Polyscience Inc., Warrington, PA, USA, 23966) was employed for transfection. The procedures followed the manufacturer's protocol.

4.7. Virus Titration

Virus-infected cells and cell culture supernatants were collected at the indicated time points, and viruses were titrated by plaque assay using A549 cells for ADV-GFP and VACV-GFP and *Ceropithecus aethiops* epithelial kidney (Vero) cells for HSV-GFP, PR8-GFP, and NDV-GFP. A549 and Vero cells were placed into 12-well plates and allowed to incubate for a duration of 12 h. Subsequently, the cells were exposed to supernatants containing the virus that had been serially diluted, using 1% DMEM, for a period of 2 h. After the incubation period, the inoculums were discarded and substituted with DMEM containing 0.1% agarose (from Sigma-Aldrich). The plates were incubated for 36 h at 37 °C and subsequently examined for plaque formation under a magnification of 200×. Determining virus titers was performed by employing the number of plaque-forming units (PFUs) and the dilution factor in the calculations.

4.8. DP96R Transcription Assay

In the ASFV experiment, primary porcine alveolar macrophages (primary PAMs) were exposed to a 0.5 multiplicity of infection (MOI) of ASFV (Korea/wild boar/Hwacheon/2020-2287). Cell pellets were collected at 0, 3, 6, 9, 12, 15, and 18 hpi. Subsequently, RNA extraction was carried out, and qRT-PCR analysis was performed using DP96R primers (Table S1). All experiments dealing with ASFV were conducted in accordance with the Standard Operating Procedure (SOP) in the biosafety level 3 (BSL-3) laboratory of the NIWDC in Korea.

4.9. RNA Interference Experiments

The specific small interfering RNA (siRNA) for targeting ASFV DP96R was designed and synthesized by Bioneer (Daejeon, Republic of Korea). Primary PAMs (cell number 1×10^6 in 24-well plates) were transfected with negative control (siControl) or DP96R (siDP96R) siRNA by RNAimax (Invivogen). The target sequence of DP96R was 256-GGAUCCCUAAUGCGCUCCA-274. Nontargeting siRNA was used as a negative control (UUCUCCGAACGUGUCACGU). At 6 hpt, the cells were left uninfected or infected with ASFV at a multiplicity of infection (MOI) of 0.5. The mRNA expression levels of the target gene, IFNs, and ISG mRNA were detected by qRT-PCR. All experiments dealing with ASFV were conducted in accordance with the Standard Operating Procedure (SOP) in the biosafety level 3 (BSL-3) laboratory of the NIWDC in Republic of Korea.

4.10. ELISA

ELISA was conducted to identify the presence of secreted interferons, pro-inflammatory cytokines, and chemokines within the cell culture supernatants. Human IL-6 (BD OptEIA, Franklin Lakes, NJ, USA, 5552220), Human IFN-β (CUSABIO, CSB-E09889h), porcine IL-6 (R&D Systems, p6000B), and porcine IFN-β (CUSABIO, CSB-E09890p), were used for analysis, and the procedures followed the manufacturer's provided protocols.

4.11. Quantitative Real-Time PCR

DP96R pIRES or pIRES stably expressing PAMs were cultivated in 6-well tissue culture plates, with a seeding density of 1×10^6 cells per well, and maintained at a temperature of 37 °C. The cells were infected with ADV-GFP and HSV-GFP at a multiplicity of infection (MOI) of 1 and then collected at 0, 12, and 24 hpi. The total RNA from the cells was extracted using the Machery Nagel Nucleospin RNA kit (790955.250), and subsequently, cDNA was synthesized using reverse transcriptase (Toyobo, Osaka, Japan). The varying levels of cDNA were quantified through real-time polymerase chain reaction (RT-PCR) employing the Smart Gene SYBR Green Q-PCR master mix (SG. SYBR.500) kit, following the manufacturer's provided instructions. The primer sequences utilized in the quantitative polymerase chain reaction (qPCR) are detailed in Table S1.

4.12. Immunoprecipitation

At 36 h post-transfection (hpt), the cells were collected, and whole-cell lysates (WCLs) were acquired following lysis using a lysis buffer containing protease inhibitor cocktail (PI), phosphatase inhibitor cocktail (Sigma), and radio-immunoprecipitation assay (RIPA) lysis buffer (comprising 50 mM Tris-HCl, 150 mM NaCl, 0.5% sodium deoxycholate, 1% IGEPAL, 1 mM NaF, and 1 mM Na3VO4). The lysates were then sonicated using a sonicator from Sonics. The whole-cell lysate (WCL) was initially subjected to a pre-clearing step by incubation with Sepharose 6B (GE Healthcare Life Science, Little Chalfont, UK) at 4 °C with rotation for 2 h. Following the pre-clearing step, for GST and Strep pull-down, the WCL was subjected to incubation with a 50% slurry of glutathione-conjugated Sepharose (GST) beads (Amersham Biosciences, Slough, UK) and Strep beads (IBA) for 12 h. In contrast, for antibody pull-down, the specified antibodies were incubated with WCL for 12 h, and protein A/G PLUS-Agarose beads (Santa Cruz, H0422) were employed for incubation, lasting 4 h. The immunoprecipitated beads were gathered following centrifugation and subjected to washing with lysis buffer using various washing conditions. Subsequently, the interaction between the relevant proteins was assessed through immunoblot analysis.

4.13. Immunoblot Analysis

The cell lysates prepared with immunoprecipitated beads were separated via SDS-PAGE and then transferred to a PVDF membrane using a Trans-Blot® semi-dry transfer cell from Bio-Rad located in Seoul, Republic of Korea. Following the transfer, the membrane was blocked for a duration of 1 h using 5% bovine serum albumin (Georgiachem, Norcross, GA, USA, BS1005) and subsequently incubated overnight at 4 °C with the primary antibody. Subsequently, the membranes were subjected to washing with either TBST or PBST, and then, the membrane was incubated with a horseradish peroxidase-conjugated (HRP) secondary antibody (GeneTex, Irvine, CA, USA, GTX213111-01) for a period of 2 h at room temperature. The membrane underwent three rounds of washing with TBST or PBST. Lastly, the reaction was visualized using an enhanced chemiluminescence detection system (ECL) (GE Healthcare, located in Little Chalfont, UK) and with Amersham ImageQuant 800 (Cytiva).

4.14. Luciferase Reporter Assay

HEK293T cells and 293-Dual™ hSTING-A162 cells were cultured in 12-well tissue culture plates (3.5×10^5 cells/well) and incubated at 37 °C. The human STING-overexpressing HEK293T cell line, 293-dual hSTING-A162, was used to examine the IFN-β luciferase activity induced by poly(dA:dT), cGAS and 2′3′cGAMP, while HEK293T cells were used to test the STING-, TBK1-, IKKε-, IRF3- and IRF3-5D-induced IFN-β luciferase reporter assay. HEK293T cells were transfected with IFN-β luminescence, TK renilla-luciferase reporter plasmid (an internal control to normalize transfection efficiency). DP96R pIRES plasmid or control vector was dose-dependently transfected. poly(dA:dT) (InvivoGen) and 2′3′cGAMP ligands (InvivoGen) were transfected with Lipofectamine 2000 (Invitrogen) and Lipofectamine RNAi max (Invitrogen), respectively. At 24 h post-transfection, cells were washed with phosphate-buffered saline (PBS) and lysed with 1× Passive Lysis buffer (Promega, Madison, WI, USA) for 15 min. Following the manufacturer's protocol, luciferase activity was measured using a dual-luciferase reporter assay system (Promega; E1980).

4.15. Nuclear Fractionation Assay

In a 6-well cell culture plate, 80–90% confluent DP96R protein stably expressing PAMs, infected with ADV-GFP 1MOI for 2 h, were harvested at specified time intervals. Then, nuclear and cytoplasmic extraction procedures were performed following the instructions provided by the manufacturer (Invent Biotechnologies Inc., Plymouth, MN, USA, SM-005). The concentration of cytosol or nuclear lysate was quantified using the Bradford assay (Bio-Rad, Hercules, CA, USA).

4.16. Immunofluorescence and Confocal Microscopy

PK-15 cells were placed into an eight-well chamber slide manufactured by Ibidi. Harvested cells were fixed by incubating them with 4% paraformaldehyde for a duration of 20 min at room temperature. The fixed cells were rinsed with 1× PBS and then made permeable by treating them with absolute methanol. This was followed by an incubation period of 20 min at −20 °C. The cells were rinsed with 1× PBS and subsequently blocked by exposing them to 2% BSA in 1× PBS for a duration of 1 h at room temperature. The primary antibodies were added and incubated at 4 °C overnight. Following this, the cells underwent three rounds of washing with 1× PBST and were then exposed to an appropriate secondary antibody for a period of 1 h at room temperature. Subsequently, the cells were subjected to three additional washes with 1× PBST and stained with DAPI (4′,6-diamidino-2-phenylindole) for a duration of 10 min at room temperature. Images were captured using a LEICA DMi8 microscope and then analyzed with LAS-X software (version 3.7.1.21655). Pearson correlation coefficient of confocal images was calculated using Fiji Image J (version 1.54c); an open-source platform for biological image analysis.

4.17. Statistical Analysis

Graphs and all statistical analyses were performed using GraphPad Prism software version 6 for Windows. Data are presented as the means and standard deviations (SDs) representing at least two independent experiments. An unpaired Student's *t*-test was performed to compare the control and treatment groups at each time point. p values of <0.05, <0.01, <0.001, or <0.0001 were considered significant.

Supplementary Materials: The following supporting information can be downloaded at: https://www.mdpi.com/article/10.3390/ijms25042099/s1.

Author Contributions: Conceptualization J.-S.L.; methodology, J.-S.L., N.D. and J.-W.C.; software, N.D; validation, N.D., J.-W.C., Y.K., W.J., K.C., W.A.G.C., A.W., L.R. and J.-S.L.; formal analysis, N.D., K.C., W.A.G.C., A.W., L.R. and J.-S.L.; resources, Y.K., W.J. and J.-S.L.; Data curation, N.D. and J.-S.L., Writing—original draft preparation, N.D. and J.-S.L.; Writing—review and editing, N.D. and J.-S.L.; Visualization, N.D. and J.-S.L.; Supervision, J.-S.L.; Project administration, J.-S.L.; Funding acquisition, J.-S.L. All authors have read and agreed to the published version of the manuscript.

Funding: This research was funded by the Ministry of Environment (Grant No. NIWDC-2021-SP-02); National Research Foundation (Grant No. 2021R1A6A1A03045495), Institute for Basic Science (IBS) Research Program (Grant No. IBS-R801-D9-A01); and the Ministry for Food, Agriculture, Forestry and Fisheries (Grant No. 119081-5), Republic of Korea.

Institutional Review Board Statement: The authors declare that porcine bone marrow-derived macrophages were isolated from the bone marrow samples sourced from ChoongAng Vaccine Laboratories, Daejeon, 34055, Republic of Korea. Bone marrow isolation was performed according to the guidelines approved by the Institutional Review Board of Choong Ang Vaccine Laboratories Co., Ltd. (protocol code 211220-06, approval date 20 December 2021), Republic of Korea.

Informed Consent Statement: Not applicable.

Data Availability Statement: All the pertinent data can be found within the manuscript and its accompanying supporting information files. The authors confirm that all data supporting the findings of this study are available at figshare (https://figshare.com/s/89a1f90c6eab8dc7d4fd) (accessed on 27 November 2023).

Conflicts of Interest: The authors declare no conflicts of interest.

References

1. Dixon, L.K.; Sun, H.; Roberts, H. African swine fever. *Antivir. Res.* **2019**, *165*, 34–41. [CrossRef]
2. Galindo, I.; Alonso, C. African Swine Fever Virus: A Review. *Viruses* **2017**, *9*, 103. [CrossRef]
3. Gaudreault, N.N.; Madden, D.W.; Wilson, W.C.; Trujillo, J.D.; Richt, J.A. African Swine Fever Virus: An Emerging DNA Arbovirus. *Front. Vet. Sci.* **2020**, *7*, 215. [CrossRef] [PubMed]
4. Alonso, C.; Borca, M.; Dixon, L.; Revilla, Y.; Rodriguez, F.; Escribano, J.M.; ICTV Report Consortium. ICTV Virus Taxonomy Profile: Asfarviridae. *J. Gen. Virol.* **2018**, *99*, 613–614. [CrossRef] [PubMed]
5. Wilkinson, P. The persistence of African swine fever in Africa and the Mediterranean. *Prev. Vet. Med.* **1984**, *2*, 71–82. [CrossRef]
6. Li, J.; Song, J.; Kang, L.; Huang, L.; Zhou, S.; Hu, L.; Zheng, J.; Li, C.; Zhang, X.; He, X. pMGF505-7R determines pathogenicity of African swine fever virus infection by inhibiting IL-1β and type I IFN production. *PLoS Pathog.* **2021**, *17*, e1009733. [CrossRef] [PubMed]
7. Sánchez-Cordón, P.; Montoya, M.; Reis, A.; Dixon, L. African swine fever: A re-emerging viral disease threatening the global pig industry. *Vet. J.* **2018**, *233*, 41–48. [CrossRef] [PubMed]
8. Alejo, A.; Matamoros, T.; Guerra, M.; Andrés, G. A Proteomic Atlas of the African Swine Fever Virus Particle. *J. Virol.* **2018**, *92*, e01293-18. [CrossRef] [PubMed]
9. Karger, A.; Pérez-Núñez, D.; Urquiza, J.; Hinojar, P.; Alonso, C.; Freitas, F.B.; Revilla, Y.; Le Potier, M.F.; Montoya, M. An Update on African Swine Fever Virology. *Viruses* **2019**, *11*, 864. [CrossRef] [PubMed]
10. McNab, F.; Mayer-Barber, K.; Sher, A.; Wack, A.; O'Garra, A. Type I interferons in infectious disease. *Nat. Rev. Immunol.* **2015**, *15*, 87–103. [CrossRef]
11. Yu, L.; Liu, P. Cytosolic DNA sensing by cGAS: Regulation, function, and human diseases. *Signal Transduct. Target. Ther.* **2021**, *6*, 170. [CrossRef] [PubMed]
12. Kato, K.; Omura, H.; Ishitani, R.; Nureki, O. Cyclic GMP-AMP as an Endogenous Second Messenger in Innate Immune Signaling by Cytosolic DNA. *Annu. Rev. Biochem.* **2017**, *86*, 541–566. [CrossRef] [PubMed]
13. Zhang, X.; Shi, H.; Wu, J.; Zhang, X.; Sun, L.; Chen, C.; Chen, Z.J. Cyclic GMP-AMP containing mixed phosphodiester linkages is an endogenous high-affinity ligand for STING. *Mol. Cell* **2013**, *51*, 226–235. [CrossRef] [PubMed]
14. Ablasser, A.; Goldeck, M.; Cavlar, T.; Deimling, T.; Witte, G.; Röhl, I.; Hopfner, K.P.; Ludwig, J.; Hornung, V. cGAS produces a 2′-5′-linked cyclic dinucleotide second messenger that activates STING. *Nature* **2013**, *498*, 380–384. [CrossRef] [PubMed]
15. Zhong, B.; Yang, Y.; Li, S.; Wang, Y.Y.; Li, Y.; Diao, F.; Lei, C.; He, X.; Zhang, L.; Tien, P.; et al. The adaptor protein MITA links virus-sensing receptors to IRF3 transcription factor activation. *Immunity* **2008**, *29*, 538–550. [CrossRef] [PubMed]
16. Tanaka, Y.; Chen, Z.J. STING specifies IRF3 phosphorylation by TBK1 in the cytosolic DNA signaling pathway. *Sci. Signal* **2012**, *5*, ra20. [CrossRef] [PubMed]
17. García-Sastre, A. Ten Strategies of Interferon Evasion by Viruses. *Cell Host Microbe* **2017**, *22*, 176–184. [CrossRef]
18. Ma, Z.; Damania, B. The cGAS-STING Defense Pathway and Its Counteraction by Viruses. *Cell Host Microbe* **2016**, *19*, 150–158. [CrossRef]
19. Deng, L.; Zeng, Q.; Wang, M.; Cheng, A.; Jia, R.; Chen, S.; Zhu, D.; Liu, M.; Yang, Q.; Wu, Y. Suppression of NF-κB activity: A viral immune evasion mechanism. *Viruses* **2018**, *10*, 409. [CrossRef]
20. Rahman, M.M.; McFadden, G. Modulation of NF-κB signalling by microbial pathogens. *Nat. Rev. Microbiol.* **2011**, *9*, 291–306. [CrossRef]
21. García-Belmonte, R.; Pérez-Núñez, D.; Pittau, M.; Richt, J.A.; Revilla, Y. African Swine Fever Virus Armenia/07 Virulent Strain Controls Interferon Beta Production through the cGAS-STING Pathway. *J. Virol.* **2019**, *93*, e02298-18. [CrossRef] [PubMed]

22. Li, Z.; Chen, W.; Qiu, Z.; Li, Y.; Fan, J.; Wu, K.; Li, X.; Zhao, M.; Ding, H.; Fan, S. African Swine Fever Virus: A Review. *Life* **2022**, *12*, 1255. [CrossRef] [PubMed]
23. Li, D.; Yang, W.; Li, L.; Li, P.; Ma, Z.; Zhang, J.; Qi, X.; Ren, J.; Ru, Y.; Niu, Q. African swine fever virus MGF-505-7R negatively regulates cGAS–STING-mediated signaling pathway. *J. Immunol.* **2021**, *206*, 1844–1857. [CrossRef] [PubMed]
24. Liu, H.; Zhu, Z.; Feng, T.; Ma, Z.; Xue, Q.; Wu, P.; Li, P.; Li, S.; Yang, F.; Cao, W. African swine fever virus E120R protein inhibits interferon beta production by interacting with IRF3 to block its activation. *J. Virol.* **2021**, *95*, e00824-21. [CrossRef] [PubMed]
25. Liu, X.; Ao, D.; Jiang, S.; Xia, N.; Xu, Y.; Shao, Q.; Luo, J.; Wang, H.; Zheng, W.; Chen, N. African swine fever virus A528R inhibits TLR8 mediated NF-κB activity by targeting p65 activation and nuclear translocation. *Viruses* **2021**, *13*, 2046. [CrossRef] [PubMed]
26. Wang, Y.; Cui, S.; Xin, T.; Wang, X.; Yu, H.; Chen, S.; Jiang, Y.; Gao, X.; Jiang, Y.; Guo, X.; et al. African Swine Fever Virus MGF360-14L Negatively Regulates Type I Interferon Signaling by Targeting IRF3. *Front. Cell Infect. Microbiol.* **2021**, *11*, 818969. [CrossRef]
27. Yang, J.; Li, S.; Feng, T.; Zhang, X.; Yang, F.; Cao, W.; Chen, H.; Liu, H.; Zhang, K.; Zhu, Z.; et al. African Swine Fever Virus F317L Protein Inhibits NF-κB Activation to Evade Host Immune Response and Promote Viral Replication. *mSphere* **2021**, *6*, e0065821. [CrossRef]
28. Yang, K.; Huang, Q.; Wang, R.; Zeng, Y.; Cheng, M.; Xue, Y.; Shi, C.; Ye, L.; Yang, W.; Jiang, Y. African swine fever virus MGF505-11R inhibits type I interferon production by negatively regulating the cGAS-STING-mediated signaling pathway. *Vet. Microbiol.* **2021**, *263*, 109265. [CrossRef]
29. Zhuo, Y.; Guo, Z.; Ba, T.; Zhang, C.; He, L.; Zeng, C.; Dai, H. African swine fever virus MGF360-12L inhibits type I interferon production by blocking the interaction of importin α and NF-κB signaling pathway. *Virol. Sin.* **2021**, *36*, 176–186. [CrossRef]
30. Chen, H.; Wang, Z.; Gao, X.; Lv, J.; Hu, Y.; Jung, Y.-S.; Zhu, S.; Wu, X.; Qian, Y.; Dai, J. ASFV pD345L protein negatively regulates NF-κB signalling by inhibiting IKK kinase activity. *Vet. Res.* **2022**, *53*, 32. [CrossRef]
31. Cheng, M.; Luo, J.; Duan, Y.; Yang, Y.; Shi, C.; Sun, Y.; Lu, Y.; Wang, J.; Li, X.; Wang, J. African swine fever virus MGF505-3R inhibits cGAS-STING-mediated IFN-β pathway activation by degrading TBK1. *Anim. Dis.* **2022**, *2*, 13. [CrossRef]
32. Dodantenna, N.; Ranathunga, L.; Chathuranga, W.G.; Weerawardhana, A.; Cha, J.-W.; Subasinghe, A.; Gamage, N.; Haluwana, D.; Kim, Y.; Jheong, W. African Swine Fever Virus EP364R and C129R Target Cyclic GMP-AMP To Inhibit the cGAS-STING Signaling Pathway. *J. Virol.* **2022**, *96*, e01022-22. [CrossRef] [PubMed]
33. Sun, M.; Yu, S.; Ge, H.; Wang, T.; Li, Y.; Zhou, P.; Pan, L.; Han, Y.; Yang, Y.; Sun, Y.; et al. The A137R Protein of African Swine Fever Virus Inhibits Type I Interferon Production via the Autophagy-Mediated Lysosomal Degradation of TBK1. *J. Virol.* **2022**, *96*, e0195721. [CrossRef] [PubMed]
34. Yang, K.; Xue, Y.; Niu, H.; Shi, C.; Cheng, M.; Wang, J.; Zou, B.; Wang, J.; Niu, T.; Bao, M. African swine fever virus MGF360-11L negatively regulates cGAS-STING-mediated inhibition of type I interferon production. *Vet. Res.* **2022**, *53*, 1–12. [CrossRef]
35. Zsak, L.; Caler, E.; Lu, Z.; Kutish, G.; Neilan, J.; Rock, D. A nonessential African swine fever virus gene UK is a significant virulence determinant in domestic swine. *J. Virol.* **1998**, *72*, 1028–1035. [CrossRef]
36. O'Donnell, V.; Risatti, G.R.; Holinka, L.G.; Krug, P.W.; Carlson, J.; Velazquez-Salinas, L.; Azzinaro, P.A.; Gladue, D.P.; Borca, M.V. Simultaneous deletion of the 9GL and UK genes from the African swine fever virus Georgia 2007 isolate offers increased safety and protection against homologous challenge. *J. Virol.* **2017**, *91*, e01760-16. [CrossRef] [PubMed]
37. Teklue, T.; Wang, T.; Luo, Y.; Hu, R.; Sun, Y.; Qiu, H.-J. Generation and evaluation of an African swine fever virus mutant with deletion of the CD2v and UK genes. *Vaccines* **2020**, *8*, 763. [CrossRef]
38. Chen, W.; Zhao, D.; He, X.; Liu, R.; Wang, Z.; Zhang, X.; Li, F.; Shan, D.; Chen, H.; Zhang, J. A seven-gene-deleted African swine fever virus is safe and effective as a live attenuated vaccine in pigs. *Sci. China Life Sci.* **2020**, *63*, 623–634. [CrossRef]
39. Zhu, M.; Fang, T.; Li, S.; Meng, K.; Guo, D. Bipartite nuclear localization signal controls nuclear import and DNA-binding activity of IFN regulatory factor 3. *J. Immunol.* **2015**, *195*, 289–297. [CrossRef]
40. Sato, M.; Suemori, H.; Hata, N.; Asagiri, M.; Ogasawara, K.; Nakao, K.; Nakaya, T.; Katsuki, M.; Noguchi, S.; Tanaka, N.; et al. Distinct and essential roles of transcription factors IRF-3 and IRF-7 in response to viruses for IFN-alpha/beta gene induction. *Immunity* **2000**, *13*, 539–548. [CrossRef]
41. Honda, K.; Taniguchi, T. IRFs: Master regulators of signalling by Toll-like receptors and cytosolic pattern-recognition receptors. *Nat. Rev. Immunol.* **2006**, *6*, 644–658. [CrossRef] [PubMed]
42. Weaver, B.K.; Kumar, K.P.; Reich, N.C. Interferon regulatory factor 3 and CREB-binding protein/p300 are subunits of double-stranded RNA-activated transcription factor DRAF1. *Mol. Cell Biol.* **1998**, *18*, 1359–1368. [CrossRef] [PubMed]
43. Holly, M.K.; Smith, J.G. Adenovirus infection of human enteroids reveals interferon sensitivity and preferential infection of goblet cells. *J. Virol.* **2018**, *92*, e00250-18. [CrossRef] [PubMed]
44. Altinkilic, B.; Brandner, G. Interferon inhibits herpes simplex virus-specific translation: A reinvestigation. *J. Gen. Virol.* **1988**, *69*, 3107–3112. [CrossRef]
45. Stewart, W.E.; Scott, W.D.; Sulkin, S.E. Relative sensitivities of viruses to different species of interferon. *J. Virol.* **1969**, *4*, 147–153. [CrossRef]
46. Shen, Q.; Wang, Y.E.; Palazzo, A.F. Crosstalk between nucleocytoplasmic trafficking and the innate immune response to viral infection. *J. Biol. Chem.* **2021**, *297*, 100856. [CrossRef]
47. Rehwinkel, J.; Gack, M.U. RIG-I-like receptors: Their regulation and roles in RNA sensing. *Nat. Rev. Immunol.* **2020**, *20*, 537–551. [CrossRef]

48. Glanz, A.; Chakravarty, S.; Varghese, M.; Kottapalli, A.; Fan, S.; Chakravarti, R.; Chattopadhyay, S. Transcriptional and non-transcriptional activation, posttranslational modifications, and antiviral functions of interferon regulatory factor 3 and viral antagonism by the SARS-coronavirus. *Viruses* **2021**, *13*, 575. [CrossRef]
49. Görlich, D.; Kraft, R.; Kostka, S.; Vogel, F.; Hartmann, E.; Laskey, R.A.; Mattaj, I.W.; Izaurralde, E. Importin provides a link between nuclear protein import and U snRNA export. *Cell* **1996**, *87*, 21–32. [CrossRef] [PubMed]
50. Nigg, E.A. Nucleocytoplasmic transport: Signals, mechanisms and regulation. *Nature* **1997**, *386*, 779–787. [CrossRef] [PubMed]
51. Ye, J.; Chen, Z.; Li, Y.; Zhao, Z.; He, W.; Zohaib, A.; Song, Y.; Deng, C.; Zhang, B.; Chen, H. Japanese encephalitis virus NS5 inhibits type I interferon (IFN) production by blocking the nuclear translocation of IFN regulatory factor 3 and NF-κB. *J. Virol.* **2017**, *91*, e00039-17. [CrossRef]
52. Khan, H.; Sumner, R.P.; Rasaiyaah, J.; Tan, C.P.; Rodriguez-Plata, M.T.; Van Tulleken, C.; Fink, D.; Zuliani-Alvarez, L.; Thorne, L.; Stirling, D. HIV-1 Vpr antagonizes innate immune activation by targeting karyopherin-mediated NF-κB/IRF3 nuclear transport. *Elife* **2020**, *9*, e60821. [CrossRef]
53. Kumar, K.P.; McBride, K.M.; Weaver, B.K.; Dingwall, C.; Reich, N.C. Regulated nuclear-cytoplasmic localization of interferon regulatory factor 3, a subunit of double-stranded RNA-activated factor 1. *Mol. Cell. Biol.* **2000**, *20*, 4159–4168. [CrossRef]
54. Li, J.; Lu, M.; Huang, B.; Lv, Y. Porcine circovirus type 2 inhibits interferon-β expression by targeting Karyopherin alpha-3 in PK-15 cells. *Virology* **2018**, *520*, 75–82. [CrossRef]
55. Zhang, L.; Qiu, S.; Lu, M.; Huang, C.; Lv, Y. Nuclear transporter karyopherin subunit alpha 3 levels modulate Porcine circovirus type 2 replication in PK-15ácells. *Virology* **2020**, *548*, 31–38. [CrossRef]
56. Cai, Z.; Zhang, M.-X.; Tang, Z.; Zhang, Q.; Ye, J.; Xiong, T.-C.; Zhang, Z.-D.; Zhong, B. USP22 promotes IRF3 nuclear translocation and antiviral responses by deubiquitinating the importin protein KPNA2. *J. Exp. Med.* **2020**, *217*, e20191174. [CrossRef]
57. Li, S.; Zhu, M.; Pan, R.; Fang, T.; Cao, Y.-Y.; Chen, S.; Zhao, X.; Lei, C.-Q.; Guo, L.; Chen, Y. The tumor suppressor PTEN has a critical role in antiviral innate immunity. *Nat. Immunol.* **2016**, *17*, 241–249. [CrossRef] [PubMed]
58. Zhang, Y.; Ke, J.; Zhang, J.; Yang, J.; Yue, H.; Zhou, X.; Qi, Y.; Zhu, R.; Miao, F.; Li, Q. African swine fever virus bearing an I226R gene deletion elicits robust immunity in pigs to African swine fever. *J. Virol.* **2021**, *95*, e02185-18. [CrossRef] [PubMed]
59. Wang, X.; Wu, J.; Wu, Y.; Chen, H.; Zhang, S.; Li, J.; Xin, T.; Jia, H.; Hou, S.; Jiang, Y. Inhibition of cGAS-STING-TBK1 signaling pathway by DP96R of ASFV China 2018/1. *Biochem. Biophys. Res. Commun.* **2018**, *506*, 437–443. [CrossRef] [PubMed]
60. Afonso, C.; Piccone, M.; Zaffuto, K.; Neilan, J.; Kutish, G.; Lu, Z.; Balinsky, C.; Gibb, T.; Bean, T.; Zsak, L. African swine fever virus multigene family 360 and 530 genes affect host interferon response. *J. Virol.* **2004**, *78*, 1858–1864. [CrossRef] [PubMed]
61. Reis, A.L.; Abrams, C.C.; Goatley, L.C.; Netherton, C.; Chapman, D.G.; Sanchez-Cordon, P.; Dixon, L.K. Deletion of African swine fever virus interferon inhibitors from the genome of a virulent isolate reduces virulence in domestic pigs and induces a protective response. *Vaccine* **2016**, *34*, 4698–4705. [CrossRef]
62. Huang, L.; Chen, W.; Liu, H.; Xue, M.; Dong, S.; Liu, X.; Feng, C.; Cao, S.; Ye, G.; Zhou, Q. African Swine Fever Virus HLJ/18 CD2v Suppresses Type I IFN Production and IFN-Stimulated Genes Expression through Negatively Regulating cGMP-AMP Synthase–STING and IFN Signaling Pathways. *J. Immunol.* **2023**, *210*, 1338–1350. [CrossRef]
63. Ran, Y.; Li, D.; Xiong, M.G.; Liu, H.N.; Feng, T.; Shi, Z.W.; Li, Y.H.; Wu, H.N.; Wang, S.Y.; Zheng, H.X.; et al. African swine fever virus I267L acts as an important virulence factor by inhibiting RNA polymerase III-RIG-I-mediated innate immunity. *PLoS Pathog.* **2022**, *18*, e1010270. [CrossRef]
64. Taniguchi, T.; Ogasawara, K.; Takaoka, A.; Tanaka, N. IRF family of transcription factors as regulators of host defense. *Annu. Rev. Immunol.* **2001**, *19*, 623. [CrossRef]
65. Lohoff, M.; Mak, T.W. Roles of interferon-regulatory factors in T-helper-cell differentiation. *Nat. Rev. Immunol.* **2005**, *5*, 125–135. [CrossRef] [PubMed]
66. Yanai, H.; Chiba, S.; Hangai, S.; Kometani, K.; Inoue, A.; Kimura, Y.; Abe, T.; Kiyonari, H.; Nishio, J.; Taguchi-Atarashi, N. Revisiting the role of IRF3 in inflammation and immunity by conditional and specifically targeted gene ablation in mice. *Proc. Natl. Acad. Sci. USA* **2018**, *115*, 5253–5258. [CrossRef] [PubMed]
67. Fitzgerald, K.A.; McWhirter, S.M.; Faia, K.L.; Rowe, D.C.; Latz, E.; Golenbock, D.T.; Coyle, A.J.; Liao, S.-M.; Maniatis, T. IKKε and TBK1 are essential components of the IRF3 signaling pathway. *Nat. Immunol.* **2003**, *4*, 491–496. [CrossRef] [PubMed]
68. Sharma, S.; TenOever, B.R.; Grandvaux, N.; Zhou, G.-P.; Lin, R.; Hiscott, J. Triggering the interferon antiviral response through an IKK-related pathway. *Science* **2003**, *300*, 1148–1151. [CrossRef]
69. Tamura, T.; Yanai, H.; Savitsky, D.; Taniguchi, T. The IRF family transcription factors in immunity and oncogenesis. *Annu. Rev. Immunol.* **2008**, *26*, 535–584. [CrossRef] [PubMed]
70. Chathuranga, K.; Weerawardhana, A.; Dodantenna, N.; Lee, J.-S. Regulation of antiviral innate immune signaling and viral evasion following viral genome sensing. *Exp. Mol. Med.* **2021**, *53*, 1647–1668. [CrossRef] [PubMed]
71. He, B.; Paterson, R.G.; Stock, N.; Durbin, J.E.; Durbin, R.K.; Goodbourn, S.; Randall, R.E.; Lamb, R.A. Recovery of paramyxovirus simian virus 5 with a V protein lacking the conserved cysteine-rich domain: The multifunctional V protein blocks both interferon-β induction and interferon signaling. *Virology* **2002**, *303*, 15–32. [CrossRef]
72. Donelan, N.R.; Dauber, B.; Wang, X.; Basler, C.F.; Wolff, T.; García-Sastre, A. The N-and C-terminal domains of the NS1 protein of influenza B virus can independently inhibit IRF-3 and beta interferon promoter activation. *J. Virol.* **2004**, *78*, 11574–11582. [CrossRef]

73. Delhaye, S.; Van Pesch, V.; Michiels, T. The leader protein of Theiler's virus interferes with nucleocytoplasmic trafficking of cellular proteins. *J. Virol.* **2004**, *78*, 4357–4362. [CrossRef]
74. Paladino, P.; Collins, S.E.; Mossman, K.L. Cellular localization of the herpes simplex virus ICP0 protein dictates its ability to block IRF3-mediated innate immune responses. *PLoS ONE* **2010**, *5*, e10428. [CrossRef]
75. Xia, H.; Cao, Z.; Xie, X.; Zhang, X.; Chen, J.Y.-C.; Wang, H.; Menachery, V.D.; Rajsbaum, R.; Shi, P.-Y. Evasion of type I interferon by SARS-CoV-2. *Cell Rep.* **2020**, *33*, 108234. [CrossRef]
76. Wang, W.; Zhou, Z.; Xiao, X.; Tian, Z.; Dong, X.; Wang, C.; Li, L.; Ren, L.; Lei, X.; Xiang, Z. SARS-CoV-2 nsp12 attenuates type I interferon production by inhibiting IRF3 nuclear translocation. *Cell. Mol. Immunol.* **2021**, *18*, 945–953. [CrossRef]
77. Fung, S.-Y.; Siu, K.-L.; Lin, H.; Yeung, M.L.; Jin, D.-Y. SARS-CoV-2 main protease suppresses type I interferon production by preventing nuclear translocation of phosphorylated IRF3. *Int. J. Biol. Sci.* **2021**, *17*, 1547. [CrossRef]
78. Oka, M.; Yoneda, Y. Importin α: Functions as a nuclear transport factor and beyond. *Proc. Jpn. Acad. Ser. B* **2018**, *94*, 259–274. [CrossRef] [PubMed]
79. Kapetanovic, R.; Fairbairn, L.; Beraldi, D.; Sester, D.P.; Archibald, A.L.; Tuggle, C.K.; Hume, D.A. Pig bone marrow-derived macrophages resemble human macrophages in their response to bacterial lipopolysaccharide. *J. Immunol.* **2012**, *188*, 3382–3394. [CrossRef] [PubMed]
80. Kapetanovic, R.; Fairbairn, L.; Downing, A.; Beraldi, D.; Sester, D.P.; Freeman, T.C.; Tuggle, C.K.; Archibald, A.L.; Hume, D.A. The impact of breed and tissue compartment on the response of pig macrophages to lipopolysaccharide. *BMC Genom.* **2013**, *14*, 1–15. [CrossRef] [PubMed]
81. Gao, J.; Scheenstra, M.R.; van Dijk, A.; Veldhuizen, E.J.; Haagsman, H.P. A new and efficient culture method for porcine bone marrow-derived M1-and M2-polarized macrophages. *Vet. Immunol. Immunopathol.* **2018**, *200*, 7–15. [CrossRef] [PubMed]
82. Kim, J.-H.; Kim, T.-H.; Lee, H.-C.; Nikapitiya, C.; Uddin, M.B.; Park, M.-E.; Pathinayake, P.; Lee, E.S.; Chathuranga, K.; Herath, T.U. Rubicon modulates antiviral type I interferon (IFN) signaling by targeting IFN regulatory factor 3 dimerization. *J. Virol.* **2017**, *91*, e00248-17. [CrossRef] [PubMed]

International Journal of
Molecular Sciences

MDPI

Article

Chicken Interferon-Alpha and -Lambda Exhibit Antiviral Effects against Fowl Adenovirus Serotype 4 in Leghorn Male Hepatocellular Cells

Jinyu Lai [1,2,†], Xingchen He [1,3,†], Rongjie Zhang [1,3], Limei Zhang [1,3], Libin Chen [2], Fengping He [1,3], Lei Li [1,3], Liangyu Yang [1,3], Tao Ren [2,*] and Bin Xiang [1,3,*]

1 College of Veterinary Medicine, Yunnan Agricultural University, Kunming 650201, China; 2018018@ynau.edu.cn (L.Z.); 1993009@ynau.edu.cn (L.Y.)
2 College of Veterinary Medicine, South China Agricultural University, Guangzhou 510642, China
3 Center for Poultry Disease Control and Prevention, Yunnan Agricultural University, Kunming 650201, China
* Correspondence: rentao6868@126.com (T.R.); xiangbin2018@126.com (B.X.)
† These authors contributed equally to this work.

Abstract: Hydropericardium hepatitis syndrome (HHS) is primarily caused by fowl adenovirus serotype 4 (FAdV-4), causing high mortality in chickens. Although vaccination strategies against FAdV-4 have been adopted, HHS still occurs sporadically. Furthermore, no effective drugs are available for controlling FAdV-4 infection. However, type I and III interferon (IFN) are crucial therapeutic agents against viral infection. The following experiments were conducted to investigate the inhibitory effect of chicken IFN against FadV-4. We expressed recombinant chicken type I IFN-α (ChIFN-α) and type III IFN-λ (ChIFN-λ) in *Escherichia coli* and systemically investigated their antiviral activity against FAdV-4 infection in Leghorn male hepatocellular (LMH) cells. ChIFN-α and ChIFN-λ dose dependently inhibited FAdV-4 replication in LMH cells. Compared with ChIFN-λ, ChIFN-α more significantly inhibited viral genome transcription but less significantly suppressed FAdV-4 release. ChIFN-α- and ChIFN-λ-induced IFN-stimulated gene (ISG) expression, such as *PKR*, *ZAP*, *IRF7*, *MX1*, *Viperin*, *IFIT5*, *OASL*, and *IFI6*, in LMH cells; however, ChIFN-α induced a stronger expression level than ChIFN-λ. Thus, our data revealed that ChIFN-α and ChIFN-λ might trigger different ISG expression levels, inhibiting FAdV-4 replication via different steps of the FAdV-4 lifecycle, which furthers the potential applications of IFN antiviral drugs in chickens.

Keywords: antiviral activity; fowl adenovirus serotype 4; hydropericardium hepatitis syndrome; interferon; interferon-stimulated genes

1. Introduction

An outbreak of fowl adenoviruses occurred in China in 2015 and rapidly disseminated across the country [1]. The subgroup C fowl adenovirus serotype 4 (FAdV-4) within group I was the most destructive among the fowl adenoviruses involved in the outbreak, responsible for inducing hydropericardium syndrome (HHS) [2,3]. Furthermore, the immunosuppressive effects induced by fowl adenoviruses contribute to secondary infections with other pathogens, increasing mortality in nearly 100% of the affected population [4–6]. Furthermore, FAdV-4 can be cross-transmitted between various hosts, such as laying hens, broilers, ducks, mandarin ducks, geese, and wild birds [7–9]. FAdV-4 is transmitted through the fecal–oral route and vertically via breeding eggs, posing challenges in adenovirus prevention and resistance [10]. Vaccination strategies against FAdV-4 have been adopted; however, HHS still occurs sporadically. In addition, immunosuppression in chickens after infection with various pathogens frequently leads to suboptimal immunization outcomes with commercial vaccines [11]. Our recent study focused on the epidemiology of fowl adenoviruses and revealed the presence of FAdV-4 in healthy flocks [12]. Thus, developing anti-FAdV-4 drugs could be a valuable adjunct to vaccine immunization, expanding the range of preventive and control measures available for managing HHS.

IFNs were categorized into three groups, type I, type II, and type III IFNs, based on their gene sequence, molecular structure, chromosomal localization, and receptor specificity [13]. The chicken interferons (IFNs) IFN-α, IFN-β, IFN-κ, IFN-ω, IFN-ζ, and IFN-τ were successively characterized in 2000, and ChIFN-λ was genetically engineered for expression in 2008 [14–16]. Previous studies revealed that the infection of chickens with FAdV-4 induces a strong innate immune response, including the upregulation of IFN expression [17–19]. Li et al. revealed that FAdV-4 induces cellular pathways in chickens to produce IFNs and antigen-presenting molecules (MHCI/II) [20]. However, research investigating the impact of IFNs on adenoviruses is lacking. Previous investigations into the antiviral properties of IFNs predominantly concentrated on type I IFNs. The presence of type I IFNs in fowl species were elucidated by characterizing its two primary members, IFN-α and IFN-β [21], and IFN-α exhibits more robust antiviral efficacy than IFN-β [22]. Nevertheless, emerging evidence suggests that IFN-λ is essential to mucosal immunity against viral pathogens [23]. Consequently, IFN-α was selected as the focus of this study. Furthermore, as a newly identified IFN, whether IFN-λ exhibits anti-FAdV-4 activity is unknown.

IFNs are versatile antiviral medications which demonstrated their efficacy in suppressing various viruses, such as encephalomyocarditis, lymphocytic choriomeningitis, herpes simplex virus type 2, and hepatitis B viruses [24,25]. Type I and type III IFNs also have inhibitory properties against severe acute respiratory syndrome-related coronavirus 2 (SARS-CoV-2), suggesting their potential as therapeutic options for SARS-CoV-2 infections. Recombinant type I IFN can effectively hinder the infection of various viruses in avian species, such as fowl leukemia, Marek's, highly pathogenic fowl influenza, and infectious bronchitis [26–29]. Nevertheless, the effects of type I and type III IFNs differ significantly among different viruses. For example, type III IFNs, specifically IFN-λ, exhibit greater efficacy against porcine epidemic diarrhea virus (PEDV) than IFN-α [30].

Type I and type III IFNs bind to different receptors on the cell membrane, with chicken type I IFNs binding to IFNAR1 and IFNAR2 and type III IFNs binding to IFNLR1 and IL10R2 [31,32]. These receptors are distributed in different abundances in different cells, and the receptors for type I IFNs are primarily distributed in fibroblasts. In contrast, the receptors for type III IFNs are mainly distributed in epithelial cells [33]. Leghorn male hepatocellular (LMH) cells were characterized and established in 1987 and belong to the epithelial cell line, which can be used to study FAdVs [34,35]. IFN induces many IFN-stimulated genes (ISGs) crucial for the antiviral response, and this induction relies on the Janus kinase/signal transducer and activator of the transcription signaling pathway [36]. Oligoadenylate synthetase (OAS) is a secondary messenger that detects foreign RNA, exerting antiviral effects [37]. Viperin hinders viral replication by impacting cellular metabolism and mitochondrial function [38]. Mx dynamin-like GTPases are crucial as antiviral effectors in type I and III IFN systems. These proteins exert their inhibitory effects on various viruses by impeding the initial stages of viral replication [39]. Initially identified as an ISG, IFI6 was initially observed to localize in mitochondria; however, subsequent investigations revealed its presence in the endoplasmic reticulum (ER). IFI6 serves a prophylactic function by safeguarding uninfected cells against virus-induced invagination formation in the ER membrane [40]. The IFN-induced protein with tetratricopeptide repeats (IFIT) family plays a crucial role in various processes that counteract viral infection, primarily by modulating translation, which consequently impedes viral replication [41]. Although the limiting effect of chicken IFN on various avian pathogens was demonstrated, the effect of chicken IFNs on FAdV-4 was not reported.

In this study, chicken type I IFN-α (ChIFN-α) and type III IFN-λ (ChIFN-λ) were prokaryotically expressed, and their ability to inhibit FAdV-4 in LMH cells was compared.

2. Results

2.1. ChIFN-α and ChIFN-λ Structure Prediction

ChIFN-α and ChIFN-λ possess signal peptides at the N-terminus. Consequently, based on the signal peptide prediction results, we amplified ChIFN-α (residues 32–193) and ChIFN-λ (residues 24–186) using polymerase chain reaction (PCR) and cloned the two genes into the pET-32α vector respectively, resulting in the creation of pET-32α-ChIFN-α and pET-32α-ChIFN-λ plasmids, both of which were fused with a His-Tag at the N-terminus.

A three-dimensional reconstruction was performed using the SWISS-MODEL online server (https://swissmodel.expasy.org/; accessed on 19 May 2023) to determine the structures of recombinant ChIFN-α and ChIFN-λ. Recombinant ChIFN-α and ChIFN-λ have similar structures, comprising five α-helices (Figure 1). The five α-helices of recombinant ChIFN-α are arranged in a barrel-like configuration, whereas the five α-helices of recombinant ChIFN-λ exhibit an irregular distribution when aggregated.

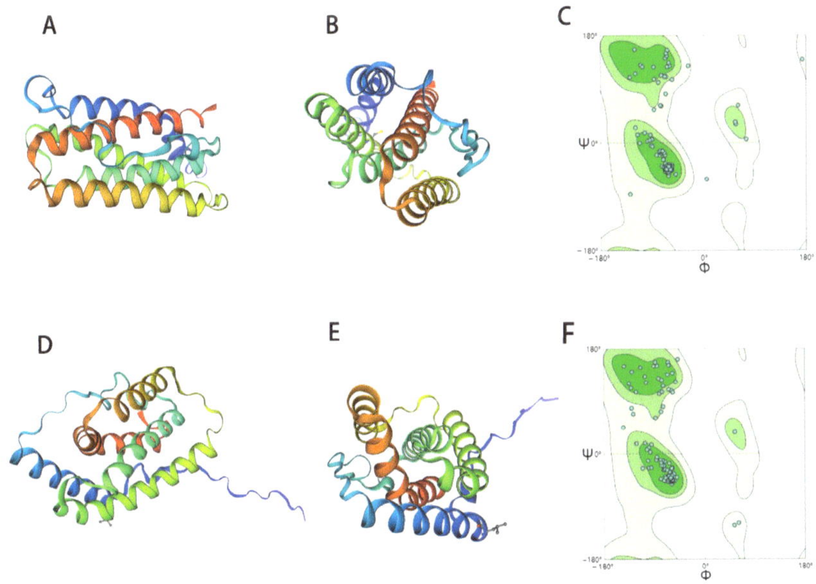

Figure 1. Structure prediction of chickem type I IFN-α (ChIFN-α) and type III IFN-λ (ChIFN-λ). The three-dimensional models of the ChIFN-α and ChIFN-λ proteins were reconstructed using the SWISS-MODEL online server (https://swissmodel.expasy.org/; accessed on 19 May 2023). Three-dimensional structure prediction of ChIFN-α (**A**,**B**). (**C**) shows the credibility of the prediction results, and the prediction results of one amino acid site are unstable. Three-dimensional structure prediction of ChIFN-λ (**D**,**E**). (**F**) shows the credibility of the prediction results.

2.2. ChIFN-α and ChIFN-λ Expression and Purification

The recombinant ChIFN-α and ChIFN-λ expression was further verified using Western blotting (Figure 2A). The expression strain *E. coli BL21 (DE3)* was used to obtain natural recombinant ChIFN-α and ChIFN-λ, and induction was achieved using isopropyl β-d-1-thiogalactopyranoside. The expression was validated, demonstrating the expression of recombinant ChIFN-α and ChIFN-λ in inclusion bodies (Figure 2B,C). Following nickel column purification, the protein purity exceeded 90% after dialysis (Figure 2B,C). The concentration of ChIFN-α was 0.0565 mg/mL, and the ChIFN-λ concentration was 0.165 mg/mL after ultrafiltration.

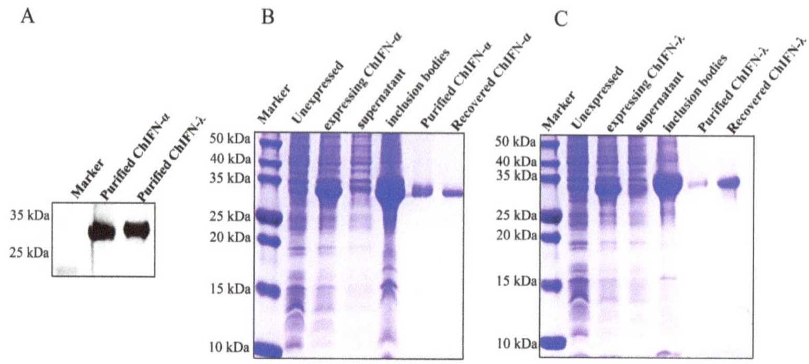

Figure 2. Purification and expression validation of ChIFN-α and ChIFN-λ. (**A**) Using Western blotting, 2 μg of the purified ChIFN-α and ChIFN-λ were validated. Validation of ChIFN-α (**B**) and ChIFN-λ (**C**) purification via sodium dodecyl-sulfate polyacrylamide gel electrophoresis (SDS-PAGE). ChIFN-α and ChIFN-λ were located in inclusion body expression and were purified using nickel columns with good purity.

2.3. Assessing the Antiviral Efficacy of ChIFN-α and ChIFN-λ

Both CHIFN-α and CHIFN-λ showed activity against Porcine vesicular stomatitis virus (VSV). The specific activity of ChIFN-α in LMH cells was 3.96×10^4 UI/mg, whereas that of ChIFN-λ was 2.3×10^4 UI/mg. These findings demonstrated a dose-dependent relationship between the anti-FAdV-4 activity of recombinant ChIFN-α and ChIFN-λ (Figure 3). Considerable inhibition of FAdV-4 hexon expression was observed with the administration of 100 ng of recombinant ChIFN-α and ChIFN-λ. Furthermore, the inhibitory effect on FAdV-4 hexon expression increased gradually with the concentration of the treatments (Figure 3A,C). However, this dose-dependent inhibitory effect became non-significant at 48 h post-infection (hpi) (Figure 3B,D).

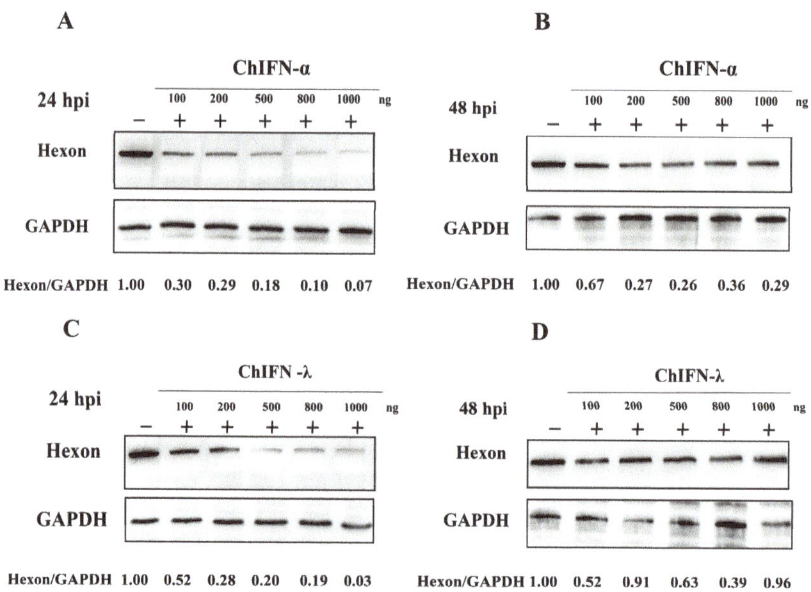

Figure 3. ChIFN-α and ChIFN-λ dose dependently inhibit fowl adenovirus serotype 4 (FAdV-4). (**A**) After pretreating Leghorn male hepatocellular (LMH) cells with different doses of ChIFN-α, the

cells were inoculated with 1 MOI FAdV-4, and the inhibitory effect measured 24 h after inoculation increased with the ChIFN-α dose increase. (**B**) The ChIFN-α dose-dependent inhibition of FAdV-4 became insignificant after 48 h of inoculation. (**C**) ChIFN-λ pretreatment inhibited FAdV-4 dose dependently within 24 h after inoculation. (**D**) This dose-dependent inhibitory effect became non-significant 48 h after inoculation.

2.4. ChIFN-α and ChIFN-λ Pretreatments Restrict FAdV-4 within a Specific Temporal Horizon

Various experiments, including Western blotting, quantitative real-time PCR, and viral titration, were conducted in LMH cells to elucidate the impact of recombinant ChIFN-α and ChIFN-λ on FAdV-4 restriction.

The inhibitory effect of recombinant ChIFN-α and ChIFN-λ on hexon expression in FAdV-4 was significant within 48 hpi (Figure 4A,B). Furthermore, when LMH cells were pretreated with recombinant ChIFN-α and ChIFN-λ, hexon expression was significantly inhibited 24 h after inoculation with FAdV-4. However, the suppressive effect of recombinant ChIFN-α and ChIFN-λ on FAdV-4 gradually diminished as the duration of viral infection increased.

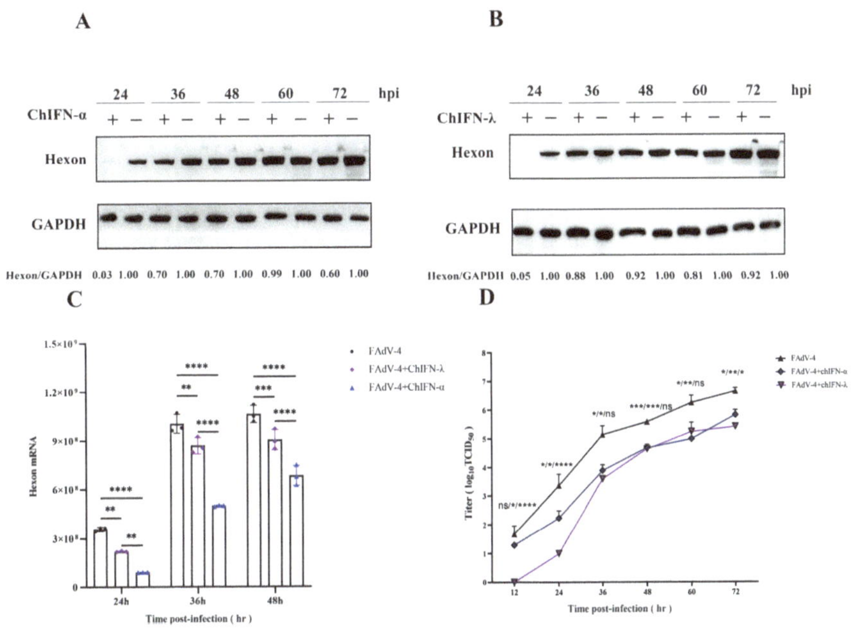

Figure 4. ChIFN-α (1000 UI) or ChIFN-λ (1000 UI) pretreatment inhibited hexon expression, transcription, and FAdV-4 viral release. After 12 h of ChIFN-α pretreatment, ChIFN-λ pretreatment, or no treatment, 1 MOI FAdV-4 was inoculated into the LMH cells. (**A**) and (**B**) Western blot experiments of protein samples collected at 12 h intervals after FAdV-4 inoculation. (**C**) Total mRNA was collected at 12 h intervals after FAdV-4 inoculation for RT-qPCR experiments. (**D**) One-step growth curves of interferon-treated and untreated groups. The first one shows the difference between FAdV-4 and ChIFN-α+FAdV-4, and the middle one shows the difference between FAdV-4 and ChIFN-λ+FAdV-4. The last one is the difference between ChIFN-α+FAdV-4 and ChIFN-λ+FAdV-4. The data above represent three independent experiments with three replicates in each experiment (the error bar represents the SEM) and were analyzed via one-way ANOVA using GraphPad Prism 9.5.0 software. (ns, not significant, * $p < 0.5$; ** $p < 0.01$; *** $p < 0.001$;**** $p < 0.0001$ compared to the FAdV-4 group).

The transcription levels of LMH cells infected with FAdV-4 decreased upon recombinant ChIFN-α and ChIFN-λ treatment. Specifically, the hexon transcription of FAdV-4 was significantly reduced in LMH cells pretreated with recombinant ChIFN-α and ChIFN-λ compared to the control group. This significant reduction was observed at 24, 36, and 48 h post-treatment (Figure 4C).

Pretreating LMH cells with ChIFN-α and ChIFN-λ significantly reduced viral release (Figure 4D). At 12 hpi after inoculation with FAdV-4, the viral titer of the control group reached $10^{1.667}$, the viral titer of the ChIFN-α-treated group was $10^{1.287}$, and the viral titer of the ChIFN-λ-treated group was 0. ChIFN-λ exhibited a substantial inhibitory effect on FAdV-4 release during the initial stages of viral infection, resulting in a 2-log10 reduction in viral titer within 24 hpi. These findings suggest that ChIFN-λ plays a crucial role in impeding the early replication process of FAdV-4. However, ChIFN-α exhibited no significant FAdV-4 release inhibition during the initial phase of FAdV-4 infection. Nonetheless, ChIFN-α demonstrated a more pronounced impact during the later stages of FAdV-4 replication, specifically between 36 and 72 hpi.

2.5. *ChIFN-α and ChIFN-λ Induce ISG Expression in LMH Cells*

LMH cells were exposed to 1000 UI of ChIFN-α or ChIFN-λ for 12 h to investigate their antiviral effects. ChIFN-α and ChIFN-λ had a stimulatory effect on the mRNA levels of ISGs, including *IFITM3, PKR, ZAP, IRF7, MX1, Viperin, IFIT5, OASL,* and *IFI6* (Figure 5). The increase in *PKR, ZAP, IRF7, MX1, OASL, IFI6, IFIT5,* and *Viperin* upregulation was significant (ChIFN-α treatment group: 4.5-, 4.5-, 6.9-, 12-,192.3-, 187.1-, 350.6-, 716-, and 771.3-fold, respectively; ChIFN-λ treatment group: 1.1-,1.5-, 2.8-, 3.8-, 34.9-, 30.8-, 191.8-, 160-, 40.7-fold, respectively). Among the ISGs examined, ChIFN-α induced significantly higher ISG levels than ChIFN-α.

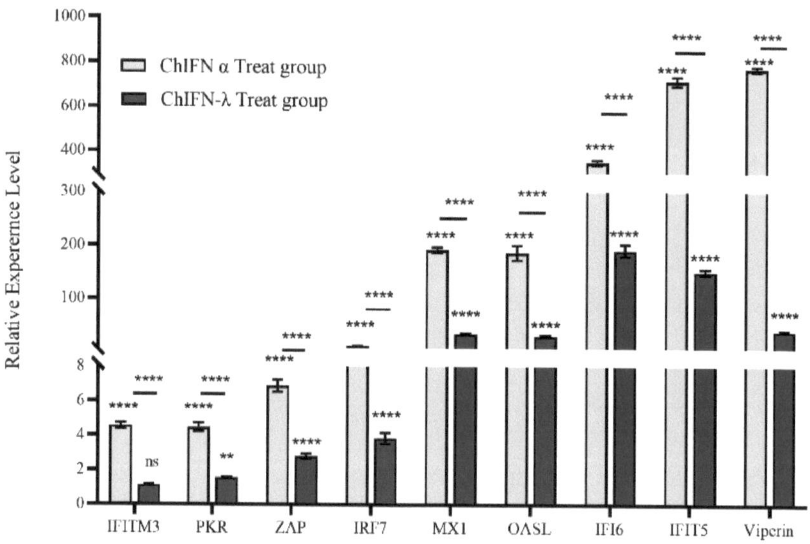

Figure 5. ChIFN-α and ChIFN-λ promoted mRNA expression of the ISGs *IFITM3, PKR, ZAP, IRF7, MX1, Viperin, IFIT5, OASL,* and *IFI6*. RT-qPCR experiments were performed on total mRNA collected 12 h after LMH cells were treated with or without ChIFN-α and ChIFN-λ. The relative quantities were determined using the $2^{-\Delta\Delta CT}$ method and normalized to β-actin. The data above represent three independent experiments with three replicates in each experiment (the error bar represents the SEM) and were analyzed via one-way ANOVA using GraphPad Prism 9.5.0 software. (ns, $p > 0.05$; ** $p < 0.01$; **** $p < 0.0001$ compared to the negative control group).

3. Discussion

FAdV-4 is a significant emerging pathogen, causing harm to poultry and infecting a diverse array of fowl species [42]. Numerous investigations substantiated the escalating prevalence of FAdV-4 infection in Chinese poultry, with evidence of cross-species transmission further complicating prevention and control efforts [7]. Consequently, other prevention and control measures should be conducted to supplement the deficiency of vaccine immunization.

Considering the specificities of the poultry industry, namely the imperative to minimize drug production costs, we opted to use cost-effective prokaryotic expression to generate recombinant ChIFN-α and ChIFN-λ. However, the complex structure of ChIFN-α and ChIFN-λ resulted in their expression as inclusion bodies during prokaryotic expression, rendering them biologically inactive. Therefore, the inclusion bodies were purified under denaturing conditions to address this issue, and optimal complexity conditions were determined to facilitate the folding of ChIFN-α and ChIFN-λ into their native conformations.

In this study, pretreatment with ChIFN-α and ChIFN-λ caused substantial FAdV-4 inhibition within 24 hpi. However, this inhibitory effect exhibited a significant decline after 48 hpi. This phenomenon could be attributed to the gradual reduction in biological activity of IFNs as treatment duration increases, particularly under elevated temperatures.

ChIFN-α and ChIFN-λ demonstrated significant FAdV-4 release inhibition within 72 hpi; however, discernible differences in their antiviral capabilities were observed. The biological activity of ChIFN-α was significantly higher than that of ChIFN-λ; however, ChIFN-λ was more effective than ChIFN-α in inhibiting viral release. Therefore, ChIFN-λ exhibited a greater ability to inhibit FAdV-4 release, particularly during the early stages of viral infection (12–24 hpi), compared to ChIFN-α. Numerous studies extensively compared the antiviral abilities of type I and III IFNs. Type I IFNs exhibit significantly superior antiviral efficacy against Crimean–Congo hemorrhagic fever virus and SARS-CoV-2 compared to type III IFNs [43,44]. However, type III IFNs also exert a markedly stronger restriction effect on PEDV than type I IFNs in intestinal epithelial cell lines [30]. Hence, the antiviral capacities of type I and III IFNs vary depending on the specific virus and cell line. Type I IFN receptors (IFNAR1 and IFNAR2) are primarily expressed in fibroblasts, whereas type III IFN receptors (IFNLR1 and IL10R2) are predominantly expressed in epithelial cells or tissues abundant in epithelial cells [45,46]. Previous studies that systematically examined the induction of ChIFN-λ-stimulated genes in LMH cells identified a total of 421 type III ISGs [47]. In contrast, only 115 ISGs of ChIFN-λ were identified in DF-1 cells [48]. This suggests that differences in receptor expression in different cell lines may lead to differences in ISG induction. Variations in receptor expression levels and the affinity for receptor binding impact the antiviral efficacy of IFNs. For instance, despite sharing the same receptor, IFN-α and IFN-β display distinct levels of antiviral activity, which is attributed to the stronger affinity of IFN-α for IFNAR1 and IFNAR2 [49]. Consequently, the dissimilarity in antiviral resistance between ChIFN-α and ChIFN-λ may be associated with the distribution of receptors and their respective affinities. ChIFN-α induced higher ISG levels than ChIFN-λ; however, excessive upregulation does not necessarily benefit the antiviral response of the host. This disparity in ISG expression levels may contribute to the stronger inhibitory effect of ChIFN-λ compared to ChIFN-α during the initial phase of FAdV-4 infection.

Many ISGs can participate in diverse viral infection processes. For instance, certain members of the *IFITM* family can hinder viral membrane fusion, impeding viral invasion [50]. Additionally, ZC3HAV1 (*ZAP*) exhibited efficacy in suppressing viral mRNA translation in mammals while concurrently preserving the integrity of host mRNAs [51]. Protein kinase R also plays a crucial role in attenuating viral mRNA translation [52], and mammalian piperin impedes viral replication and restricts viral outgrowth [53]. ISG expression was detected after ChIFN-α and ChIFN-λ treatment to investigate the mechanism of ChIFN-α and ChIFN-λ FAdV-4 inhibition. ChIFN-α and ChIFN-λ promoted the upregulation of mRNAs associated with ISGs, including *PKR*, *ZAP*, *IRF7*, *MX1*, *Viperin*, *IFIT5*, *OASL*, and *IFI6*. *PKR*, *ZAP*, *IRF7 MX1*, *Viperin*, *IFIT5*, *OASL*, and *IFI6* upregulation exhibited significant increases. However, ChIFN-α-induced transcriptional upregulation was significantly higher in all ISGs compared with the ChIFN-λ treatment group.

The present study demonstrates the pronounced inhibitory effects of ChIFN-α and ChIFN-λ on FAdV-4. These effects include suppressing viral transcription in LMH cells, restricting viral protein synthesis, and limiting viral release. These findings strongly indicate that ChIFN-α- and ChIFN-λ-induced ISG expression play a pivotal role in various stages of viral infection. Further studies are underway to identify the effective host restriction factors of FAdV-4 with transcriptome sequencing. In relation to practical implementation, we posit that it holds potential as a therapeutic approach for addressing the immune prevention and control insufficiency associated with FAdV-4, particularly within the precious chicken population. Nevertheless, regarding the precise execution strategy, we contend that additional comprehensive experimentation is imperative. This aspect constitutes our research team's primary area of interest, which we aim to explore further in subsequent investigations.

4. Materials and Methods

4.1. Cells, Viruses, and Antibodies

LMH cells were obtained from the American Type Culture Collection (Manassas, VA, USA). Cells were cultured in Dulbecco's Modified Eagle's Medium/F12 (Gibco, Waltham, MA, USA) supplemented with 10% fetal bovine serum (Gibco, Waltham, MA, USA) in a 5% CO_2 incubator at 37 °C. The FAdV-4 KM strain was isolated and preserved in our laboratory and propagated in LMH cells. Vesicular stomatitis virus was obtained from the Department of Avian Diseases, South China Agricultural University. Prof. Shijun J. Zheng from the China Agricultural University kindly donated the hexon monoclonal FAdV-4 antibody [54]. His-Tag Monoclonal antibody (66005-1-Ig) and GAPDH monoclonal rabbit antibody (60004-1-Ig) were purchased from Wuhan Proteintech Group (Proteintech, Wuhan, China), and horseradish peroxidase-labeled goat anti-rabbit antibody (CW0103) and goat anti-mouse antibody (CW0102) were purchased from CWBIO (CWBIO, Beijing, China).

4.2. ChIFN-α and ChIFN-λ Structure Prediction and Epression

Signal peptide prediction was performed via the SignalP website using the ChIFN-α (assembly: EU937528.1) and ChIFN-λ (assembly: EF587763) sequences published on the National Library of Medicine website. The three-dimensional models of the ChIFN-α and ChIFN-λ proteins were reconstructed using the SWISS-MODEL online server (https://swissmodel.expasy.org/; accessed on 19 May 2023). ChIFN-α and ChIFN-λ were amplified using polymerase chain reaction (PCR) from chicken embryo fibroblasts. The primers used during PCR are presented in Table 1. The amplified gene was cloned into the pET32α vector (Novagen; EMD Millipore, Billerica, MA, USA) and expressed in *Escherichia coli BL21 (DE3)*. The specific procedure involved transforming the plasmid into *BL21 (DE3)* and subsequently culturing the *BL21 (DE3)* at 37 °C and 220 rpm until its optical density at 600 nm (OD600) reached 0.6~0.8. Following this, 1 mM isopropyl-β-d-thiogalactoside (IPTG) was introduced for induction, and the culture was sustained for a duration of 10 h. Subsequently, the bacterial solution was collected.

Table 1. Primers for gene cloning and reverse transcription qPCR.

Name	Sequence(5′-3′)
Primers for gene cloning	
ChIFN-α-F	*CCGGAATT*CTGCAACCACCTTC
ChIFN-α-R	*CCCAAGCTT*CTAAGTGCGCGTGTTGCC
ChIFN-λ-F	*CCGGAATT*CCAGGTCACCCCGAAGAA
ChIFN-λ-R	*CCCAAGCTT*CTAAGTGCAATCCTCGCGCTGGGC
Primers for RT-qPCR	
Q-Hexon-F	CGAGGACTACGACGATTA
Q-Hexon-R	CGTGATACAGCAGGTTAATG
Q-MX1-F	AAGCCTGAGCATGAGCAGAA

Table 1. *Cont.*

Name	Sequence(5′-3′)
Q-MX1-R	TCTCAGGCTGTCAACAAGATCAA
Q-OASL-F	ACATCCTCGCCATCATCGA
Q-OASL-R	GCGGACTGGTGATGCTGACT
Q-IFIT5-F	TGCTCTGAGGGAAGAACCCAACA
Q-IFIT5-R	AGGCTCCAGGGATGAGTCCACTT
Q-Viperin-F	AACGGTGGTTCAAGAAGTATGG
Q-Viperin-R	ACAGCATAATCTCGGCACCA
Q-IFITM3-F	TGGTGACGGTGGAGACG
Q-IFITM3-R	GGCAACCAGGGCGATGA
Q-ZAP-F	TTCCAAGTCAAGCCTGTCCC
Q-ZAP-R	CTCCGCTCTGCCTCTTCATC
Q-PKR-F	TGACTTCTGTGACATACAACCCTC
Q-PKR-R	TTTCAAACCAAATCAATCCC
Q-IRF7-F	AACGACGACCCGCACAAG
Q-IRF7-R	GCAGCAGGTCCAAATCCA
Q-IFI6-F	TCAACACACTCCTCAGGCTTTACC
Q-IFI6-R	GAACTCCGCCTCCGCAAGAG
Q-β-actin-F	CAACACAGTGCTGTCTGGTGGTA
Q-β-actin-R	ATCGTACTCCTGCTTGCTGATCC

4.3. IFN Purification and Renaturation

After harvesting the expression strains, the organisms were lysed using an ultrasonic crusher (Jinxing, Shanghai, China), the inclusion bodies were lysed with 8 M urea, and ChIFN-α and ChIFN-λ were purified using the His-Tagged Protein Purification Kit (CWBIO, Beijing, China). The protein was renatured by gradually removing urea from the purified protein through dialysis. The synthesized proteins were confirmed through 15% sodium dodecyl-sulfate polyacrylamide gel electrophoresis (SDS-PAGE) analysis with Kaumas Brilliant Blue staining and Western blotting with anti-His antibody (1:8000 dilution with Tris-buffered saline [TBS]).

4.4. Antiviral Assay

The biological activities of recombinant ChIFN-α and ChIFN-λ were measured using VSV according to previously described methods [55]. The biological activity of ChIFN-α and ChIFN-λ in LMH was analyzed by inhibiting FAdV-4-induced LMH cytopathic effects as described previously [47]. Recombinant IFN was diluted with F12 at a 4-fold dilution and inoculated into 96-well plates with 100 μL per well. The supernatant was incubated for 12 h and discarded, and the 96-well plates were inoculated with FAdV-4 at 100 $TCID_{50}$. The cytopathic lesions were observed 24 h after infection, and the assay was calculated using the Reed–Muench method and expressed as UI/mg.

LMH cells were treated with 100–1000 ng of ChIFN-α or ChIFN-λ for 12 h and inoculated with 1 MOI of FAdV-4 to further compare the anti-FAdV-4 activity of ChIFN-α and ChIFN-λ. Cell samples were collected for Western blotting at 24 h and 48 h post-infection (hpi).

LMH cells were treated with 1000 UI of ChIFN-α or ChIFN-λ for 12 h and then inoculated with 1 MOI of FAdV-4 to elucidate the time frame of viral inhibition caused by ChIFN-α and ChIFN-λ. Cell and supernatant samples were collected at 12, 24, 36, 48, 60, and 72 hpi for further use.

4.5. Western Blot Analysis

Proteins (20 μg) were resolved using SDS-PAGE in Tris-Gly buffer at 150V and transferred to a polyvinylidene difluoride membrane (Sigma-Aldrich, St. Louis, Missouri, USA) at 250 mA. Specific mouse anti-hexon (1:8000 dilution with TBS) and rabbit anti-GAPDH monoclonal antibodies (1:10,000 dilution with TBS) were used for Western blotting. Protein expression was analyzed using a Tanon 5200 instrument (Tanon, Shanghai, China).

4.6. Virus Titration

LMH cells were inoculated into 96-well plates and allowed to attain 80% confluence to conduct the $TCID_{50}$ assay. The samples underwent a 10-fold serial dilution at 100 μL/well, with three repetitions per sample. Following the manifestation of a noticeable cytopathic effect (CPE) in the cells, the number of wells exhibiting CPE at each dilution was documented, and the $TCID_{50}$ value was determined using the Reed–Muench method.

4.7. Quantitative Real-Time PCR

mRNA was extracted from the cells using RNAfast2000 (Feijie, Shanghai, China) according to the manufacturer's guidelines, and 100 ng of RNA was used for qRT-PCR experiments. Subsequently, HiScript II QRT SuperMix for quantitative PCR (Vazyme, Nanjing, China) was used for quantitative real-time PCR (qRT-PCR) analysis. The primer sequences used to quantify cDNA through qRT-PCR experiments were documented in Table 1. The relative quantities were determined using the $2^{-\Delta\Delta CT}$ method and normalized to β-actin.

4.8. Statistical Analyses

Statistical analyses were conducted using GraphPad Prism 9.5.0 (GraphPad Software, San Diego, CA, USA), using paired *t*-tests and one-way ANOVA to assess group differences. p-values of <0.05 indicated statistical significance. The intensity of the protein bands was analyzed using Image J software (https://imagej.net/ij/; accessed on 11 December 2023; National Institutes of Health, Bethesda, MD, USA), and image layout and cropping were performed using Adobe Photoshop 2020 (Adobe, San Jose, CA, USA).

5. Conclusions

In conclusion, this study successfully generated ChIFN-α and ChIFN-λ through prokaryotic expression and demonstrated their strong anti-FAdV-4 activity in LMH cells. Furthermore, ChIFN-α and ChIFN-λ significantly enhanced *PKR*, *ZAP*, *IRF7*, *MX1*, *OASL*, *IFI6*, *IFIT5*, and *Viperin* expression levels. These findings suggest that ChIFN-α has stronger overall anti-FadV-4 activity than ChIFN-λ; however, ChIFN-λ exhibited greater efficacy in inhibiting FAdV-4 release during the early stages of infection. Our findings indicate that ChIFN-α and ChIFN-λ hold promise as potential therapeutic agents for treating HHS.

Author Contributions: B.X. was responsible for fund acquisition, conceptualization, and project management; T.R. and L.Y. were responsible for supervision; J.L., X.H. and R.Z. completed the experiment; J.L. was responsible for data management and formal analysis; L.C., L.Z., L.L. and F.H. provided guidance on the methodology; J.L. completed writing the manuscript and replying to the review comments; B.X. revised the manuscript. All authors have read and agreed to the published version of the manuscript.

Funding: This research was funded by the Academician (Expert) Workstation of Yunnan Province Program (202305AF150127), Yunnan Ten Thousand Talents Plan Leading Talents of Industrial Technology Project of China (YNWR-CYJS-2019-020), National Key Research and Development Program of China (2022YFD1601900), and Yunnan Fundamental Research Projects (202301AT070488).

Institutional Review Board Statement: Not applicable.

Informed Consent Statement: Not applicable.

Data Availability Statement: Data are contained within the article.

Acknowledgments: We would like to thank Shijun Zheng of China Agricultural University for providing monoclonal antibodies against Hexon for this study, which greatly helped our research.

Conflicts of Interest: The authors declare no conflicts of interest.

References

1. Xia, J.; Yao, K.C.; Liu, Y.Y.; You, G.J.; Li, S.Y.; Liu, P.; Zhao, Q.; Wen Rui Wu, Y.P.; Huang, X.B.; Cao, S.J.; et al. Isolation and molecular characterization of prevalent Fowl adenovirus strains in southwestern China during 2015–2016 for the development of a control strategy. *Emerg. Microbes Infect.* **2017**, *6*, e103. [CrossRef]
2. Ren, G.; Wang, H.; Yan, Y.; Liu, F.; Huang, M.; Chen, R. Pathogenicity of a fowl adenovirus serotype 4 isolated from chickens associated with hydropericardium-hepatitis syndrome in China. *Poult. Sci.* **2019**, *98*, 2765–2771. [CrossRef] [PubMed]
3. Li, H.; Wang, J.; Qiu, L.; Han, Z.; Liu, S. Fowl adenovirus species C serotype 4 is attributed to the emergence of hepatitis-hydropericardium syndrome in chickens in China. *Infect. Genet. Evol.* **2016**, *45*, 230–241. [CrossRef] [PubMed]
4. Xu, A.H.; Sun, L.; Tu, K.H.; Teng, Q.Y.; Xue, J.; Zhang, G.Z. Experimental co-infection of variant infectious bursal disease virus and fowl adenovirus serotype 4 increases mortality and reduces immune response in chickens. *Vet. Res.* **2021**, *52*, 61. [CrossRef] [PubMed]
5. Niu, Y.; Sun, Q.; Shi, Y.; Ding, Y.; Li, Z.; Sun, Y.; Li, M.; Liu, S. Immunosuppressive potential of fowl adenovirus serotype 4. *Poult. Sci.* **2019**, *98*, 3514–3522. [CrossRef] [PubMed]
6. Ganesh, K.; Raghavan, R. Hydropericardium hepatitis syndrome of broiler poultry: Current status of research. *Res. Vet. Sci.* **2000**, *68*, 201–206. [CrossRef]
7. Shen, Z.; Xiang, B.; Li, S.; Ren, X.; Hong, Y.; Liao, J.; Yu, D.; Ren, T.; Liao, M.; Xu, C. Genetic characterization of fowl adenovirus serotype 4 isolates in Southern China reveals potential cross-species transmission. *Infect. Genet. Evol.* **2019**, *75*, 103928. [CrossRef]
8. Jiang, Z.; Liu, M.; Wang, C.; Zhou, X.; Li, F.; Song, J.; Pu, J.; Sun, Y.; Wang, M.; Shahid, M.; et al. Characterization of fowl adenovirus serotype 4 circulating in chickens in China. *Vet. Microbiol.* **2019**, *238*, 108427. [CrossRef] [PubMed]
9. Niczyporuk, J.S.; Kozdrun, W.; Czekaj, H.; Piekarska, K.; Stys-Fijol, N. Isolation and molecular characterization of Fowl adenovirus strains in Black grouse: First reported case in Poland. *PLoS ONE* **2020**, *15*, e0234532. [CrossRef]
10. Li, P.H.; Zheng, P.P.; Zhang, T.F.; Wen, G.Y.; Shao, H.B.; Luo, Q.P. Fowl adenovirus serotype 4: Epidemiology, pathogenesis, diagnostic detection, and vaccine strategies. *Poult. Sci.* **2017**, *96*, 2630–2640. [CrossRef]
11. Meng, F.; Dong, G.; Zhang, Y.; Tian, S.; Cui, Z.; Chang, S.; Zhao, P. Co-infection of fowl adenovirus with different immunosuppressive viruses in a chicken flock. *Poult. Sci.* **2018**, *97*, 1699–1705. [CrossRef] [PubMed]
12. Lai, J.; Yang, L.; Chen, F.; He, X.; Zhang, R.; Zhao, Y.; Gao, G.; Mu, W.; Chen, X.; Luo, S.; et al. Prevalence and Molecular Characteristics of FAdV-4 from Indigenous Chicken Breeds in Yunnan Province, Southwestern China. *Microorganisms* **2023**, *11*, 2631. [CrossRef] [PubMed]
13. Pestka, S.; Krause, C.D.; Walter, M.R. Interferons, interferon-like cytokines, and their receptors. *Immunol. Rev.* **2004**, *202*, 8–32. [CrossRef]
14. Sick, C.; Schultz, U.; Staeheli, P. A family of genes coding for two serologically distinct chicken interferons. *J. Biol. Chem.* **1996**, *271*, 7635–7639. [CrossRef]
15. Karpala, A.J.; Morris, K.R.; Broadway, M.M.; McWaters, P.G.; O'Neil, T.E.; Goossens, K.E.; Lowenthal, J.W.; Bean, A.G. Molecular cloning, expression, and characterization of chicken IFN -lambda. *J. Interferon Cytokine Res.* **2008**, *28*, 341–350. [CrossRef] [PubMed]
16. Sekellick, M.J.; Ferrandino, A.F.; Hopkins, D.A.; Marcus, P.I. Chicken interferon gene: Cloning, expression, and analysis. *J. Interferon Res.* **1994**, *14*, 71–79. [CrossRef]
17. Zhao, W.; Li, X.; Li, H.; Han, Z.; Wang, F.; Liu, C.; Shao, Y.; Ma, D. Fowl adenoviruse-4 infection induces strong innate immune responses in chicken. *Comp. Immunol. Microbiol. Infect. Dis.* **2020**, *68*, 101404. [CrossRef]
18. Wu, N.; Yang, B.; Wen, B.; Li, W.; Guo, J.; Qi, X.; Wang, J. Pathogenicity and Immune Responses in Specific-Pathogen-Free Chickens During Fowl Adenovirus Serotype 4 Infection. *Avian Dis.* **2020**, *64*, 315–323. [CrossRef] [PubMed]
19. Wang, J.; Ba, G.; Han, Y.Q.; Ming, S.L.; Wang, M.D.; Fu, P.F.; Zhao, Q.Q.; Zhang, S.; Wu, Y.N.; Yang, G.Y.; et al. Cyclic GMP-AMP synthase is essential for cytosolic double-stranded DNA and fowl adenovirus serotype 4 triggered innate immune responses in chickens. *Int. J. Biol. Macromol.* **2020**, *146*, 497–507. [CrossRef] [PubMed]
20. Li, M.; Raheem, M.A.; Han, C.; Yu, F.; Dai, Y.; Imran, M.; Hong, Q.; Zhang, J.; Tan, Y.; Zha, L.; et al. The fowl adenovirus serotype 4 (FAdV-4) induce cellular pathway in chickens to produce interferon and antigen-presented molecules (MHCI/II). *Poult. Sci.* **2021**, *100*, 101406. [CrossRef] [PubMed]
21. Zhou, H.; Chen, S.; Wang, M.; Cheng, A. Interferons and their receptors in birds: A comparison of gene structure, phylogenetic analysis, and cross modulation. *Int. J. Mol. Sci.* **2014**, *15*, 21045–21068. [CrossRef] [PubMed]
22. Sick, C.; Schultz, U.; Munster, U.; Meier, J.; Kaspers, B.; Staeheli, P. Promoter structures and differential responses to viral and nonviral inducers of chicken type I interferon genes. *J. Biol. Chem.* **1998**, *273*, 9749–9754. [CrossRef]
23. Ye, L.; Schnepf, D.; Staeheli, P. Interferon-lambda orchestrates innate and adaptive mucosal immune responses. *Nat. Rev. Immunol.* **2019**, *19*, 614–625. [CrossRef]
24. Ank, N.; West, H.; Bartholdy, C.; Eriksson, K.; Thomsen, A.R.; Paludan, S.R. Lambda interferon (IFN-lambda), a type III IFN, is induced by viruses and IFNs and displays potent antiviral activity against select virus infections in vivo. *J. Virol.* **2006**, *80*, 4501–4509. [CrossRef] [PubMed]
25. Li, Q.; Sun, B.; Zhuo, Y.; Jiang, Z.; Li, R.; Lin, C.; Jin, Y.; Gao, Y.; Wang, D. Interferon and interferon-stimulated genes in HBV treatment. *Front. Immunol.* **2022**, *13*, 1034968. [CrossRef] [PubMed]

26. Dai, M.; Wu, S.; Feng, M.; Feng, S.; Sun, C.; Bai, D.; Gu, M.; Liao, M.; Cao, W. Recombinant chicken interferon-alpha inhibits the replication of exogenous avian leukosis virus (ALV) in DF-1 cells. *Mol. Immunol.* **2016**, *76*, 62–69. [CrossRef]
27. Xia, C.; Liu, J.; Wu, Z.G.; Lin, C.Y.; Wang, M. The interferon-alpha genes from three chicken lines and its effects on H9N2 influenza viruses. *Anim. Biotechnol.* **2004**, *15*, 77–88. [CrossRef]
28. Mo, C.W.; Cao, Y.C.; Lim, B.L. The in vivo and in vitro effects of chicken interferon alpha on infectious bursal disease virus and Newcastle disease virus infection. *Avian Dis.* **2001**, *45*, 389–399. [CrossRef]
29. Levy, A.M.; Heller, E.D.; Leitner, G.; Davidson, I. Effect of native chicken interferon on MDV replication. *Acta Virol.* **1999**, *43*, 121–127.
30. Li, L.; Fu, F.; Xue, M.; Chen, W.; Liu, J.; Shi, H.; Chen, J.; Bu, Z.; Feng, L.; Liu, P. IFN-lambda preferably inhibits PEDV infection of porcine intestinal epithelial cells compared with IFN-alpha. *Antivir. Res.* **2017**, *140*, 76–82. [CrossRef]
31. Reboul, J.; Gardiner, K.; Monneron, D.; Uze, G.; Lutfalla, G. Comparative genomic analysis of the interferon/interleukin-10 receptor gene cluster. *Genome Res.* **1999**, *9*, 242–250. [CrossRef] [PubMed]
32. International Chicken Genome Sequencing, C. Sequence and comparative analysis of the chicken genome provide unique perspectives on vertebrate evolution. *Nature* **2004**, *432*, 695–716. [CrossRef] [PubMed]
33. Santhakumar, D.; Rubbenstroth, D.; Martinez-Sobrido, L.; Munir, M. Avian Interferons and Their Antiviral Effectors. *Front. Immunol.* **2017**, *8*, 49. [CrossRef] [PubMed]
34. Kawaguchi, T.; Nomura, K.; Hirayama, Y.; Kitagawa, T. Establishment and characterization of a chicken hepatocellular carcinoma cell line, LMH. *Cancer Res.* **1987**, *47*, 4460–4464. [PubMed]
35. Zhang, J.; Zou, Z.; Huang, K.; Lin, X.; Chen, H.; Jin, M. Insights into leghorn male hepatocellular cells response to fowl adenovirus serotype 4 infection by transcriptome analysis. *Vet. Microbiol.* **2018**, *214*, 65–74. [CrossRef] [PubMed]
36. Darnell, J.E., Jr.; Kerr, I.M.; Stark, G.R. Jak-STAT pathways and transcriptional activation in response to IFNs and other extracellular signaling proteins. *Science* **1994**, *264*, 1415–1421. [CrossRef]
37. Hornung, V.; Hartmann, R.; Ablasser, A.; Hopfner, K.P. OAS proteins and cGAS: Unifying concepts in sensing and responding to cytosolic nucleic acids. *Nat. Rev. Immunol.* **2014**, *14*, 521–528. [CrossRef]
38. Seo, J.Y.; Yaneva, R.; Cresswell, P. Viperin: A multifunctional, interferon-inducible protein that regulates virus replication. *Cell Host Microbe* **2011**, *10*, 534–539. [CrossRef]
39. Haller, O.; Staeheli, P.; Schwemmle, M.; Kochs, G. Mx GTPases: Dynamin-like antiviral machines of innate immunity. *Trends Microbiol.* **2015**, *23*, 154–163. [CrossRef]
40. Richardson, R.B.; Ohlson, M.B.; Eitson, J.L.; Kumar, A.; McDougal, M.B.; Boys, I.N.; Mar, K.B.; De La Cruz-Rivera, P.C.; Douglas, C.; Konopka, G.; et al. A CRISPR screen identifies IFI6 as an ER-resident interferon effector that blocks flavivirus replication. *Nat. Microbiol.* **2018**, *3*, 1214–1223. [CrossRef]
41. Zhou, X.; Michal, J.J.; Zhang, L.; Ding, B.; Lunney, J.K.; Liu, B.; Jiang, Z. Interferon induced IFIT family genes in host antiviral defense. *Int. J. Biol. Sci.* **2013**, *9*, 200–208. [CrossRef] [PubMed]
42. Schachner, A.; Matos, M.; Grafl, B.; Hess, M. Fowl adenovirus-induced diseases and strategies for their control—A review on the current global situation. *Avian Pathol.* **2018**, *47*, 111–126. [CrossRef]
43. Bordi, L.; Lalle, E.; Caglioti, C.; Travaglini, D.; Lapa, D.; Marsella, P.; Quartu, S.; Kis, Z.; Arien, K.K.; Huemer, H.P.; et al. Antagonistic antiviral activity between IFN-lambda and IFN-alpha against lethal Crimean-Congo hemorrhagic fever virus in vitro. *PLoS ONE* **2015**, *10*, e0116816. [CrossRef]
44. Felgenhauer, U.; Schoen, A.; Gad, H.H.; Hartmann, R.; Schaubmar, A.R.; Failing, K.; Drosten, C.; Weber, F. Inhibition of SARS-CoV-2 by type I and type III interferons. *J. Biol. Chem.* **2020**, *295*, 13958–13964. [CrossRef] [PubMed]
45. Kumaran, J.; Colamonici, O.R.; Fish, E.N. Structure-function study of the extracellular domain of the human type I interferon receptor (IFNAR)-1 subunit. *J. Interferon Cytokine Res.* **2000**, *20*, 479–485. [CrossRef] [PubMed]
46. Reuter, A.; Soubies, S.; Hartle, S.; Schusser, B.; Kaspers, B.; Staeheli, P.; Rubbenstroth, D. Antiviral activity of lambda interferon in chickens. *J. Virol.* **2014**, *88*, 2835–2843. [CrossRef] [PubMed]
47. Dai, M.; Xie, T.; Liao, M.; Zhang, X.; Feng, M. Systematic identification of chicken type I, II and III interferon-stimulated genes. *Vet. Res.* **2020**, *51*, 70. [CrossRef]
48. Santhakumar, D.; Rohaim, M.A.; Munir, M. Genome-Wide Classification of Type I, Type II and Type III Interferon-Stimulated Genes in Chicken Fibroblasts. *Vaccines* **2019**, *7*, 160. [CrossRef]
49. Pei, J.; Sekellick, M.J.; Marcus, P.I.; Choi, I.S.; Collisson, E.W. Chicken interferon type I inhibits infectious bronchitis virus replication and associated respiratory illness. *J. Interferon Cytokine Res.* **2001**, *21*, 1071–1077. [CrossRef]
50. Yount, J.S.; Moltedo, B.; Yang, Y.Y.; Charron, G.; Moran, T.M.; Lopez, C.B.; Hang, H.C. Palmitoylome profiling reveals S-palmitoylation-dependent antiviral activity of IFITM3. *Nat. Chem. Biol.* **2010**, *6*, 610–614. [CrossRef]
51. Gao, G.; Guo, X.; Goff, S.P. Inhibition of retroviral RNA production by ZAP, a CCCH-type zinc finger protein. *Science* **2002**, *297*, 1703–1706. [CrossRef]
52. Munir, M.; Berg, M. The multiple faces of proteinkinase R in antiviral defense. *Virulence* **2013**, *4*, 85–89. [CrossRef]
53. Tan, K.S.; Olfat, F.; Phoon, M.C.; Hsu, J.P.; Howe, J.L.C.; Seet, J.E.; Chin, K.C.; Chow, V.T.K. In vivo and in vitro studies on the antiviral activities of viperin against influenza H1N1 virus infection. *J. Gen. Virol.* **2012**, *93*, 1269–1277. [CrossRef] [PubMed]

54. Li, W.; You, G.; Haiyilati, A.; Wang, H.; Jiao, H.; Wang, Y.; Gao, L.; Cao, H.; Li, X.; Zheng, S.J. Critical Role of Viral Protein Hexon in Hypervirulent Fowl Adenovirus Serotype-4-Induced Autophagy by Interaction with BAG3 and Promotion of Viral Replication in LMH Cells. *J. Virol.* **2023**, *97*, e0028423. [CrossRef] [PubMed]
55. Cai, M.; Zhu, F.; Shen, P. Expression and purification of chicken beta interferon and its antivirus immunological activity. *Protein Expr. Purif.* **2012**, *84*, 123–129. [CrossRef] [PubMed]

MDPI AG
Grosspeteranlage 5
4052 Basel
Switzerland
Tel.: +41 61 683 77 34

International Journal of Molecular Sciences Editorial Office
E-mail: ijms@mdpi.com
www.mdpi.com/journal/ijms

www.ingramcontent.com/pod-product-compliance
Lightning Source LLC
Chambersburg PA
CBHW041349210926
51850CB00013B/3032
9783725840908